ROYAL HORTICULTURAL SOCIETY

PESTS&
DISEASES

ROYA **SOCIETY**

P
D

ES

PI ALSTEAD

with the RHS Plant Health team

DK

Canterbury College

156177

Contents

Introduction 6

How to Use this Book 8

GALLERY OF SYMPTOMS

GARDEN HEALTH AND PROBLEMS

A–Z OF PESTS, DISEASES, AND DISORDERS

INDIVIDUAL PLANT PROBLEMS

Introduction

Since it was first published more than 20 years ago,
RHS Pests & Diseases has established itself as the classic
gardener's reference on plant health. For this fifth edition we
have undertaken the most extensive revision in its history,
with over 25 new pests and diseases added and all entries
comprehensively updated with new information on
diagnosis, methods of control, and resistant plant choices.

Laying an emphasis on organic and biological treatments, and with an in-depth
look at how to prevent problems through good garden practice, this book will
help you work in harmony with nature to keep your garden thriving.

Whatever the size of your growing space, it is impossible to garden without
encountering pests, diseases, and disorders. At the very least they spoil a plant's
appearance and, if left to spread unchecked, the more serious problems may
dramatically affect a plant's vigour, causing a reduction in growth, flowering
potential, or yield. Some can create wounds that may appear to be insignificant,
but which allow the entry of secondary organisms that can then wreak havoc.
And the correct identification of a problem is essential: if you are uncertain
about what is attacking a plant, you won't be able to tell if it is something to
worry about and what, if anything, can be done to remedy the situation.

If your garden is to flourish, it is not only important to identify and treat pests
and diseases promptly, but also to understand how to keep plants healthy in the
first place. Many common problems develop or become more significant
because of how plants are grown. A plant can become stressed if conditions are
less than ideal or its care requirements are being neglected, and this stress will
make it more susceptible to attack and more likely to show serious symptoms.
Once a cultural problem has been identified and understood, it is often quite
simple to overcome it and so avoid a whole host of other issues.

This new edition of *RHS Pests & Diseases* provides a wealth of up-to-date
information on all aspects of the more problematic side of gardening,
whether you are a home gardener, a student, or a professional horticulturist.
By helping you both to pinpoint specific problems and take a broad view of
garden health, this book will help you stay one step ahead of pests,
diseases, and disorders in the garden.

How to Use this Book

The easy-to-use format provides three ways to help you identify a plant pest, disease, or disorder: by symptom, by plant, and by problem. Symptoms of pests and diseases are shown in colour photographs in the **Gallery of Symptoms**, which is cross-referenced to the **A–Z of Pests, Diseases, and Disorders**, a dictionary of most possible garden problems. **Individual Plant Problems** is a plant-by-plant listing, enabling you to look up the problems your particular plant could be suffering from and find their reference in the A–Z and Gallery. A section on **Garden Health and Problems** looks at the garden and plant cultivation as a whole, discussing the background to problems that might occur, and includes tables with information on the use and correct application of all the most common commercially available garden pesticides, while at the back of the book a **Glossary** of specialist terms and an **Index** can be found.

IDENTIFYING AND CONTROLLING PROBLEMS:

Colour photograph of affected plant part with symptoms clearly visible

By Symptom

See GALLERY OF SYMPTOMS, *pp.10–71*

The Gallery section contains hundreds of photographs, enabling you to identify the cause of the symptoms on your plant and then referring you to the A–Z section. It is divided into six parts: Fruits, Berries, Nuts, and Vegetables; Soil, Tubers, Roots, and Bulbs; Leaves; Stems and Buds; Flowers; and Lawns. At the end are four pages enabling the identification of Beneficial Garden Creatures.

Name of problem, corresponding with detailed A–Z entry

Whether problem is a pest, a disease, or a disorder

Page reference in A–Z

Apple Scald   DISORDER • *See p.110*

Reddish-brown patches of discoloration develop on the skin of the fruit. The flesh beneath is usually undamaged, but it may be discoloured. Exposed surfaces are most likely to be badly affected.

PLANTS AFFECTED Apples
SEASON Mainly summer

Plants affected by that particular problem

Brief description of visible and other symptoms

Season when problem is most likely to occur

---

 12   GALLERY OF SYMPTOMS • **Fruit, Berry, Nut, and Vegetable Problems**

Fruits, Berries, Nuts, and Vegetables

Fruit Problems

If the edible parts of a plant are attacked by a pest or pathogen, the whole reason for growing the plant may well be lost. Fruits and berries are particularly prone to attack, as they usually have a high sugar and moisture content. Nuts, peas, and beans are also valuable food sources for many creatures.

In addition, some ornamental trees and shrubs are commonly grown mainly for their attractive show of berries, and when these are eaten, removed, or spoiled by disease, the plant's appearance is ruined.

In many cases prolonged protection of the plant may be needed, perhaps by using barriers or traps to keep pests away from the flowers or the developing and ripening fruits. Pesticides can give protection, but these may sometimes need to be applied before the first signs of damage are noticed.

Apple Sawfly   PEST • *See p.109*
Most infested fruits drop off in early summer (*see p.14*). Some do develop as ripe fruit but have distinctive ribbon scars where the young larvae have fed earlier in the summer.

PLANTS AFFECTED Apples
SEASON Mid-spring to early autumn

Apple Scald   DISORDER • *See p.110*
Reddish-brown patches of discoloration develop on the skin of the fruit. The flesh beneath is usually undamaged, but it may be discoloured. Exposed surfaces are most likely to be badly affected.

PLANTS AFFECTED Apples
SEASON Mainly summer

Bird Damage   PEST • *See p.115*
Blackbirds, starlings, and other birds will peck holes in ripening fruits. Wasps may enlarge this damage, and brown rot frequently develops once the skin has been broken.

PLANTS AFFECTED Apples, pears, plums
SEASON Midsummer to early autumn

Apple Capsid   PEST • *See p.109*
This pest sucks sap from the fruitlets and causes yellowish-brown, raised, corky swellings. The damage is only skin deep and does not affect the fruits' eating or keeping qualities.

PLANTS AFFECTED Apples
SEASON Late spring to early autumn

By Plant

See INDIVIDUAL PLANT PROBLEMS, *pp.200–213*

A comprehensive listing of garden plants by type, with listings of the general and specific ailments to which each is liable. The problem can then be identified by referring to the *Gallery of Symptoms* or to the A–Z section.

Pests, diseases, or disorders that the plant might suffer from, listed alphabetically

Apple Canker **52**, 108
Apple Capsid **12**, 109
Apple Leaf Miner **32**, 109
Apple Sawfly **12**, **14**, 109
Apple Scab **28**, 110
Apple Scald **12**, 110
Apple Sooty Blotch 110

Page reference for colour photograph in Gallery of Symptoms

Page reference for entry in A–Z of Pests, Diseases and Disorders

Page reference in bold type indicates that symptoms are illustrated in the Gallery section

---

Fruits and Nuts

Fruit and Nut Problems ▪ INDIVIDUAL PLANT PROBLEMS   **201**

Many gardeners are happy to use chemicals on ornamentals, but when it comes to cropping plants, most prefer to avoid them wherever possible. Good cultivation, preventive measures taken at the correct time, and the use of resistant cultivars will all help. Some problems affect many fruits (*see right*). Others are more host-specific; plants so affected are listed below. More resistant species or cultivars may be suggested in the entry for the problem in question in the *A–Z of Pests, Diseases, and Disorders.*

COMMON PROBLEMS

The most common problems affecting plants and trees bearing edible fruits and nuts are:

Aphid 108	Powdery Mildews 173
Bird Damage on Fruits 115	Rabbits 179
Coral Spot 128	Squirrels 186
Grey Mould 142	Wasps 195
Honey Fungus 145	Winter Moth 198
Magnesium Deficiency 154	

Page numbers in **bold** refer to colour illustrations

A

Apple
Apple Bitter Pit **13**, 108
Apple Blossom Weevil **65**, 108
Apple Canker **52**, 108
Apple Capsid **12**, 109
Apple Leaf Miner **32**, 109
Apple Sawfly **12**, **14**, 109
Apple Scab **28**, 109
Apple Scald **12**, 110
Apple Sooty Blotch 110
Apple Water Core 110
Blossom Wilt **56**, 117
Brown Rot **13**, 120
Codling Moth **13**, 127
Coral Spot **56**, 128
Crown Gall **54**, 129
Fireblight **51**, 135
Fruit Tree Red Spider Mite **34**, **57**, 138
High Temperature Injury 144
Honey Fungus **50**, 145
Iron Deficiency **39**, 148
Leopard Moth 153
Magnesium Deficiency **37**, 154

Crown Gall **54**, 129
Frost Damage (Buds and Flowers) **62**, 137
Frost Damage (Foliage and Stems) **43**, 138
Grey Mould **15**, 142
Leafhoppers 151
Leaf Spot (Fungal) **25**, 152
Powdery Mildews **15**, 173
Raspberry Beetle **15**, 175
Red Berry Mite **14**, 176
Rusts 180
Spotted Wing Fruit Fly **15**, 185
Viruses **38**, 194

C

Cherry
Apple Leaf Miner **13**, 108
Bacterial Canker **51**, 112
Blossom Wilt **56**, 117
Brown Rot **13**, 120
Cherry Blackfly **39**, 125
Crown Gall **54**, 129
Frost Damage (Buds and Flowers) **62**, 137
Frost Damage (Foliage and Stems) **43**, 138
Honey Fungus **50**, 145
Magnesium Deficiency **37**, 154

Citrus *see* **Greenhouse and House Plants 212–13**

Currant, black
American Gooseberry Mildew **15**, 40, 107
Big Bud Mites **59**, 115
Blackcurrant Gall Midge **40**, 116
Capsid Bugs **45**, **65**, 123
Coral Spot **56**, 128
Currant Blister Aphid 129
Currant Clearwing Moth **49**, 130
Glasshouse Red Spider Mite **62**, 32, 141
Gooseberry Dieback 142
Grey Mould **15**, 142
Honey Fungus **50**, 145
Leaf Spot (Bacterial) **26**, 44, 151
Leaf Spot (Fungal) **25**, 152
Spotted Wing Fruit Fly **15**, 185
Viruses **38**, 194
Woolly Vine Scale **58**, 199

Currant, red/white
Brown Scale **58**, 120
Capsid Bugs **45**, **65**, 123
Coral Spot **56**, 128
Currant and Gooseberry Leaf Spot 129
Currant Blister Aphid 129
Currant Clearwing Moth **49**, 130
Dieback 132
Gooseberry Sawflies **47**, 142
Grey Mould **15**, 142
Honey Fungus **50**, 145
Leaf Spot (Fungal) **25**, 152
Spotted Wing Fruit Fly **15**, 185
Viruses **38**, 194
Woolly Vine Scale **58**, 199

Fig
Coral Spot **56**, 128
Dieback **132**
Frost Damage, **62**, 137
Glasshouse Red Spider Mite **38**, 62, 141
Grey Mould **15**, 142

Leaf Spot (Fungal) **25**, 152
Mealybugs **48**, **58**, 155
Soft Scale 184
Viruses **38**, 194

G

Gooseberry
American Gooseberry Mildew **15**, 40, 107
Aphids 108
Brown Scale **58**, 120
Bullfinches **65**, 121
Crown Gall **54**, 129
Currant and Gooseberry Leaf Spot 129
Dieback 132
Frost Damage (Buds and Flowers) **62**, 137
Frost Damage (Foliage and Stems) **43**, 138
Gooseberry Sawflies **47**, 142
Grey Mould **63**, 142
Honey Fungus **50**, 145
Iron Deficiency and Lime-induced Chlorosis **39**, 148
Rusts 180
Viruses **38**, 194

Grape *see* **Vine**

H

Hazel nut
Aphids 108
Big Bud Mites **59**, 115
Leaf Spot (Fungal) **25**, 152
Nut Weevil **18**, 159
Powdery Mildews **15**, 173
Squirrels **18**, 186
Viruses **38**, 194
Winter Moth **47**, 198

M

Mulberry
"Bleeding" After Pruning 116
Mulberry Bacterial Blight 157
Mulberry Canker **54**, 157

P

Peach, nectarine, apricot
Bacterial Canker **51**, 112
Blossom Wilt **56**, 117
Brown Scale **58**, 120
Crown Gall **54**, 129
Frost Damage (Buds and Flowers) **62**, 147
Frost Damage (Foliage and Stems) **43**, 138
Glasshouse Red Spider Mite **38**, 62, 141
Honey Fungus **50**, 145
Iron Deficiency and Lime-induced Chlorosis **39**, 148
Nitrogen Deficiency 154
Peach Aphids 164
Peach Leaf Curl **40**, 164
Phosphorus Deficiency 167
Phytophthora 167
Plum Rust **36**, 169
Powdery Mildews **15**, 173
Shothole **31**, **45**, 182
Silver Leaf 182
Spotted Wing Fruit Fly **15**, 185

Pear
Apple and Pear Canker **52**, 108
Apple Powdery Mildew **40**, 109
Apple Scab **28**, 109
Blossom Wilt **56**, 117
Boron Deficiency 117
Brown Rot **13**, **14**, 120
Bullfinches **65**, 121
Codling Moth **13**, 127
Coral Spot **56**, 128
Fireblight **51**, 135
Frost Damage (Buds and Flowers) **62**, 147
Honey Fungus **50**, 145
Magnesium Deficiency **37**, 154
Nitrogen Deficiency 154
Pear and Cherry Slugworm **32**, 164
Pear Bedstraw Aphid 164
Pear Leaf Blister Mite **29**, 164

---

110   A–Z OF PESTS, DISEASES, AND DISORDERS ▪ Apple Scald

Apple Scald

SYMPTOMS Reddish or brown areas develop on the skin of the fruit (*see p.12*), on surfaces that are subjected to the full impact of the sun. In extreme cases the flesh beneath may be damaged, but the discoloration is usually limited to the skin.

CAUSE The scorching or scalding is caused by strong sunlight on the tender skin of young apples.

CONTROL In temperate zones, scald is generally only seen in unusually hot, sunny summers, so controls are not really justified.

Apple Sooty Blotch

SYMPTOMS Sooty, greenish-black spots develop on the fruit skin, usually on apples but occasionally on plums. They do not penetrate into the flesh, however, and can be scraped off. The fruit's development is not affected and it is not distorted.

CAUSE The fungus *Gloeodes pomigena*, which overwinters on young stems and which is most common in trees growing in shade, and in cool, damp conditions.

CONTROL Although common in some seasons, it is rarely widespread and usually only occasional. Increasing air circulation around the fruits will decrease the likelihood of sooty blotch developing, so fruit thinning and regular pruning are advisable.

Apple Sucker

SYMPTOMS Flattened, pale green insects, up to 2mm long, occur on apple blossom trusses in the spring (*see p.61*). Heavy infestations kill the flowers and prevent fruit setting.

CAUSE Sap-sucking, aphid-like insects, *Psylla mali*, that hatch from overwintering eggs as apples come into flower. Most of the damage is caused by the nymphs during the blossom

period. The adult stage resembles a winged aphid and is present after petal fall.

CONTROL Spray with an insecticide approved for use on apples (*see p.95*) before flowering, at the green bud stage. Do not spray when the blossom is open.

Apple Water Core

SYMPTOMS This disorder is also known as "glassiness" because affected fruits develop a glassy, almost wet appearance to the flesh. This is usually most apparent close to the core, but may occur throughout. Symptoms may decrease when apples are stored, but more often affected fruits become brown and start to disintegrate.

CAUSE Can be encouraged by unsuitable cultural conditions, including excessive pruning and thinning, high temperature, high light intensity and extensive damage or removal of bark on the trunk or branches.

CONTROL Avoid excessive thinning or pruning. Other factors are usually impossible to control.

Aquilegia Downy Mildew

SYMPTOMS Angular yellowish patches develop on the foliage of columbines (*Aquilegia* spp.) and increase in size and number until the entire leaf begins to curl and then dies. A delicate white, slightly purple fungal growth develops in patches on the lower leaf surface (*see p.28*), each corresponding to an area of yellowing on the upper surface. Purplish blotches may also develop on the plant stem, while infected flowers take on a water-soaked appearance and become distorted. Seed pods may still form if the plant is infected relatively late in the season and after flowering,

but these invariably develop brown blotches and are unlikely to set seed.

CAUSE The fungus-like *Peronospora* species, which was first reported in Britain in 2013. It thrives in cool, damp weather in spring and autumn, and is spread on air currents, rain, or splash from watering. It may also be seed-borne on home-saved seed.

CONTROL There is no chemical available to control this infection. Avoid overhead watering and reduce the risk of infection by growing plants spaced well apart, for improved air circulation. Do not save seeds from infected or suspect plants, or any grown in the vicinity of an outbreak.

Aquilegia Gall Midge

SYMPTOMS Flower buds of columbines (*Aquilegia* spp.) are distorted and fail to open fully, with the petals turning brown (*see p.64*). Yellowish-orange maggots up to 3mm long may be found inside affected buds.

CAUSE A tiny fly, *Macrolabis aquilegiae*, which lays eggs in the developing flower buds in late spring. After feeding, the larvae go into the soil to pupate and overwinter.

CONTROL Remove and destroy infested flower buds before the larvae escape into the soil to help reduce infestations. Insecticide controls are likely to be ineffective since it is difficult to prevent the females from laying eggs, and the larvae are well protected inside the buds.

Aquilegia Leaf Miners

SYMPTOMS During the summer, the foliage of Aquilegia is disfigured with large, whitish-purple blotches or white sinuous lines (*see p.32*).

CAUSE The larvae of two species of leaf-mining flies. The large blotch mines are caused by larvae of *Phytomyza aquilegiae*, while the linear mines are made by larvae of *P. minuscula*.

CONTROL Aquilegia will still flower and grow, so control is not essential, particularly with the less damaging linear type of mine. If required, pick off infested leaves to prevent the larvae completing their feeding.

Aquilegia Sawfly

SYMPTOMS Pale green caterpillar-like larvae, up to 18mm long, feed on the leaf margins of aquilegias. They can quickly defoliate plants, leaving just the leaf stalks and flower stems. Several generations occur between mid-spring and early autumn.

CAUSE A sawfly, *Pristiphora rufipes*, which lays small batches of eggs on the undersides of leaves. The adults have black bodies 6mm long, two pairs of transparent wings and reddish-brown legs.

CONTROL Check for signs of leaf damage on aquilegias, especially in spring and early summer, when defoliation will have a more severe effect on the plant's health. Either remove the larvae by hand or spray with a contact insecticide (*see p.94*).

Arabis Mosaic Virus

SYMPTOMS Yellow mosaic-like flecking and discoloration on leaves, not only of arabis but of many plants. Infected plants are often stunted and distorted, and commonly fail to reach maturity.

CAUSE Arabis mosaic virus, which is one of the most common and widespread viruses infecting garden plants, both edible and ornamental.

CONTROL There is no control measure available. Affected plants should be removed and disposed of promptly.

By Problem

See A–Z OF PESTS, DISEASES, AND DISORDERS, *pp.105–199*

Comprehensive entries for the pests, diseases, and disorders that may be found on garden plants clearly describe the symptoms, cause, and control of each problem. Preventive and control measures include organic, biological, and pesticide methods as appropriate. In addition, suggestions are given for techniques for avoiding and minimizing further damage, and for resistant cultivars, if they are available. It is cross-referenced to the gallery section and to other associated entries.

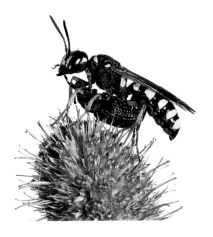

Gallery of Symptoms

Even for experienced gardeners, the correct
identification of what is causing a plant
problem, be it a pest, disease, or disorder,
can be difficult. For ease of comparison with
unhealthy plants, the photographs in this
section are arranged by the parts of plants
that are affected, and show similar types
of symptoms together so that differences
can be seen. Key diagnostic features
of each problem are described.

Fruits, Berries, Nuts, and Vegetables

If the edible parts of a plant are attacked by a pest or pathogen, the whole reason for growing the plant may well be lost. Fruits and berries are particularly prone to attack, as they usually have a high sugar and moisture content. Nuts, peas, and beans are also valuable food sources for many creatures.

In addition, some ornamental trees and shrubs are commonly grown mainly for their attractive show of berries, and when these are eaten, removed, or spoiled by disease, the plant's appearance is ruined.

In many cases prolonged protection of the plant may be needed, perhaps by using barriers or traps to keep pests away from the flowers or the developing and ripening fruits. Pesticides can give protection, but these may sometimes need to be applied before the first signs of damage are noticed.

Fruit Problems

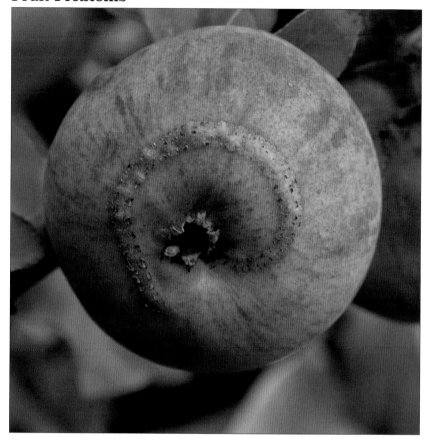

Apple Sawfly
PEST • *See p.109*

Most infested fruits drop off in early summer (*see p.14*). Some do develop as ripe fruit but have distinctive ribbon scars where the young larvae have fed earlier in the summer.

PLANTS AFFECTED Apples
SEASON Mid-spring to early autumn

Apple Scald
DISORDER • *See p.110*

Reddish-brown patches of discoloration develop on the skin of the fruit. The flesh beneath is usually undamaged, but it may be discoloured. Exposed surfaces are most likely to be badly affected.

PLANTS AFFECTED Apples
SEASON Mainly summer

Bird Damage
PEST • *See p.115*

Blackbirds, starlings, and other birds will peck holes in ripening fruits. Wasps may enlarge this damage, and brown rot frequently develops once the skin has been broken.

PLANTS AFFECTED Apples, pears, plums
SEASON Midsummer to early autumn

Apple Capsid
PEST • *See p.109*

This pest sucks sap from the fruitlets and causes yellowish-brown, raised, corky swellings. The damage is only skin deep and does not affect the fruits' eating or keeping qualities.

PLANTS AFFECTED Apples
SEASON Late spring to early autumn

Fruit Problems *CONTINUED*

Codling Moth
PEST • *See p.127*

This is the cause of maggoty apples in late summer. The caterpillar feeds at the core and tunnels out through the side or the eye end of the ripening fruit when it is fully developed.

PLANTS AFFECTED Apples, pears
SEASON Midsummer to early autumn

Wasps
PEST • *See p.195*

These familiar insects enlarge damage started by birds, but can also initiate feeding on softer-skinned fruits (*see p.14*). Their presence can make fruit picking hazardous.

PLANTS AFFECTED Tree fruits, grapes
SEASON Midsummer to early autumn

Brown Rot
DISEASE • *See p.120*

Soft, brown rot develops on fruit (here on apple) and soon becomes studded with raised, creamy-white pustules (*see also p.14*). These are often arranged in concentric rings.

PLANTS AFFECTED Tree fruits
SEASON Late summer and autumn

Rosy Apple Aphid
PEST • *See p.180*

Small greyish-pink aphids infest the foliage and fruitlets in spring. They cause leaf curling and yellowing, and affected fruits remain small and distorted with a pinched appearance at the eye end.

PLANTS AFFECTED Apples
SEASON Early spring to late summer

Pear Midge
PEST • *See p.165*

Pear fruitlets become abnormally swollen, turn black and drop off the tree in early summer. These fruitlets contain many small, whitish-orange maggots. The whole crop may be lost.

PLANTS AFFECTED Pears
SEASON Late spring to early summer

Pear Stony Pit Virus
DISEASE • *See p.165*

Fruits become knobbly and pitted, with hard patches throughout the flesh. Affected fruits usually only appear on isolated branches.

PLANTS AFFECTED Pears
SEASON Late summer to early autumn

Pear Scab
DISEASE • *See p.165*

Blackish-brown, scabby patches appear on the skin of the fruits. The fruits are often small and misshapen. They may be spoiled by splits that allow entry of secondary organisms, which then cause rotting.

PLANTS AFFECTED Pears
SEASON Late summer to early autumn

Quince Leaf Blight
DISEASE • *See p.174*

Numerous small, dark red or blackish-brown spots are found on the skin. Fruits may be distorted and under-sized, and occasionally split.

PLANTS AFFECTED Quinces
SEASON Late summer or early autumn

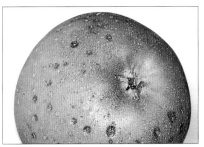

Apple Bitter Pit
DISORDER • *See p.108*

Pale brown spots of discoloration appear within the flesh, or sometimes on the fruit's skin. Those on the skin are usually slightly sunken. The fruit is otherwise perfectly normal in shape and size.

PLANTS AFFECTED Apples
SEASON Mainly late summer and in store

Fruit and Berry Problems

Apple Sawfly   PEST • *See p.109*

After initially tunnelling under the skin, apple sawfly larvae bore into the core of fruitlets. Wet, blackish-brown excrement comes out of the entry hole. For apple sawfly damage to mature fruit see *p.12*.

PLANTS AFFECTED Apples
SEASON Mid-spring to midsummer

Fruit Split (Apple)   DISORDER • *See p.109*

Straight or branched splits develop in the skin. The splits usually occur singly, but several may appear on one fruit. Secondary organisms, such as brown rot, may invade and cause rotting.

PLANTS AFFECTED Most fruits, especially apples
SEASON Mid- to late summer

Brown Rot   DISEASE • *See p.120*

Fruits (here a cherry) turn brown and rot and rapidly develop numerous raised, creamy-white pustules (*see also p.13*). The fruits then dry out, but may remain hanging on the tree or fall to the ground.

PLANTS AFFECTED Tree fruits
SEASON Late summer to autumn

Plum Moth   PEST • *See p.169*

Infested plum fruits tend to ripen prematurely. They contain pinkish caterpillars, which eat the flesh around the stone, leaving behind a trail of brown excrement pellets.

PLANTS AFFECTED Plums, damsons, gages
SEASON Late summer to early autumn

Wasps   PEST • *See p.195*

These familiar yellow and black stinging insects are attracted to ripe fruits. They can eat out extensive cavities in some softer-skinned fruits. For wasp damage to apples and pears *see p.13*.

PLANTS AFFECTED Tree fruits, grapes
SEASON Midsummer to mid-autumn

Red Berry Mite   PEST • *See p.176*

Blackberries fail to ripen completely, with parts or all of the fruit remaining red. Ripening becomes progressively unsatisfactory as the season develops. The cause is microscopic mites.

PLANTS AFFECTED Blackberries
SEASON Late summer to mid-autumn

Fruit and Berry Problems *CONTINUED*

Raspberry Beetle
PEST • *See p.175*

The stalk end of the fruit dries up and becomes greyish-brown. An elongate, whitish-brown beetle grub, up to 8mm long, may be found feeding in the core or plug of the fruit.

PLANTS AFFECTED Cane fruits
SEASON Midsummer to early autumn

Spotted-wing Fruit Fly
PEST • *See p.185*

Larvae of fly feed inside ripening fruits, causing sunken patches and holes to appear in the skin, with white maggots visible within.

PLANTS AFFECTED Most soft fruits, cherry, peach
SEASON Spring to autumn

Strawberry Grey Mould
DISEASE • *See p.186*

Soft, pale brown areas develop on the fruit and rapidly increase in size. Fuzzy grey fungal growth develops on the surface and the fruit rapidly rots.

PLANTS AFFECTED Strawberries
SEASON Late spring to early autumn

Powdery Mildew
DISEASE • *See p.173*

Fruits (here grapes) develop a white, off-white, or greyish-beige fungal coating in patches or all over their surface. They fail to swell fully, may be small or distorted, and may crack or split.

PLANTS AFFECTED Many fruits
SEASON Summer

American Gooseberry Mildew
DISEASE • *See p.107*

Powdery or felty white fungal growth develops on the skin, soon turning buff or pale brown. The fungus may be rubbed off. Fruits may be under-sized.

PLANTS AFFECTED Gooseberries, blackcurrants
SEASON Summer

Pyracantha Scab
DISEASE • *See p.173*

Greenish-black, scabby fungal growth develops on the berries, which may cause them to split and die off. In extreme cases the whole surface of the berry may be obscured.

PLANTS AFFECTED *Pyracantha*
SEASON Summer to autumn

Vegetable Problems

Tomato Blight
DISEASE • *See p.189*

Brown, discoloured areas develop within the fruits and are visible on the surface of both ripe and unripe fruits. Discoloured areas may shrink inwards and suffer secondary rotting.

PLANTS AFFECTED Tomatoes
SEASON Summer and early autumn

Tomato Greenback
DISORDER • *See p.189*

A partial or complete ring of unripened flesh appears around the stalk end of a ripening fruit. The affected flesh remains either green or yellow.

PLANTS AFFECTED Tomatoes
SEASON Summer

Tomato Blotchy Ripening
DISORDER • *See p.189*

Randomly distributed, hard, yellow or green patches of flesh remain unripened. The rest of the fruit develops and ripens normally.

PLANTS AFFECTED Tomatoes
SEASON Summer and early autumn

Blossom End Rot
DISEASE • *See p.117*

A sunken, leathery, dark brown-to-black patch develops at the bottom end of the developing fruit (here tomatoes). The remainder of the fruit ripens normally, but remains distorted.

PLANTS AFFECTED Tomatoes, peppers
SEASON Summer

Tomato Fruit Splitting
DISORDER • *See p.189*

Fruits develop normally and then split shortly before they are ready to pick. The split may dry out or may become infected with secondary fungi.

PLANTS AFFECTED Tomatoes
SEASON Summer or early autumn

Grey Mould
DISEASE • *See p.142*

Fruits (here a tomato) turn soft and discolour and become covered in fuzzy grey fungal growth. On tomatoes, ghost spotting symptoms (yellow circles) on otherwise normal fruits may appear.

PLANTS AFFECTED Many fruits
SEASON Summer and early autumn

Bean Seed Fly
PEST • *See p.114*

White fly maggots feed on the germinating seeds, making holes in the cotyledons and stems. Seedlings may be killed or emerge with the growing point killed.

PLANTS AFFECTED Runner and French beans
SEASON Mid-spring to early summer

Vegetable Problems *CONTINUED*

Bean Seed Beetles
PEST • *See p.113*

Beans, especially broad beans, kept for seed may have circular holes where adult seed beetles have emerged. The grubs feed inside on the cotyledons, but damaged seeds will usually germinate.

PLANTS AFFECTED Beans and peas
SEASON Midsummer to mid-autumn

Pea Moth
PEST • *See p.163*

Small cream-coloured caterpillars with dark spots feed inside pea pods on the developing seeds. Piles of excrement pellets indicate where the caterpillars have been feeding.

PLANTS AFFECTED Peas
SEASON Early to late summer

Pea Leaf and Pod Spot
DISEASE • *See p.163*

Yellow or brown, often sunken, patches develop on the pods. Dark brown or black pycnidia (fungal fruiting bodies) may be visible on affected areas.

PLANTS AFFECTED Peas
SEASON Summer

Sweetcorn Smut
DISEASE • *See p.188*

Individual kernels on sweetcorn cobs become enlarged and distorted. Initially pale grey in colour, they rupture to release masses of powdery black spores. The rest of the plant develops normally.

PLANTS AFFECTED Sweetcorn
SEASON Summer

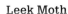

Carrot Fly
PEST • *See p.124*

Elongate, yellow-white fly maggots tunnel in the tap roots, causing orange-brown discoloured lines on the outside of the roots. Secondary rots may extend the damage.

PLANTS AFFECTED Carrots, parsnips, parsley
SEASON Early summer to late autumn

Leek Moth
PEST • *See p.152*

Whitish-brown patches develop on the leaves where the leaf-mining caterpillars have been feeding. Stems and bulbs are also tunnelled. Small, oval, net-like cocoons may be found on the foliage.

PLANTS AFFECTED Leeks and onions
SEASON Early summer to mid-autumn

Vegetable Problems *CONTINUED*

Onion Neck Rot
DISEASE • *See p.161*

The neck area of the onion turns pale brown and soft and appears semi-transparent. Fuzzy grey fungal growth and black sclerotia may develop. The onion can become mummified.

PLANTS AFFECTED Onions, shallots
SEASON Late summer and while in store

Onion Eelworm
PEST • *See p.161*

Microscopic, worm-like animals live within the bulb and leaves. Infested plants are swollen and distorted, with softened tissues that are susceptible to secondary rots.

PLANTS AFFECTED Onions, leeks
SEASON Late spring to early autumn

Onion White Rot
DISEASE • *See p.161*

Dense, fluffy, white fungal growth appears around the basal plate. Small black sclerotia may be visible in amongst this and the bulb starts to rot.

PLANTS AFFECTED Onions, leeks, shallots, garlic
SEASON Summer

Nut Problems

Nut Weevil
PEST • *See p.159*

The kernel is eaten by white grubs with brown heads. The fully-fed grubs bore circular holes in ripe nut shells (here of a hazelnut) when they leave to pupate.

PLANTS AFFECTED Hazel and cob nuts
SEASON Midsummer to early autumn

Squirrels
PEST • *See p.186*

Grey squirrels often strip trees before the nuts are fully ripe and the ground becomes littered with broken nut shells. Nuts are buried, sometimes in lawns, for consumption during the winter.

PLANTS AFFECTED Hazel and cob nuts, walnuts
SEASON Late summer to mid-autumn

Acorn Gall Wasp
PEST • *See p.106*

Oak acorns are converted into ridged, yellow-brown structures, which are sometimes called knopper galls. The wasp grubs feed in a seed-like capsule inside the gall.

PLANTS AFFECTED Oaks (*Quercus*)
SEASON Late summer to mid-autumn

Soil, Roots, Tubers, and Bulbs

Pests and pathogens that attack below ground are common, but because the symptoms they cause are usually first seen above ground, they may well be misidentified. Often the only way to identify a soil-borne problem conclusively is to dig up all or part of the plant, a process which in itself may obviously cause considerable and long-term damage.

Most soil-borne problems, or pests or pathogens which attack roots, tubers, bulbs, or corms directly, are potentially very damaging, simply because the root system or the base of the plant is so fundamental to the survival of the plant as a whole. Early investigation, identification, and, where possible, control is therefore essential. Control is often difficult, since locating and targeting an underground pest or pathogen with a pesticide is not easy.

Soil Problems

Vine Weevil (Damage by Larvae)
PEST • *See p.193*

Plump, white, legless larvae with pale brown heads feed on roots, corms, and tubers, particularly on plants grown in containers. The plant lacks vigour.

PLANTS AFFECTED Many plants
SEASON Late summer to mid-spring

Leatherjackets
PEST • *See p.152*

Leatherjackets are the larvae of daddy-long-legs. They are greyish-brown, legless maggots with no obvious head. They eat roots and seedling plants and also cause problems in lawns (*see p.66*).

PLANTS AFFECTED Mostly annuals and vegetables
SEASON Mid-spring to late summer

Woodlice
PEST • *See p.199*

Woodlice may be grey or pinkish-brown and up to 15mm long. They live mainly in the surface layers of the soil, where they feed on decaying vegetation and occasionally seedlings.

PLANTS AFFECTED Seedlings
SEASON All year

Fungus Gnats (Larvae)
PEST • *See p.138*

The larvae of fungus gnats, or sciarid flies, are white, legless maggots with black heads. They feed on decaying vegetation and seedlings.

PLANTS AFFECTED Greenhouse and house plants
SEASON All year

Chafer Grubs
PEST • *See p.125*

Chafer grubs have plump, white bodies, curved like the letter C, with brown heads and three pairs of legs. They eat roots and make cavities in root vegetables and potato tubers.

PLANTS AFFECTED Lawns and garden plants
SEASON Mostly mid-spring to mid-autumn

Cutworms
PEST • *See p.130*

Cutworms are the earth-coloured caterpillars of several species of moths. They kill small plants by severing the stems or by gnawing away the outer bark at soil level.

PLANTS AFFECTED Mostly annuals and vegetables
SEASON Mainly mid-spring to mid-autumn

Swift Moths (Larvae)
PEST • *See p.188*

These are slender, white-bodied caterpillars with brown heads, which feed on roots, corms, and rhizomes, especially in herbaceous borders.

PLANTS AFFECTED Many garden plants
SEASON Mainly mid-spring to mid-autumn

Soil Problems *CONTINUED*

These slow-moving larvae have tough bodies

Wireworms are up to 25mm long

Wireworms
PEST • *See p.198*

Yellowish-brown grubs, with three pairs of short legs at their front, eat roots and bore into root vegetables. They cause most damage where grass has been dug to create flower or vegetable gardens.

PLANTS AFFECTED Seedlings, potato tubers, and others
SEASON All year

Millipedes
PEST • *See p.156*

These spotted snake millipedes are creamy-white spotted with red, with legs on every body segment. They feed on rotting vegetation and enlarge damage initiated by other pests and diseases.

PLANTS AFFECTED Seedlings, potato tubers
SEASON Spring to autumn

Springtails
PEST • *See p.185*

Small, white animals come to the surface of pot plants or are washed out of the pot when the plant is watered. They cause no harm to the plant as they feed on dead plant material.

PLANTS AFFECTED Pot plants
SEASON All year

Bulb Problems

Blue Mould on Bulbs
DISEASE • *See p.117*

Reddish-brown lesions develop on the flesh of the bulb and may be sunken. Fluffy, white or blue-green fungal growth may appear on top of lesions.

PLANTS AFFECTED Most bulbs in store
SEASON Mainly late summer to autumn

Narcissus Eelworm
PEST • *See p.158*

Plants are stunted and distorted. If the bulb is cut in half transversely, brown concentric rings can be seen where the microscopic eelworms are feeding. Infested bulbs rot and die.

PLANTS AFFECTED Daffodils
SEASON All year

Narcissus Bulb Fly
PEST • *See p.158*

Infested bulbs rot in the soil or produce a few distorted leaves in the spring. One large maggot feeds in the centre of the bulb and fills it with its muddy-brown excrement.

PLANTS AFFECTED Daffodils, amaryllis, and others
SEASON Late summer to early spring

Bulb Problems *CONTINUED*

Tulip Fire
DISEASE • *See p.190*

Tiny, irregularly shaped, black sclerotia (fungal resting bodies) develop on the bulb, clustered around the neck. The bulb may rot or appear firm. Foliage that appears above ground may be withered, discoloured, or dead.

PLANTS AFFECTED Tulips
SEASON Late summer

Gladiolus Corm Rot
DISEASE • *See p.140*

Brown, ridged areas of discoloration appear on the corm. The corm dries out and becomes mummified. No fungal growths or sclerotia are visible.

PLANTS AFFECTED Gladioli, crocuses, bulbous iris
SEASON In storage

Gladiolus Corm Rot – Foliage Symptoms
DISEASE
See p.140

Yellow flecking and, later, striping develops towards the tips of the outermost leaves. The discoloration spreads and the leaves turn brown and die.

PLANTS AFFECTED Gladioli, crocuses, bulbous iris
SEASON Late spring and summer

Root and Tuber Problems

Foot and Root Rots
DISEASE • *See p.137*

Roots discolour (here on *Daphne*), often turning black or brown and shrinking inwards. Deterioration usually starts at the root tips and roots may disintegrate.

PLANTS AFFECTED Most plants
SEASON All year

Iris Rhizome Rot
DISEASE • *See p.148*

Roots and rhizomes deteriorate rapidly, turning soft and slimy. The infection usually starts just beneath the leaf, but spreads quickly. Foliage may fail to develop, or yellow and die back.

PLANTS AFFECTED Rhizomatous irises
SEASON Mainly spring and summer

Clubroot
DISEASE • *See p.127*

Numerous irregularly shaped swellings develop on roots of cabbage and related plants. Swellings do not contain any insect larvae or grubs. The entire root system may be distorted and swollen.

PLANTS AFFECTED Any in the Brassica family
SEASON All year

Root Aphids
PEST • *See p.177*

Infested plants grow slowly and tend to wilt in sunny weather. Aphids, usually globular and creamy-brown, feed on the roots. They often secrete a white, waxy powder from their bodies.

PLANTS AFFECTED Lettuce, beans and others
SEASON Mid- to late summer

Potato Cyst Eelworms
PEST • *See p.171*

Pinhead-sized, spherical cysts, which may be white, pale yellow, or brown, form on the roots. These are the swollen bodies of the female eelworms.

PLANTS AFFECTED Potatoes, tomatoes
SEASON Mid- to late summer

Root Mealybugs
PEST • *See p.178*

These sap-feeding insects have white, elongate bodies up to 2mm long. They secrete a white, waxy powder from their bodies, which coats roots and soil particles.

PLANTS AFFECTED Pot plants and *Meconopsis*
SEASON All year

Root and Tuber Problems CONTINUED

Potato Common Scab
DISEASE • *See p.171*

Raised, roughened, scabby patches develop on the skin of the tuber. Scabs have ragged edges. The damage is usually superficial.

PLANTS AFFECTED Potatoes
SEASON Late spring onwards

Potato Spraing
DISEASE • *See p.172*

Tan-coloured arcs of discoloration appear in the flesh. Corky tissue may be present around the discoloured areas. The tubers are occasionally distorted. Foliage and stems may be mottled yellow.

PLANTS AFFECTED Potatoes
SEASON Late spring onwards

Potato Silver Scurf
DISEASE • *See p.172*

Pale and inconspicuous silvery-grey markings develop on the tuber skin. The size, shape, flesh, and eating quality of the potato are not affected.

PLANTS AFFECTED Potatoes
SEASON Mainly in storage

Potato Blight
DISEASE • *See p.170*

Slightly sunken, darkened areas develop on the tuber skin. Internally, the flesh is discoloured by a dry reddish-brown rot. Secondary organisms cause soft rotting and an unpleasant smell.

PLANTS AFFECTED Potatoes
SEASON Late summer onwards

Potato Powdery Scab
DISEASE • *See p.172*

Small, near-circular scabby areas appear on the tuber skin. The scabs have raised margins and burst to release masses of brown spores.

PLANTS AFFECTED Potatoes
SEASON Summer onwards

Wireworms
PEST • *See p.198*

Elongate, yellowish-brown grubs with three pairs of legs at the head end bore into the tubers. They are mainly a problem where grass has recently been dug over to make a vegetable garden.

PLANTS AFFECTED Root vegetables, potatoes
SEASON Early summer to mid-autumn

Root and Tuber Problems *CONTINUED*

Violet Root Rot
DISEASE • *See p.193*

The surface of the root (here of a carrot) or tuber is covered in a network of dark purple fungal strands. Sclerotia may be present. Roots and tubers may rot.

PLANTS AFFECTED Various plants
SEASON Mainly spring to autumn

Slugs
PEST • *See p.183*

Slugs make small, round holes in the skin of potato tubers but hollow out extensive cavities once they are inside. Some potato cultivars are particularly susceptible to slug damage.

PLANTS AFFECTED Potatoes, root vegetables
SEASON Midsummer to mid-autumn

Potato Gangrene
DISEASE • *See p.172*

Slightly sunken, round-edged areas develop on the tuber shortly after lifting. Lesions may remain defined or enlarge. The tuber rots, turning pale pink.

PLANTS AFFECTED Potatoes
SEASON Summer and autumn

Potato Dry Rot
DISEASE • *See p.171*

One end of the tuber shows wrinkled skin and then shrinks inwards rapidly, forming concentric rings of wrinkles and discoloration. Pink, white, or blue-green fungal pustules may develop.

PLANTS AFFECTED Potatoes in store
SEASON Any time of year

Secondary infection causes rotting of shoulder area

Deep fissure running the length of root

Parsnip Canker
DISEASE • *See p.162*

Roughened, reddish-brown, orange or black cankered areas develop on the root, most commonly around the shoulders. Affected tissues may develop secondary rotting.

PLANTS AFFECTED Parsnips
SEASON Mainly autumn and winter

Irregular Watering
DISORDER • *See p.148*

Longitudinal cracks appear in roots (here of a carrot). Top growth appears perfectly normal. Secondary organisms may cause deterioration of root.

PLANTS AFFECTED Many root crops
SEASON Mainly summer and autumn

Leaves

Damage to the foliage can spoil a plant's appearance and, if any part of the plant is edible, the crop may be reduced. Leaf markings may take the form of spots, either flat (*pp.24–26*) or raised to the touch (*pp.27–30*), often with the pycnidia (fruiting bodies) of fungi visible; blotches (*pp.31–34*); mottling (*pp.34–36*); or discoloration (*pp.37–39*). Leaves may also be distorted (*pp.40–42*), dying (*pp.43–44*), or damaged (*pp.45–47*). Pests may be visible (*p.48*).

Leaf markings are often caused by diseases. Weaker fungal diseases are only likely to cause extensive damage to a plant already lacking in vigour. Plants are particularly vulnerable if growing conditions are unsuitable, they have not been well cared for, or have suffered some other serious attack. Both cultural and pesticide remedies may be necessary to improve the general health of the plant and to control the specific problem.

Leaf marking can occur at any stage, but older foliage is usually worst affected, especially in the autumn. Variegated leaf areas are especially prone to damage.

Flat Markings on Leaves

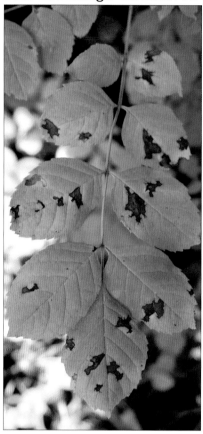

Ash Dieback
DISEASE • *See p.111*

Black blotching develops on leaves, particularly on the mid-vein and leaf base, causing them to wilt. Patches of dead tissue appear on stems and when cankers develop, the stem, branch, or whole tree may die.

PLANTS AFFECTED Many ash species (*Fraxinus*)
SEASON Mainly autumn

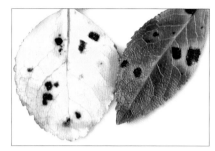

Escallonia Leaf Spot
DISEASE • *See p.134*

Purplish-black spots, usually with pale grey centres, develop on leaves, which then yellow and drop prematurely. The plant's vigour may be reduced.

PLANTS AFFECTED Escallonias
SEASON Mainly summer

Orchid Viruses
DISEASE • *See p.162*

Dark brown or black, or occasionally yellow, markings appear on the leaf. These are usually seen as streaks or spotting. Flowering is often affected and the plant's overall vigour will decline.

PLANTS AFFECTED Most orchids
SEASON All year

Strawberry Leaf Spot
DISEASE • *See p.186*

Reddish-purple spots are found on the leaves, which may later enlarge (*see p.31*). Off-white fungal growth may develop over the surface. Flowers can be affected.

PLANTS AFFECTED Strawberries
SEASON Mainly midsummer to autumn

Broad Bean Chocolate Spot
DISEASE • *See p.120*

Chocolate-brown spots are seen on the leaves and there are similar coloured streaks on the stems, pods, and flowers. The spots may join together.

PLANTS AFFECTED Broad beans
SEASON Mainly late spring to autumn

Delphinium Black Blotch
DISEASE • *See p.131*

Black blotches appear on the leaves, which enlarge, and often spread to the leaf stalks and stems. The leaves may eventually die off.

PLANTS AFFECTED Delphiniums
SEASON Late spring to autumn

Flat Markings on Leaves *CONTINUED*

Mahonia Rust
DISEASE • *See p.155*

Dark brown, spore-filled pustules are found on the lower leaf surface, with bright orange spots on the upper surface (*see p.24*). Leaves may fall early, but overall plant health is rarely affected.

PLANTS AFFECTED Mahonias
SEASON Late spring to early autumn

Leaf Spots (Fungal)
DISEASE • *See p.152*

Rounded, grey or brown spots develop on the leaves (here of a castor oil plant) and may join together. Numerous pycnidia are often apparent on the spots.

PLANTS AFFECTED Many plants
SEASON Varies with host plant

Celery Leaf Spot
DISEASE • *See p.125*

Dark brown or black, pinprick-sized spots appear on the leaves, which may discolour and die off. The leaf stalks can also be affected and the crop reduced.

PLANTS AFFECTED Celery and celeriac
SEASON All season

Rose Black Spot
DISEASE • *See p.178*

Purple-black spots are seen on leaves. The spots may join together or remain separate. Leaf yellowing and premature fall are common.

PLANTS AFFECTED Roses
SEASON Mainly early summer to autumn

Blotches appear shiny and may be up to 1.5cm in diameter

Main part of leaf tissue remains healthy green colour

Surface of spot is rough and slightly raised

Leaf Spots (Fungal)
DISEASE • *See p.152*

Brown spots are seen on the leaves (here of a primula), often with a grey centre and sometimes accompanied by leaf yellowing. Pycnidia may be visible.

PLANTS AFFECTED Many plants
SEASON Varies with host plant

Tar Spot of Acer
DISEASE • *See p.188*

Large, black blotches with bright yellow "halos" develop on upper leaf surfaces. Leaves may fall early and the tree can look disfigured, but overall vigour is not usually affected.

PLANTS AFFECTED Acers (usually sycamore)
SEASON Late spring to autumn

Flat Markings on Leaves *CONTINUED*

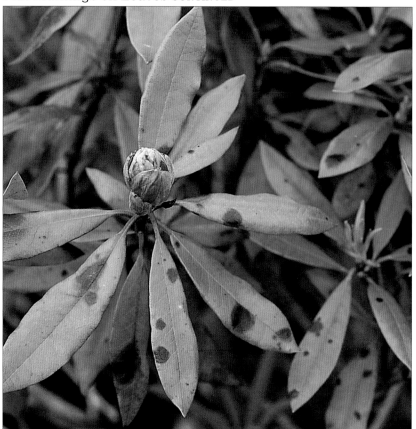

Rhododendron Leaf Spot
DISEASE • *See p.176*

Brownish-purple spots on leaves often have a pink or purple ring around the edge of each one. Concentric ringing is also seen, with pycnidia in the centre.

PLANTS AFFECTED Rhododendrons
SEASON All year (mainly late summer onwards)

Arbutus Leaf Spot
DISEASE • *See p.111*

Small spots appear, each with a dry, pale grey centre and a purple edge. These may enlarge but will rarely join together. Leaf fall may be premature.

PLANTS AFFECTED Strawberry trees (*Arbutus*)
SEASON Mainly midsummer to autumn

Pea Leaf and Pod Spot
DISEASE • *See p.163*

Yellow or brown, often sunken, spots are found on the leaves and pods. There is a general yellowing of adjacent tissues. Pycnidia are usually apparent.

PLANTS AFFECTED Peas
SEASON All season (mainly summer)

Brassica Dark Leaf Spot
DISEASE • *See p.119*

Grey-brown, round spots are seen on the leaf surface, often with concentric rings of pycnidia on them. Affected areas may drop away, leaving a ragged leaf.

PLANTS AFFECTED Brussels sprouts, cauliflowers
SEASON Mainly midsummer to late autumn

Leaf Spots (Bacterial)
DISEASE • *See p.151*

Circular or angular spots of dead tissue are found on the leaves (here of hibiscus), usually surrounded by a bright yellow "halo". No fungal bodies present.

PLANTS AFFECTED Many plants
SEASON Varies with host plant

Pansy Leaf Spot
DISEASE • *See p.162*

Purple-brown or buff-coloured spots appear on leaves and may join together. The affected areas may fall away, leaving holes. Older plants are usually the most badly affected.

PLANTS AFFECTED Pansies
SEASON Mainly midsummer to late autumn

Raised Markings on Leaves

Chrysanthemum White Rust
DISEASE • *See p.127*

Raised, warty, buff-coloured pustules appear beneath the leaf surface, corresponding with yellow, sunken areas above. Leaves may die back.

PLANTS AFFECTED Chrysanthemums
SEASON All season (mainly summer and autumn)

White Blister
DISEASE • *See p.196*

Shiny, white, raised spots are seen on the lower leaf surface of brassicas (here on a cabbage), often arranged in concentric rings. The upper leaf surface is yellowed and usually distorted.

PLANTS AFFECTED Edible brassicas
SEASON Mainly early summer to late autumn

Lime Nail Gall Mite
PEST • *See p.154*

Red or yellowish-green, tubular structures containing microscopic mites grow on the upper leaf surfaces. Heavily infested leaves may be distorted.

PLANTS AFFECTED Lime trees (*Tilia*)
SEASON Late spring to autumn

Viburnum Whitefly
PEST • *See p.193*

Black, oval pupae encrusted with white wax are present on the lower leaf surface for much of the year. White-winged adults are found during the summer.

PLANTS AFFECTED *Viburnum tinus*; also *Arbutus*
SEASON All year

Cushion Scale
PEST • *See p.130*

Honeydew and sooty mould develop on the upper leaf surfaces. Oval, yellow or brown scales live on the underside (here of a camellia leaf) and in early summer white egg bands are deposited.

PLANTS AFFECTED Camellias, rhododendrons, hollies
SEASON All year (mainly winter to spring)

Periwinkle Rust
DISEASE • *See p.166*

Numerous small, dark brown, spore-filled pustules are found on the underside of the leaf. Leaves become pitted and distorted. The whole plant may become distorted and die back.

PLANTS AFFECTED Periwinkles (*Vinca*)
SEASON All season (mainly summer and autumn)

Iris Rust
DISEASE • *See p.148*

Numerous raised, orange pustules are found on leaves. Leaves yellow and may die back prematurely. Older leaves are usually the worst affected. The plant's vigour is rarely reduced.

PLANTS AFFECTED Irises
SEASON Mainly late summer to autumn

Antirrhinum Rust
DISEASE • *See p.107*

Concentric rings of tiny, dark brown, spore-filled pustules are seen on lower leaf surfaces. A yellow "halo" appears around the edge of each spot.

PLANTS AFFECTED Snapdragons (*Antirrhinum*)
SEASON All season (mainly spring and summer)

Broad Bean and Pea Rust
DISEASE • *See p.120*

Discrete, dark brown patches of spores appear on leaves, often with a yellowish halo. Plants may defoliate and cropping may be reduced.

PLANTS AFFECTED Broad beans, peas, and related crops
SEASON Early summer to autumn

Raised Markings on Leaves *CONTINUED*

Pelargonium Rust
DISEASE • *See p.165*

Dark brown, fungal pustules are seen on the lower leaf surfaces, usually arranged in concentric rings. Yellowish discoloration develops on the upper surfaces. Leaves die.

PLANTS AFFECTED Zonal pelargoniums
SEASON All season

Hydrangea Scale
PEST • *See p.147*

Oval, white, waxy egg masses appear on the leaves and stems. Flat, oval, yellowish-brown scale insects are also present, and live on the underside of the leaves next to the leaf veins.

PLANTS AFFECTED Mainly hydrangeas
SEASON All year; eggs late spring to midsummer

Aquilegia Downy Mildew
DISEASE • *See p.110*

Angular, yellowish patches develop on leaves. Infected foliage then curls and dies. A purplish-white fungal growth may appear on the lower leaf surface. Stems and flowers may also be affected.

PLANTS AFFECTED Columbines *(Aquilegia)*
SEASON Mid-spring to mid-autumn

Vine Erinose Mite
PEST • *See p.193*

Parts of the leaf bulge upwards; the underside of affected areas is covered in creamy-white hairs which turn brown. Microscopic mites live among the hairs.

PLANTS AFFECTED Grape vines *(Vitis)*
SEASON Late spring to autumn

Rhododendron Powdery Mildew
DISEASE • *See p.177*

Pale brown to buff, felty patches of fungal growth develop under leaves, with yellow blotches above. Leaves fall early and the plant may be weakened.

PLANTS AFFECTED Rhododendrons
SEASON All year on evergreen rhododendrons

Apple Scab
DISEASE • *See p.109*

Greyish-green or khaki, scabby spots develop on the leaf surface. Slight blistering or puckering may also occur. Leaves yellow and fall prematurely and the tree is weakened.

PLANTS AFFECTED Edible and ornamental *Malus*
SEASON Mainly late spring and summer

Raised Markings on Leaves *CONTINUED*

Pear Leaf Blister Mite
PEST • *See p.164*

The foliage develops many yellowish-green or pink, slightly raised patches, which later turn black. These patches often form a band on each side of the midrib.

PLANTS AFFECTED Pear trees
SEASON Mid-spring to autumn

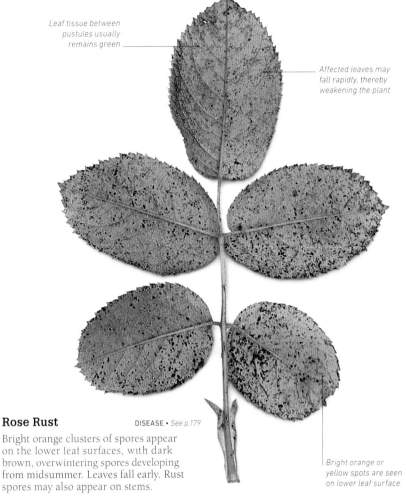

Leaf tissue between pustules usually remains green

Affected leaves may fall rapidly, thereby weakening the plant

Bright orange or yellow spots are seen on lower leaf surface

Willow Bean Gall Sawflies
PEST • *See p.197*

Coffee bean-sized swellings, which are red or yellowish-green, form on willow leaves. Each gall contains a small, caterpillar-like larva.

PLANTS AFFECTED Willows (*Salix*)
SEASON Early summer to early autumn

Rose Rust
DISEASE • *See p.179*

Bright orange clusters of spores appear on the lower leaf surfaces, with dark brown, overwintering spores developing from midsummer. Leaves fall early. Rust spores may also appear on stems.

PLANTS AFFECTED Roses
SEASON Mainly summer and autumn

Hemispherical Scale
PEST • *See p.144*

Convex, dark brown, hemispherical scale insects infest leaves (here of a fern) and stems. They are often accompanied by sticky honeydew and sooty mould.

PLANTS AFFECTED Many greenhouse plants
SEASON All year

Plum Leaf Gall Mite
PEST • *See p.169*

Whitish-green structures containing microscopic mites develop on the upper leaf surface, especially around the leaf margins.

PLANTS AFFECTED Wild and cultivated plums
SEASON Late spring to autumn

Elm Gall Mite
PEST • *See p.134*

Hard, pale green swellings containing microscopic mites develop on the upper surface of elm leaves. Almost every leaf may be affected, but the tree's vigour is not usually reduced.

PLANTS AFFECTED Elms (*Ulmus*)
SEASON Mid-spring to autumn

Raised Markings on Leaves *CONTINUED*

Leaf has turned brown and dropped

Patches of spores cover the underside

Leek Rust
DISEASE • *See p.152*

Bright orange, oval pustules are found on the leaves, the outer leaves being the worst affected. The leaves may yellow and die off, usually becoming papery and brown.

PLANTS AFFECTED Leeks, chives,garlic onions, shallots
SEASON Mainly midsummer to late autumn

Hollyhock Rust
DISEASE • *See p.144*

Orange, raised pustules appear on the lower leaf surface (here of a mallow), later turning buff-brown. The upper leaf surface has corresponding yellow or orange spots. Leaves may die.

PLANTS AFFECTED Hollyhocks and mallows
SEASON All season

Heuchera Rust
DISEASE • *See p.144*

Discrete raised areas or depressions develop on upper leaf surfaces with patches of dark brown or bright orange fungal spores beneath. Leaves may become distorted, turn brown, and die.

PLANTS AFFECTED *Heuchera*
SEASON Summer

Acer Gall Mites
PEST • *See p.106*

Numerous small, round, reddish or yellowish-green structures (galls) form on the upper leaf surface, or felty hairs develop on the undersurface. Heavily infested leaves may become distorted.

PLANTS AFFECTED Maples and sycamores (*Acer*)
SEASON Late spring to autumn

Oak Gall Wasp (Spangle Galls)
PEST • *See p.160*

Flat, circular discs or galls containing tiny grubs form on the lower leaf surfaces. They are reddish at first, but yellowish-brown when mature.

PLANTS AFFECTED Deciduous oaks (*Quercus*)
SEASON Late summer to mid-autumn

Oak Gall Wasp (Silk Button Galls)
PEST • *See p.160*

Golden-brown discs with central depressions develop on the underside of oak leaves. Each disc contains a single grub, which will become a gall wasp.

PLANTS AFFECTED Deciduous oaks (*Quercus*)
SEASON Late summer to autumn

Blotchy Markings on Leaves

Beet Leaf Miner
PEST • *See p.115*

Pale green blotches later dry up and become brown. White maggots live within the mined areas and later turn into brown pupae, from which adult flies will emerge.

PLANTS AFFECTED Beetroot, chard, and spinach beet
SEASON Late spring to late summer

Strawberry Leaf Spot
DISEASE • *See p.187*

Early spotting (*see p.24*) develops into large blotches, each with a grey centre. These may appear slightly raised. Off-white fungal growth may develop.

PLANTS AFFECTED Strawberries
SEASON Mainly midsummer to autumn

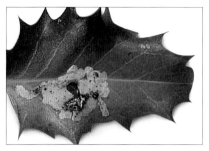

Holly Leaf Miner
PEST • *See p.144*

Irregular, yellowish-purple blotches occur on the upper leaf surfaces. There may also be linear tunnels in the leaf coming from the central blotch.

PLANTS AFFECTED Hollies (*Ilex*)
SEASON All year

Willow Leaf Beetles
PEST • *See p.197*

Whitish or brown dried-up areas form where the adult beetles or their larvae have grazed away part of the leaves. The larvae often feed in groups of up to 20.

PLANTS AFFECTED Willows, poplars, and aspens
SEASON Early summer to early autumn

Yucca Leaf Spot
DISEASE • *See p.199*

Circular, brown or grey-brown spots appear on leaf surfaces with concentric rings and central pycnidia. The spots may join together.

PLANTS AFFECTED Yuccas
SEASON All year (mainly midsummer onwards)

Shothole
DISEASE • *See p.182*

Spots or blotches of discoloured tissues develop. The damaged areas then fall out, leaving holes (*see p.45*). No insects are present, but fungal growth is sometimes visible.

PLANTS AFFECTED Mainly *Prunus* spp.
SEASON Mainly mid- to late summer

Blotchy Markings on Leaves *CONTINUED*

Astrantia Leaf Miner
PEST • *See p.112*

Foliage becomes discoloured, with brownish-white areas where the larval stage has been feeding inside leaves. White spots caused by adult feeding also occur.

PLANTS AFFECTED *Astrantia*
SEASON Spring and late summer

Delphinium Leaf Miner
PEST • *See p.131*

Brown, dried-up areas extend from the leaf tips to the leaf stalk. Small, white grubs live within the mined areas. Adult flies cause pale spots on the leaves.

PLANTS AFFECTED Delphiniums and monkshoods
SEASON Early to late summer

Laburnum Leaf-mining Moth
PEST • *See p.149*

Roughly spherical, whitish-brown leaf mines are caused by small caterpillars. In heavy infestations the mines join together to form irregular blotches.

PLANTS AFFECTED Laburnums
SEASON Early summer to early autumn

Aquilegia Leaf Miner
PEST • *See p.110*

Two species of flies have larvae that feed inside the foliage of aquilegias: one makes large blotch mines, the other winding linear mines. Both show as whitish discoloured areas.

PLANTS AFFECTED Columbine (*Aquilegia*)
SEASON Late spring-summer

Apple Leaf Miner
PEST • *See p.109*

Small caterpillars make white or brown lines where they have bored through the leaves. Pupation occurs in white silk cocoons spun on the outside of the leaf.

PLANTS AFFECTED Apples and cherries
SEASON Early to late summer

Fuchsia Rust
DISEASE • *See p.138*

Numerous tiny, pale orange pustules are found on the lower leaf surface, with the upper surface showing yellow-orange or purple discoloration. Leaves may wither and fall prematurely.

PLANTS AFFECTED Fuchsias and willowherbs
SEASON Mainly early summer to autumn

Rose Slugworm
PEST • *See p.180*

Pale green caterpillar-like larvae with brown heads graze the leaves, causing white or brown dried-up areas. They usually feed on the lower leaf surface, but may occur on the upper surface in shaded areas.

PLANTS AFFECTED Roses
SEASON Early summer to early autumn

Pear and Cherry Slugworm
PEST • *See p.164*

Small larvae covered in black slime graze the leaf surface, causing it to dry up and turn brown. The larvae often feed on the upper leaf surface.

PLANTS AFFECTED Cherries, pears, plums, hawthorns
SEASON Early summer to mid-autumn

Lilac Leaf Miner
PEST • *See p.153*

Irregular, brown blotch mines develop where whitish-green caterpillars have fed inside the foliage. The caterpillars then roll up the leaf tips with silk threads and complete their feeding there.

PLANTS AFFECTED Lilacs, privet, and ash
SEASON Early summer to early autumn

Blotchy Markings on Leaves *CONTINUED*

Galls may form between the leaf veins or along them

Spots are usually creamy-white, but on purple-tinged leaves they are often red or pink

Felt galls disfigure but cause little real damage to plants

Felt Gall Mites
PEST • *See p.135*

Microscopic mites cause the abnormal growth of dense patches of hairs (here on beech), usually on the underside of the foliage. In late summer the hairs dry up and become brown.

PLANTS AFFECTED Some deciduous trees
SEASON Late spring to autumn

Raspberry Viruses
DISEASE • *See p.175*

Yellow-green flecks, mosaics, mottles, or ring spots are seen on the leaves. The leaves may become small and distorted and the plant will have a reduced crop. The plant's overall growth is very poor.

PLANTS AFFECTED Cane fruits
SEASON Spring to autumn

Raspberry Leaf and Bud Mite
PEST • *See p.175*

Pale yellow blotches form on the upper leaf surface, with reduced hairiness of the lower leaf surfaces. Leaves at the shoot tips are often distorted.

PLANTS AFFECTED Raspberries
SEASON Late spring to early autumn

Rose Powdery Mildew
DISEASE • *See p.179*

White, powdery fungal growth is found on the leaves. They may become distorted, generally yellowed, and fall early. Stems and buds can also be affected.

PLANTS AFFECTED Roses
SEASON All season (mainly summer to autumn)

Oedema
DISORDER • *See p.160*

Raised, pale green and, later, brown spots are found on the lower leaf surface (here of a pelargonium). The upper leaf surface may show corresponding yellow spotting.

PLANTS AFFECTED Many plants
SEASON All year

Celery Leaf Miner
PEST • *See p.125*

Brown, dried-up patches develop on the leaves. White maggots live within the mined areas and later become brown pupae.

PLANTS AFFECTED Celery, parsnip, parsley, lovage
SEASON Early summer to early autumn

Blotchy Markings on Leaves *CONTINUED*

Camellia Leaf Blight
DISEASE • *See p.123*

Brownish blotches develop on leaves, which fall early; this may be followed by twig dieback. Black pycnidia may be seen and may exude black spore tendrils.

PLANTS AFFECTED Camellias
SEASON All year

Hellebore Black Death
DISEASE • *See p.143*

Black streaking and mottling of the tissues along or between the leaf veins. Leaves and stems become stunted and distorted. Flowers may also be affected.

PLANTS AFFECTED Hellebores
SEASON All year

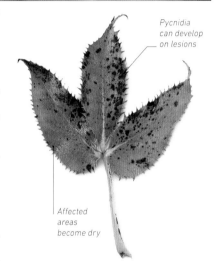

Pycnidia can develop on lesions

Affected areas become dry

Hellebore Leaf Blotch
DISEASE • *See p.143*

Slate-grey or grey-brown blotches are seen on the leaves. The lesions may eventually join together. Leaf and flower stems and flowers may also be affected.

PLANTS AFFECTED Hellebores
SEASON Mainly early summer to autumn

Horse Chestnut Leaf Blotch
DISEASE • *See p.146*

Tan-brown, irregular blotches appear, each with a bright yellow margin, on the leaves. The leaf margins and tips are most severely affected.

PLANTS AFFECTED Horse chestnuts (*Aesculus*)
SEASON Midsummer to autumn

Mottled Markings on Leaves

Glasshouse Leafhopper
PEST • *See p.141*

Coarse, pale spots are found on the upper leaf surface, caused by small, active insects feeding from the lower leaf surface.

PLANTS AFFECTED Greenhouse and garden plants
SEASON All year

Fruit Tree Red Spider Mite
PEST • *See p.138*

The foliage of apple and plum trees becomes increasingly dull and yellowish, usually with fine, pale mottling, and may fall prematurely.

PLANTS AFFECTED Apples and plums
SEASON Late spring to early autumn

Carnation Tortrix Moth
PEST • *See p.123*

Pale green caterpillars with brown heads bind leaves together with silk threads and graze away the inner surfaces. The leaves will then dry up and turn brown.

PLANTS AFFECTED Greenhouse and garden plants
SEASON All year

Mottled Markings on Leaves *CONTINUED*

Pelargonium Viruses
DISEASE • *See p.166*

Yellow flecks, spots, mottling, or sometimes ring spots are seen on the leaves. Leaves may be small and distorted, and overall growth is poor.

PLANTS AFFECTED Pelargoniums
SEASON All season

Euonymus Scale
PEST • *See p.135*

Pale yellow mottling and elongate white scale insects are found on the foliage. Stems become encrusted with the dark brown, pear-shaped female scales. Heavy infestations may cause dieback.

PLANTS AFFECTED Evergreen spindles (*Euonymus*)
SEASON All year

Rose Leafhopper
PEST • *See p.179*

A coarse, pale mottling develops on the upper leaf surface. Pale yellow insects live on the underside of the leaves. Roses growing in warm, sheltered places can become heavily infested by summer.

PLANTS AFFECTED Roses
SEASON Mid-spring to mid-autumn

Rhododendron Lacebug
PEST • *See p.176*

A coarse, pale yellow mottling develops on the upper leaf surface. Insects with mottled wings feed on the lower leaf surface and leave rusty brown deposits.

PLANTS AFFECTED Rhododendrons
SEASON Early spring to late summer

Rose Viruses
DISEASE • *See p.180*

Yellow flecks, spots, mottling, vein-banding, or ring spots are found on the leaves. Flowers may show slight colour-breaking symptoms and growth may be affected, but these are both rare.

PLANTS AFFECTED Roses
SEASON All season

Cucumber Mosaic Virus
DISEASE • *See p.129*

Yellow mottling, mosaic patterning, and blotches are seen on leaves (here of a marrow), which may be distorted (*see p.37*). The whole plant is stunted and dies early.

PLANTS AFFECTED Wide host range (not just cucurbits)
SEASON All season

Mottled Markings on Leaves *CONTINUED*

Beech Red Spider Mite
PEST • *See p.114*

Foliage develops a fine yellowish mottling, with tiny mites present on the underside of leaves, especially along the veins. Beech hedges are more susceptible than trees.

PLANTS AFFECTED Beech (*Fagus*)
SEASON Late summer to autumn

Glasshouse Red Spider Mite
PEST • *See p.141*

Early stages of spider mite attack show as a fine, pale mottling of the upper leaf surface. Tiny, yellow-green mites are found on the underside of the leaves.

PLANTS AFFECTED Greenhouse and garden plants
SEASON Mid-spring to late autumn

Plum Rust
DISEASE • *See p.169*

Numerous dark brown, spore-filled pustules are found on the underside of the leaf. Corresponding bright yellow or orange spots are seen on the upper surface. Leaves fall prematurely.

PLANTS AFFECTED Plums and anemones
SEASON Mainly late summer and autumn

Banded Palm Thrips
PEST • *See p.113*

Black, white-banded, narrow-bodied, elongate insects cause a silvery-brown discoloration of the upper leaf surfaces of the younger foliage.

PLANTS AFFECTED Some house plants
SEASON All year

Camellia Yellow Mottle
DISEASE • *See p.123*

Yellow or white mottled markings and patches of discoloration form on leaves. General growth is not affected. Usually only isolated branches are affected.

PLANTS AFFECTED Camellias
SEASON All year

Tulip Fire
DISEASE • *See p.190*

Buff-coloured or bleached flecks develop on the leaves. Early infection causes extensive leaf distortion and the development of fuzzy, grey fungal growth. Flowers may also be affected.

PLANTS AFFECTED Tulips
SEASON Mid- to late spring

Privet Thrips
PEST • *See p.173*

Elongate, narrow-bodied insects cause green leaves (*above right*) to develop a silvery-brown discoloration (*above left*), especially during late summer. The thrips are found on the upper leaf surfaces.

PLANTS AFFECTED Privet and lilac
SEASON Early summer to early autumn

Discoloured Leaves

Magnesium Deficiency
DISORDER • *See p.154*

Yellow, brown, or purple discoloration between the leaf veins and around the leaf edge (here on a tomato). Older leaves are the worst affected.

PLANTS AFFECTED Many plants
SEASON Mainly late spring to autumn

Lettuce Downy Mildew
DISEASE • *See p.153*

Angular, yellow discoloured areas on the leaves rapidly turn pale brown and papery or soft. Off-white fungal growth may develop beneath the leaf.

PLANTS AFFECTED Lettuces
SEASON All season (mainly summer and autumn)

Cactus Corky Scab
DISORDER • *See p.122*

Pale, roughened areas appear on the surface and may become brown and sunken. In extreme cases holes develop as the affected tissue dies.

PLANTS AFFECTED Cacti, succulents
SEASON All year

Geranium Downy Mildew
DISEASE • *See p.153*

Pale green or light brown, angular patches of discoloration are seen on the leaves. Off-white, slightly fuzzy fungal patches develop beneath.

PLANTS AFFECTED *Geranium* species and cultivars
SEASON All season

Cucumber Mosaic Virus
DISEASE • *See p.129*

Leaves may be distorted and reduced in size, and also discoloured, with blotches and mosaicing (*see p.35*). On some hosts, flowers show colour-breaking.

PLANTS AFFECTED Wide host range
SEASON All season

Chrysanthemum Leaf Miner
PEST • *See p.126*

White or brown sinuous lines develop where the leaf-mining grubs have been feeding. In heavy attacks most of the leaf area becomes discoloured.

PLANTS AFFECTED Chrysanthemum, cineraria
SEASON Mid-spring to late autumn

Powdery Mildew
DISEASE • *See p.173*

White, powdery fungal growth develops on the upper leaf surface (here of a geranium). The leaf can become distorted and the leaf stalk may also be affected.

PLANTS AFFECTED Wide range of plants
SEASON Varies with host plant

Leaf and Bud Eelworms
PEST • *See p.151*

Brown areas enclosed by the larger leaf veins form islands or wedges in the leaves. Eventually the whole leaf will become discoloured.

PLANTS AFFECTED Many plants
SEASON Mainly late summer to late autumn

Carrot Motley Dwarf Virus
DISEASE • *See p.124*

Foliage turns yellow and pink and may become twisted and stunted. Leaf stalks are also twisted. Foliage dies off prematurely and the crop is reduced.

PLANTS AFFECTED Carrots and parsley
SEASON Mainly summer

Discoloured Leaves *CONTINUED*

Viruses
DISEASE • *See p.194*

Virus symptoms include yellow vein-clearing (here on a sweet pea), mottling, mosaicing, spotting, or ring spotting seen on the leaves. Leaves may be distorted (*see p.40*) and general growth is stunted.

PLANTS AFFECTED Wide host range
SEASON Mainly late spring to late summer

Begonia Powdery Mildew
DISEASE • *See p.115*

White or pale grey, powdery fungal growth is found mainly on upper leaf surfaces. Affected leaves turn papery or soft. May affect stems and buds.

PLANTS AFFECTED Begonias
SEASON All year

Horse Chestnut Bleeding Canker
DISEASE • *See p.146*

Symptoms include yellowing foliage, premature leaf drop, and eventually crown death. Leaves may be smaller than those on healthy trees.

PLANTS AFFECTED Horse chestnuts (*Aesculus*)
SEASON Mainly spring to autumn

Narcissus Viruses
DISEASE • *See p.159*

Yellow streaking or flecking is seen on leaves. Plants can become stunted and occasionally symptoms are apparent on flower stems or petals. Growth and performance may be unaffected.

PLANTS AFFECTED Daffodils, narcissi
SEASON Mainly early to late spring

Glasshouse Red Spider Mite
PEST • *See p.141*

If heavily infested, leaves lose much of their green colour and dry up. Fine silk webbing may appear between the leaves and the mites swarm over the plant.

PLANTS AFFECTED Greenhouse and garden plants
SEASON Mid-spring to late autumn

Weedkiller Damage – Contact
DISORDER • *See p.196*

Yellow spots and blotches appear (here on a rose), which may turn brown as the leaf tissue dies. Plant growth can be stunted and the plant may die.

PLANTS AFFECTED Any plant
SEASON Mainly spring to summer

Discoloured Leaves *CONTINUED*

Iron and Manganese Deficiency
DISORDER
See p.148

Yellowing between the veins on leaves (here of a rhododendron) is seen. The youngest leaves are affected most severely. Overall growth may be affected.

PLANTS AFFECTED Many lime-hating plants
SEASON Varies with host plant

Oak Powdery Mildew
DISEASE • *See p.160*

Powdery, white fungal growth appears on the leaf surface. Leaves, particularly young ones, may be distorted and die off prematurely.

PLANTS AFFECTED Oaks (*Quercus*)
SEASON Spring to autumn

Cherry Blackfly
PEST • *See p.125*

Leaves at the shoot tips become curled, with black aphids on the underside. Affected leaves often turn brown in midsummer.

PLANTS AFFECTED Cherry trees
SEASON Late spring to late summer

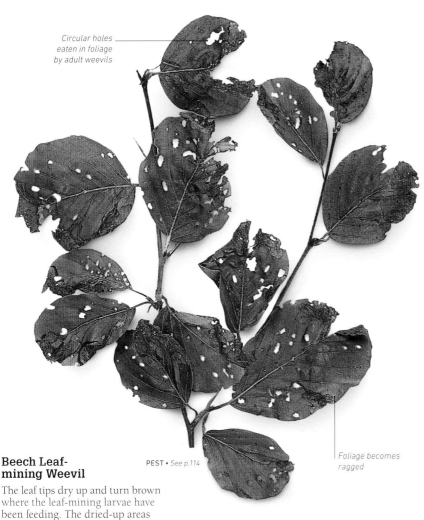

Circular holes eaten in foliage by adult weevils

Foliage becomes ragged

Beech Leaf-mining Weevil
PEST • *See p.114*

The leaf tips dry up and turn brown where the leaf-mining larvae have been feeding. The dried-up areas will then break off.

PLANTS AFFECTED Beeches (*Fagus*)
SEASON Mid-spring to late summer

Potassium Deficiency
DISORDER • *See p.170*

Leaf tips and edges are scorched and may show brownish-purple spotting on the underside (here on a pelargonium). Flowering and fruiting will be poor.

PLANTS AFFECTED Many plants
SEASON Mainly midsummer to autumn

Mahonia Rust
DISEASE • *See p.155*

Bright, discoloured spots appear on the upper leaf surface, with pustules on the underside (*see p.25*). Leaves may fall prematurely, but the overall health of the plant is rarely affected.

PLANTS AFFECTED Mahonias
SEASON Late spring to early autumn

Distorted Leaves

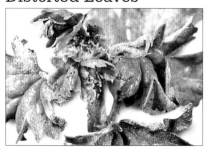

American Gooseberry Mildew
DISEASE • *See p.107*

Powdery fungal growth is seen on puckered leaves, usually on the upper surfaces. Young leaves wither and die. Stems and fruits may be infected.

PLANTS AFFECTED Gooseberries and blackcurrants
SEASON All season

Apple Powdery Mildew
DISEASE • *See p.109*

Powdery, white fungal growth develops on leaves. On young leaves the upper and lower surfaces may be attacked, and the foliage withers and dies.

PLANTS AFFECTED Edible and ornamental apples
SEASON Spring to autumn

Virus
DISEASE • *See p.194*

Growth may be distorted and stunted (as on this tobacco plant) and the plant usually dies off early. Other symptoms include yellow mottling, spotting, or mosaicing on the leaves.

PLANTS AFFECTED Most plants
SEASON Mainly summer and autumn

Blackcurrant Gall Midge
PEST • *See p.116*

Leaves fail to expand, remaining tightly crumpled and later drying up. Small, white maggots may be found between the folds of infested leaves.

PLANTS AFFECTED Blackcurrant
SEASON Late spring to late summer

Box Sucker
PEST • *See p.118*

Shoot extension is severely restricted and the cupped leaves form cabbage-like structures. Flattened, pale green, aphid-like insects may be present and excrete a white, runny liquid.

PLANTS AFFECTED Box (*Buxus*)
SEASON Mid-spring to midsummer

Bay Sucker
PEST • *See p.113*

Leaves at the shoot tips become yellow, with thickened, down-curled margins. Later, the damaged areas dry up and turn brown. Fluffy, white sucker nymphs may be seen on the lower leaf surface.

PLANTS AFFECTED Bay (*Laurus nobilis*)
SEASON Late spring to mid-autumn

Irregular Watering
DISORDER • *See p.148*

Leaves (here of a rhododendron) become puckered and distorted, but retain their normal colour. Not all leaves will be affected.

PLANTS AFFECTED Many plants
SEASON All season (mainly spring and summer)

Peach Leaf Curl
DISEASE • *See p.164*

Leaves become puckered; they are pale green at first, but soon turn red and purple. A powdery, white spore layer develops on the surface.

PLANTS AFFECTED Peaches, nectarines, almonds
SEASON Spring and early summer

Distorted Leaves *CONTINUED*

Texture is firm but compressible and becomes softer with age

Withered remains of leaf still attached to gall

Gall is rounded, but may also be forked with finger-like protrusions

Camellia Gall   DISEASE • *See p.122*

Large pale green galls, up to 18cm long, develop in place of leaves. The surface becomes covered in spores and so may appear white. The plant's overall vigour is rarely affected.

PLANTS AFFECTED Camellias
SEASON Late summer to autumn

Lupin Anthracnose   DISEASE • *See p.154*

Brown patches develop on stems and leaves; these may be covered with orange spores. The stems may develop a distinct corkscrew-like distortion. Plants may collapse.

PLANTS AFFECTED Lupins
SEASON Mid-spring to late summer

Violet Gall Midge   PEST • *See p.193*

Infested leaves have greatly swollen and curled leaf margins. Pale orange maggots or white silk cocoons can be found inside. Galled leaves stay on the plant over the winter.

PLANTS AFFECTED Violets
SEASON All year

Plum Leaf-curling Aphid   PEST • *See p.169*

In the spring, leaves become tightly curled and crinkled, due to greenfly feeding on the underside of the leaves. Leaves remain distorted all summer.

PLANTS AFFECTED Plums, damsons, and gages
SEASON Early to late spring

Currant Blister Aphid   PEST • *See p.129*

Leaves at the shoot tips develop raised, red or yellowish-green, puckered areas. Pale yellow aphids may be found on the underside of affected leaves.

PLANTS AFFECTED Red, white, and blackcurrants
SEASON Mid-spring to early summer

Rose Leaf-rolling Sawfly   PEST • *See p.178*

The margins of affected leaflets are rolled downwards to form slender tubes in which small, pale green larvae feed. Rolled leaflets often hang down.

PLANTS AFFECTED Cultivated and wild roses
SEASON Late spring to midsummer

Distorted Leaves *CONTINUED*

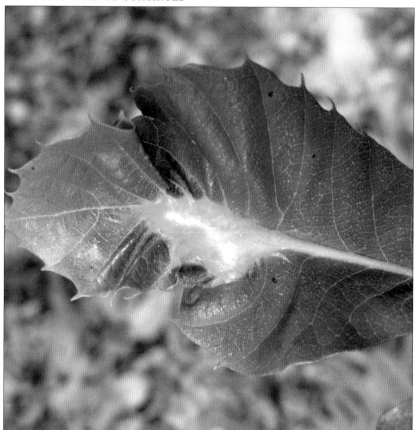

Oriental (Sweet) Chestnut Gall Wasp
PEST • *See p.162*

Hard, pale green swellings containing gall wasp larvae develop in the foliage, petioles, and buds. Affected leaves are often undersized and distorted.

PLANTS AFFECTED Sweet chestnut
SEASON Develop late spring, visible to leaf fall

Viruses
DISEASE • *See p.194*

The plant (here a *Passiflora*) may be distorted and stunted and will usually die off early. Leaves and entire plant often also display yellow mottling, spotting, mosaicing, or veining.

PLANTS AFFECTED Many plants
SEASON Mainly summer and autumn

Weedkiller Damage – Broadleaved
DISORDER
See p.196

Leaves (here of a pelargonium) become cup-shaped, often with the veins close together. Stems and petioles may be curled or distorted (*see below*).

PLANTS AFFECTED Any plant
SEASON Mainly spring and summer

Tarsonemid Mites
PEST • *See p.188*

Microscopic mites live and feed at shoot tips and in flower buds. Their feeding causes scarring on the stems and progressive distortion and stunting of the foliage. Growing tips may be killed.

PLANTS AFFECTED Many greenhouse plants
SEASON All year

Gleditsia Gall Midge
PEST • *See p.141*

Leaves at the shoot tip fail to expand and resemble small seed pods. Each galled leaflet contains several whitish-orange maggots.

PLANTS AFFECTED Honey locust (*Gleditsia*)
SEASON Early to late summer

Weedkiller Damage – Broadleaved
DISORDER
See p.196

Stems and petioles (here of a tomato plant) may become curled and distorted. Leaves narrow or curl up (*see above*) and stems may be roughened.

PLANTS AFFECTED Most plants
SEASON Mainly summer and autumn

Leaves Dying Off

Thuja Blight
DISEASE • See p.189

Tiny, black, slightly raised spots are embedded in the foliage. These are fungal fruiting bodies that leave pits or holes as the spores are released. Foliage turns brown and dies back.

PLANTS AFFECTED *Thuja plicata*
SEASON Mainly late summer and early autumn

Frost Damage
DISORDER • See p.138

The foliage (here of a hydrangea) becomes withered and hangs limply. It may become blackened or turn pale green or brown. The exposed parts of the plant are most severely affected.

PLANTS AFFECTED Many plants
SEASON Mainly winter and spring

Potato and Tomato Blight
DISEASE • See pp.170, 189

Brown patches develop on leaves, mainly around the tips and edges. Fluffy white growth can develop. Leaves eventually wither and die. Top growth may collapse.

PLANTS AFFECTED Potatoes and tomatoes
SEASON Mid- to late summer

Drought
DISORDER • See p.133

Leaves (here of a hosta) wilt and later dry out or shrivel, often turning brown and papery. In the early stages, lowered temperatures and watering may revive the plant; later, dieback or death occurs.

PLANTS AFFECTED Many plants
SEASON Mainly summer

Pea Wilt
DISEASE • See p.164

Leaves turn yellow and die back. Internally the stem's vascular system is stained brown, but this is only visible if the outer stem is stripped away. The whole plant wilts and is killed rapidly.

PLANTS AFFECTED Peas and sweet peas
SEASON Mainly summer

Lettuce Grey Mould
DISEASE • See p.153

Fluffy, grey fungal growth appears on the leaves. This is often combined with brown or orange slimy rotting of the stem and base of the plant.

PLANTS AFFECTED Lettuces
SEASON Mainly spring to autumn

Leaves Dying Off *CONTINUED*

Sirococcus Blight
DISEASE • *See p.183*

Shoot tips die back and needles discolour pink, turn brown, and drop. Cankers appear on stems, often oozing resin, and these may girdle stems, causing dieback.

PLANTS AFFECTED Species of *Cedrus* and *Tsuga*
SEASON All year

Narcissus Leaf Scorch
DISEASE • *See p.158*

Leaves are red-brown and scorched as they emerge. Brown spots with pycnidia are found on the leaves. Bulbs do not rot or show discoloration.

PLANTS AFFECTED Narcissi, amaryllis, crinums
SEASON Spring

Verticillium Wilt
DISEASE • *See p.192*

Foliage wilts and dies and soft stems wilt. Usually only part of the plant is affected initially. The leaf veins and vascular tissue are discoloured.

PLANTS AFFECTED Many plants
SEASON All year

Leaf Spot (Bacterial)
DISEASE • *See p.151*

Spots develop on the foliage (here of a *Calathea*), often with a water-soaked appearance. A distinct yellow "halo" may develop around each spot (*see p.26*).

PLANTS AFFECTED Depends on bacterium involved
SEASON Varies with host plant

Phyllosticta Leaf Spot and Dieback
DISEASE • *See p.167*

Leaves develop brown blotches or spots, and on conifers needles may turn brown. Foliage falls prematurely and there may be associated dieback of stems or twigs.

PLANTS AFFECTED Holly, rhododendron, ivy, yew
SEASON All year

Impatiens Downy Mildew
DISEASE • *See p.147*

Leaves yellow and are quickly shed. A pale, fungal-like growth may be seen beneath the leaves. The flowers may be similarly affected. Plants are killed rapidly.

PLANTS AFFECTED Busy Lizzie (*Impatiens walleriana*)
SEASON Spring to summer

Kerria Blight
DISEASE • *See p.149*

Leaves develop reddish-brown spots with purple borders and, in damp weather, white spores in the centres. Leaves eventually turn brown and drop. Brownish-purple cankers may girdle stems and cause dieback.

PLANTS AFFECTED *Kerria* spp.
SEASON Stems all year; leaves late spring to summer

Holes in Leaves

Iris Sawfly
PEST • *See p.148*

Brownish-grey, caterpillar-like larvae eat V-shaped notches from the leaf margins. Older larvae cause extensive defoliation, especially of the upper parts of the leaves.

PLANTS AFFECTED Waterside irises only
SEASON Early to midsummer

Slugs and Snails
PESTS • *See pp.183, 184*

These common pests feed mainly at night and eat irregular holes in leaves or graze away the surface tissues. A silvery slime trail is sometimes left on plants. Snails are dormant in winter, but slugs remain active.

PLANTS AFFECTED Many plants
SEASON All year

Alder Leaf Beetle
PEST • *See p.106*

Dark blue beetles, 6–7mm long, and their black larvae, up to 11mm long, make holes on alder leaves and feeding can lead to dried up patches. Heavy infestations can cause extensive defoliation.

PLANTS AFFECTED Alder
SEASON Late spring to midsummer

Water Lily Beetle
PEST • *See p.195*

Both the adult beetles and their larvae feed on the upper leaf surface, grazing elongate strips which rot away to form holes. Damaged leaves turn yellow and decay prematurely.

PLANTS AFFECTED Water lilies
SEASON Late spring to late summer

China Mark Moth
PEST • *See p.126*

Elliptical pieces of various sizes are cut from the leaf margins of water lilies and used to form protective cases for the caterpillars. The larval cases can be found on the underside of the leaves.

PLANTS AFFECTED Water lilies
SEASON Early to late summer

Shothole
DISEASE • *See p.182*

Leaves develop spots (*see p.31*) where the leaf tissue dies and falls away, leaving holes, usually with a slight browning around their edges, in the leaves (here of a *Prunus*).

PLANTS AFFECTED Many plants
SEASON Mainly summer and autumn

Capsid Bug
PEST • *See p.123*

Leaves at the shoot tip (here of a *Phygelius*) develop many small, brown-edged holes. The leaves are often distorted and flower buds may be killed or damaged.

PLANTS AFFECTED Many plants
SEASON Late spring to late summer

Flea Beetles
PEST • *See p.136*

Small, rounded holes are scalloped out of the upper leaf surface (here of a wallflower). Small, blue-black or black, yellow-striped beetles, which jump off the foliage, may be seen.

PLANTS AFFECTED Brassicas and related plants
SEASON Mid-spring to late summer

Vine Weevil
(Damage by Adult)
PEST • *See p.193*

Adult vine weevils eat irregular notches from the leaf margins (here of a rhododendron). On shrubs, the leaves closest to the ground are attacked.

PLANTS AFFECTED Many plants
SEASON Mid-spring to mid-autumn

Holes in Leaves *CONTINUED*

Box Tree Moth
PEST • *See p.118*

Yellowish-green caterpillars with dark coloured flanks feed on the leaves of box plants and spin silk webbing over the stems and foliage where they are feeding. Plants can suffer severe defoliation.

PLANTS AFFECTED Box (*Buxus*)
SEASON Spring to autumn

Elephant Hawk Moth
PEST • *See p.134*

Large holes are eaten in the leaves by blackish-brown or green caterpillars. The caterpillar's eye markings give them a snake-like appearance.

PLANTS AFFECTED Mainly fuchsias in gardens
SEASON Mid- to late summer

Leaf-cutting Bees
PEST • *See p.151*

Semi-circular or elliptical pieces are removed from the leaf margins (here of a rose). The smooth outline of the missing pieces distinguishes leaf-cutting bee damage from other leaf-eating pests.

PLANTS AFFECTED Many plants, especially roses
SEASON Early to late summer

Cabbage Moth
PEST • *See p.122*

Infested plants have many holes of various sizes in their foliage. On cabbages the caterpillars bore into the heart leaves and soil the edible parts with their droppings.

PLANTS AFFECTED Brassicas and onions
SEASON Early summer to early autumn

Aruncus Sawfly
PEST • *See p.111*

Pale green caterpillar-like larvae initially eat small holes between the leaf veins. However, by the time they are fully grown, they may have eaten everything except the larger veins.

PLANTS AFFECTED Goatsbeard (*Aruncus dioicus*)
SEASON Early to late summer

Lackey Moth
PEST • *See p.150*

Hairy caterpillars, with red, blue, and white stripes running along their bodies, cause extensive defoliation (here of a *Malus*). The feeding area is covered by silk webbing spun by the caterpillars.

PLANTS AFFECTED Many trees and shrubs
SEASON Mid-spring to early summer

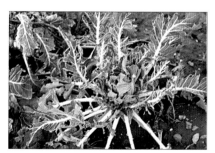

Pigeons
PEST • *See p.168*

Pigeons rip pieces off leaves with their beaks, causing ragged tears on the foliage (here of a cabbage). In heavy attacks most of the soft parts of the leaves will be removed.

PLANTS AFFECTED Brassicas and some other plants
SEASON All year

Holes in Leaves *CONTINUED*

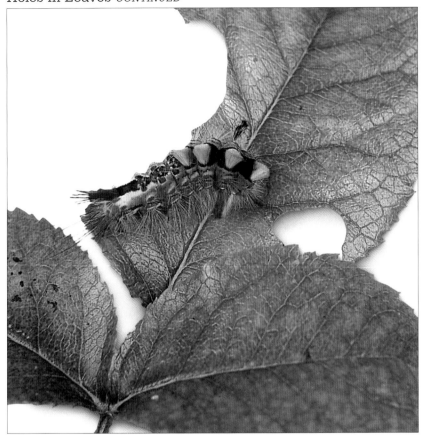

Vapourer Moth
PEST • *See p.192*

Irregular holes are eaten in the leaf margins by very hairy caterpillars. Plumes of black hairs project from their head and rear ends, with four tufts of buff hairs on the upper surface.

PLANTS AFFECTED Mainly trees and shrubs
SEASON Late spring to midsummer

Solomon's Seal Sawfly
PEST • *See p.185*

Whitish-grey caterpillars with black heads eat slots in the middle of the leaves. By the time they are fully grown, they will have caused extensive defoliation.

PLANTS AFFECTED Solomon's seal (*Polygonatum*)
SEASON Late spring to midsummer

Mint Beetles
PEST • *See p.156*

Irregular holes are eaten in the foliage by iridescent green or dark blue beetles and their rotund, black larvae. Infestations are rarely heavy enough to cause serious damage to the plant.

PLANTS AFFECTED Wild and cultivated mints
SEASON Mid- to late summer

Gooseberry Sawflies
PEST • *See p.142*

The soft parts of the foliage are rapidly devoured, leaving just the stalks and larger veins. The larvae are pale green with many black spots or uniformly green in colour.

PLANTS AFFECTED Gooseberries, red/white currants
SEASON Mid-spring to late summer

Winter Moth
PEST • *See p.198*

Small holes are eaten in newly-emerged spring foliage; these holes increase in size as the leaves expand. The pale green caterpillars loosely bind leaves together with silk threads.

PLANTS AFFECTED Fruit and other trees
SEASON Early to late spring

Pea and Bean Weevil
PEST • *See p.163*

Uniform, U-shaped notches are eaten from the leaf margins (here of pea seedlings). The greyish-brown beetles do not usually cause extensive damage.

PLANTS AFFECTED Broad beans and peas
SEASON Late spring to late summer

Insects on Leaves

Beech Woolly Aphid
PEST • *See p.114*

Pale yellow aphids covered with a fluffy, white wax form dense colonies on the underside of beech leaves. Foliage at the shoot tips may be distorted.

PLANTS AFFECTED Beech (*Fagus*)
SEASON Late spring to late summer

Black Bean Aphid
PEST • *See p.116*

Dense colonies of black aphids with small, white markings on their upper surface infest the foliage, and sometimes the shoot tips and flowers. Ants often collect the honeydew they excrete.

PLANTS AFFECTED Beans, nasturtiums, dahlias
SEASON Late spring to late summer

Cedar Aphid
PEST • *See p.124*

Greyish-brown aphids feed at the base of the leaves. Honeydew excreted by the aphids makes the foliage sticky and allows the growth of sooty mould. Leaves may turn yellow and fall.

PLANTS AFFECTED Cedars (*Cedrus*)
SEASON Late spring to midsummer

Cabbage Whitefly
PEST • *See p.122*

Small, white-winged insects live on the underside of leaves and readily fly up when disturbed. Flat, oval, scale-like immature nymphs also occur on the lower leaf surface.

PLANTS AFFECTED Cabbage and other brassicas
SEASON All year

Mealybugs
PEST • *See p.155*

Flattened, oval insects covered in a white, mealy, or fluffy wax live on the underside of leaves and in the leaf axils. These sap-feeding insects excrete honeydew, which makes the foliage sticky.

PLANTS AFFECTED Many greenhouse plants
SEASON All year

Mealy Cabbage Aphid
PEST • *See p.155*

Dense colonies of grey-white aphids live on the underside of brassica leaves and on the shoot tips of young plants. Mottled yellow patches develop on leaves.

PLANTS AFFECTED Brassicas and swedes
SEASON Late spring to early autumn

Alder Sucker
PEST • *See p.106*

Flattened, greenish-black insects covered in a fluffy, white wax live on the stems and undersides of alder leaves during the spring. The fluffiness disappears in early summer when the nymphs mature.

PLANTS AFFECTED Alders (*Alnus*)
SEASON Mid-spring to midsummer

Juniper Webber Moth
PEST • *See p.149*

Dead patches develop in juniper bushes as the foliage turns brown and dries up. The patches are bound together by silk webbing produced by the caterpillars.

PLANTS AFFECTED Junipers (*Juniperus*)
SEASON Late spring to early summer

Larch Adelgid
PEST • *See p.150*

Small, black, aphid-like insects concealed under fluffy, white wax feed by sucking sap from the foliage. Heavy attacks make the foliage sticky with the pests' excrement.

PLANTS AFFECTED Larches (*Larix*)
SEASON Mid-spring to late summer

Stems and Buds

The stems of a plant can be compared to the skeleton of an animal, their support giving the plant its characteristic shape and holding the leaves, flowers, and buds in the position in which they function best. Since the stems contain the plant's vascular, or transport, system they are also fundamental to its function. When stems are damaged by a pest or by disease, the buds, flowers, and leaves carried on them may show symptoms too.

Stems may be attacked in a variety of ways, by pests that gnaw the outer tissues (sometimes ringing the whole stem), feed from them by sucking sap or even tunnel through them. Pest damage may then allow pathogens to enter and so cause further infections, which in turn may spread down the stem and into the crown of the plant; if this occurs, the whole plant may be killed. The tissues of the stem may also be prevented from working by the presence of fungal wilts, which clog up the vascular system, so preventing it from functioning. Fungal infections may also damage the stem from the outside.

Stem Problems

Currant Clearwing Moth
PEST • *See p.130*

White caterpillars with brown heads bore into the stems of currants, making them liable to snap. Emergence holes may be seen on stems in early summer.

PLANTS AFFECTED Currants
SEASON Late spring to late summer

Shothole Borers
PEST • *See p.182*

Small holes in the bark, resembling damage by shotgun pellets, are the emergence holes of small, brownish-black beetles. Attacks mainly occur on trees growing under stressful conditions.

PLANTS AFFECTED *Prunus* and other trees
SEASON Late spring to midsummer

Sirococcus Blight
DISEASE • *See p.183*

Cankers appear on stems, often oozing resin, and these may girdle stems, causing dieback. Shoot tips die back and needles discolour pink, turn brown, and drop.

PLANTS AFFECTED Species of *Cedrus* and *Tsuga*
SEASON All year

Phlox Eelworm
PEST • *See p.166*

Stems are stunted and abnormally swollen and there may be splitting at the base. Leaves at shoot tips are narrow and may consist of little more than the midribs (as on the top stem, above).

PLANTS AFFECTED Perennial and annual phlox
SEASON Mid-spring to midsummer

Didymella Stem Rot
DISEASE • *See p.132*

Blackish-brown, sunken blotches appear on the stem base (here of a tomato). Adventitious roots may form above soil level. Older leaves turn yellow.

PLANTS AFFECTED Tomatoes and aubergines
SEASON Mainly summer

Bark Beetles
PEST • *See p.113*

The larvae of several species of beetle feed underneath the bark of recently dead or dying trees and shrubs. This network of tunnels was made by elm bark beetle grubs.

PLANTS AFFECTED Many trees and shrubs
SEASON All year

Stem Problems *CONTINUED*

Compost is sodden

Flowers, leaves, and stems have lost all turgidity

Damping Off
DISEASE • *See p.131*

Seedlings (here of lettuce) collapse. The stem base is discoloured and may shrink inwards, often appearing water-soaked initially. Fluffy white fungal growth may be present.

PLANTS AFFECTED All seedlings
SEASON Mainly late winter and spring

Waterlogging
DISORDER • *See p.195*

Foliage wilts and yellows, as if suffering from drought, but the condition does not improve if more water is applied. The entire top growth may flop over (as with the cyclamen above).

PLANTS AFFECTED Any plant
SEASON All year

Snowdrop Grey Mould
DISEASE • *See p.184*

Dense, fuzzy, grey mould growth appears on foliage, often as soon as leaves emerge. Bulbs may rot. Small, black sclerotia may develop.

PLANTS AFFECTED Snowdrops (*Galanthus*)
SEASON Late winter to early spring.

Verticillium Wilt
DISEASE • *See p.192*

Soft stems and leaves wilt (*see p.44*). Conducting tissue is stained brown, causing longitudinal streaking which is only visible if the bark is removed.

PLANTS AFFECTED Many trees and shrubs
SEASON Symptoms most apparent in summer

Honey Fungus
DISEASE • *See p.145*

A creamy white mycelium (fungal sheet), smelling strongly of mushrooms, develops beneath the bark on infected stem bases, trunks, and roots. The woody material beneath discolours.

PLANTS AFFECTED Most trees, shrubs, and climbers
SEASON All year

Raspberry Cane Spot
DISEASE • *See p.175*

Purple spots with silvery-white centres develop on the canes. The spots enlarge rapidly and the canes may split and die off. Leaves and flowers may be affected.

PLANTS AFFECTED Raspberries and other cane fruits
SEASON Early summer to autumn

Stem Problems *CONTINUED*

Bacterial Canker   DISEASE • *See p.112*

Patches of bark flatten and sink inwards. Amber-coloured, resin-like ooze may appear close by. Foliage withers and dies off or buds fail to open. Branches may be killed by girdling cankers.

PLANTS AFFECTED Ornamental and edible *Prunus*
SEASON Mainly spring and autumn

Seiridium Canker   DISEASE • *See p.181*

Foliage (here on a cypress) yellows and dies. Bark becomes roughened as cankers develop. Resin oozes from the infected areas and tiny, black pycnidia appear. Cankers may ring stems.

PLANTS AFFECTED Several conifers
SEASON All year (mainly autumn and winter)

Smuts   DISEASE • *See p.184*

Raised, blister-like areas develop on stems (here on *Trollius*). These erupt to produce black, powdery spore masses. Affected stems may be killed if the pustules are sufficiently large.

PLANTS AFFECTED Many plants
SEASON Mainly summer

Rose Root Aphid – Eggs   PEST • *See p.179*

Large, dark brown aphids come up from the roots in autumn and lay clusters of eggs on the stems. The eggs are brown when freshly laid but turn shiny black.

PLANTS AFFECTED Roses
SEASON Eggs seen autumn to spring

Fireblight   DISEASE • *See p.135*

Flattened areas develop on the bark and may exude bacterial ooze. The wood beneath is stained foxy-red brown. Flowers wither and die, followed by adjacent leaves (here of a pear tree).

PLANTS AFFECTED Pome fruits
SEASON Late spring to summer

Lilac Blight   DISEASE • *See p.153*

Young shoots wilt and die, with the older stems developing cankers. Angular brown spots appear on leaves which then die off. Grey mould growth is often present, masking the symptoms.

PLANTS AFFECTED Lilacs (*Syringa*)
SEASON Late spring and early summer

Beech Bark Scale   PEST • *See p.114*

An off-white, powdery wax accumulates in crevices in the bark of beech trees. It is secreted from tiny, sap-feeding scale insects. Heavy infestations are linked with beech bark disease (*see p.54*).

PLANTS AFFECTED Beech (*Fagus*)
SEASON All year

Grey Mould   DISEASE • *See p.142*

Discoloured patches develop on stems (here of a tomato) and may girdle them, causing dieback. Fluffy grey fungal growth and black fungal resting bodies can develop on affected areas.

PLANTS AFFECTED Soft and woody-stemmed plants
SEASON All year

Stem Problems *CONTINUED*

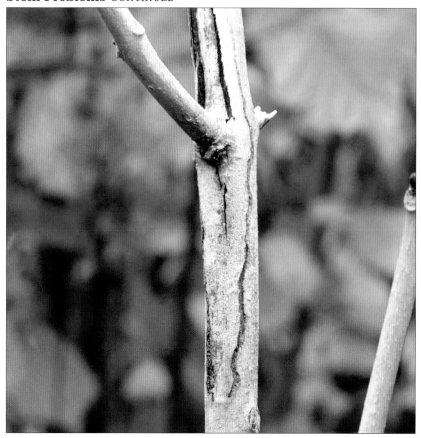

Ash Dieback
DISEASE • *See p.111*

Patches of dead tissue appear on stems and when cankers develop, the stem, branch, or whole tree may die. Black blotching develops on leaves, particularly on the mid-vein and leaf base, causing them to wilt.

PLANTS AFFECTED Many ash species (*Fraxinus*)
SEASON Spring to mid-autumn

Sclerotinia
DISEASE • *See p.181*

Stems (here of a Jerusalem artichoke) become discoloured and covered in dense, white, fluffy fungal growth. Large black sclerotia (fungal resting bodies) are embedded in the fungal growth.

PLANTS AFFECTED Wide range of plants
SEASON All year (mainly early autumn)

Apple and Pear Canker
DISEASE • *See p.108*

Flattened areas of discoloured bark crack and split, forming concentric rings of flaky bark. White or red fungal bodies may develop on the cankered areas.

PLANTS AFFECTED Mainly apples and pears
SEASON All year

Sycamore Sooty Bark Disease
DISEASE • *See p.188*

Green or yellow staining develops under bark. Later, black spore masses or blister-like outgrowths develop on the stems. Large areas of bark die and break off.

PLANTS AFFECTED Sycamores (*Acer pseudoplatanus*)
SEASON All year

Irregular Watering
DISORDER • *See p.148*

Longitudinal cracks or splits in bark may penetrate into the woody tissue. Older splits have swollen edges due to the formation of scar tissue.

PLANTS AFFECTED All woody plants
SEASON All year

Voles
PEST • *See p.194*

Branches or whole plants die as a result of these small rodents gnawing away the bark on the stems or roots. Damage often occurs in winter, but the results may not be apparent until summer.

PLANTS AFFECTED Many trees and shrubs
SEASON All year

Stem Problems *CONTINUED*

Cotoneaster Webber Moth
PEST • *See p.128*

The foliage turns brown and dries up where small, dark brown caterpillars have been feeding. They cover the infested branches with silk webbing.

PLANTS AFFECTED Cotoneaster, hawthorn
SEASON Late spring to late summer

Juniper Scale
PEST • *See p.149*

Flat, yellow-white, rounded scales encrust the stems, often causing yellowing of the green parts. Heavy infestations of this sap-feeder can cause stems to die back.

PLANTS AFFECTED Juniper, cypress, thuya
SEASON All year

Rose Canker and Dieback
DISEASE • *See p.178*

Stems die back and discoloration occurs, either in patches or all over the affected stem. Tissues may dry out. Growth above the affected areas dies back.

PLANTS AFFECTED Roses
SEASON All year

Bracket Fungus
DISEASE • *See p.119*

Bracket- or shelf-shaped, fungal fruiting bodies of variable size grow out from the tree (here a *Malus*), most commonly at the base of the trunk, but sometimes high in the crown.

PLANTS AFFECTED Most trees and some shrubs
SEASON All year (particularly autumn)

Fomes Root and Butt Rot
DISEASE • *See p.137*

Dark red-brown, woody-textured bracket fungus grows on the trunk or butt of trees. The lower surface is creamy-white and covered in minute pores.

PLANTS AFFECTED Mainly conifers
SEASON All year (mainly autumn)

Daedaleopsis confragosa
DISEASE • *See p.130*

Semi-circular, rubbery, annual bracket fungus develops on trunks or branches, reddish-brown and smooth or slightly ridged with a paler undersurface.

PLANTS AFFECTED Willow and other deciduous trees
SEASON Mainly autumn

Stem Problems *CONTINUED*

Crown Gall – Aerial Form
DISEASE • *See p.129*

Roughened, rounded growths break out from within the stem (here of a loganberry), rupturing it. Galls become woody and if extensive, stems may die back.

PLANTS AFFECTED Many woody plants
SEASON All year (mainly summer)

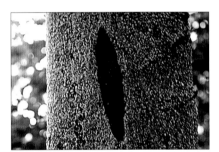

Beech Bark Disease
DISEASE • *See p.114*

Small areas of bark die off and sink inwards. Black, tar-like exudations appear and as the damage increases foliage may yellow and die.

PLANTS AFFECTED Beeches (*Fagus*)
SEASON All year

Horse Chestnut Bleeding Canker
DISEASE • *See p.146*

Drops of brown gummy ooze appear in patches on the trunk and main stems. Affected bark often cracks, allowing other decay organisms to enter.

PLANTS AFFECTED Horse chestnuts (*Aesculus*)
SEASON All year

Laetiporus sulphureus
DISEASE • *See p.150*

Overlapping, annual, bright yellowish-orange fungal brackets are found on the trunk or large limbs of trees. The brackets have wavy edges.

PLANTS AFFECTED Many trees
SEASON Mainly autumn

Bracket Fungus
DISEASE • *See p.119*

Single or overlapping, small bracket fungi grow on stems or trunks (here on dead wood). The upper surface may be concentrically zoned in shades of tan, brown, cream, and grey.

PLANTS AFFECTED Deciduous trees and shrubs
SEASON All year (particularly autumn)

Poor Pruning
See pp.102–103

This pruning cut – here on birch – has removed a branch too close to the trunk and the wound has failed to heal successfully. Infection has entered the wound, causing the wood to deteriorate.

PLANTS AFFECTED Any tree or shrub
SEASON All year

Mulberry Canker
DISEASE • *See p.157*

These are small cankers that may girdle and so kill stems. Tiny red-brown pustules appear around the cankered area. Leaves and shoots may die back.

PLANTS AFFECTED Mulberries
SEASON All year (most obvious in summer)

Stem Problems *CONTINUED*

Phellinus pomaceus
DISEASE • *See p.166*

Hoof-shaped or rounded, fungal fruiting bodies appear on the bark of dead or living stems (here of an apple tree). The upper surface is pale grey-brown.

PLANTS AFFECTED Various deciduous trees
SEASON All year (particularly autumn)

Meripilus giganteus
DISEASE • *See p.155*

Large, fleshy, golden-brown fungal brackets are found around the base of trees or in the near vicinity of the roots. The undersurface is creamy white.

PLANTS AFFECTED Oaks, beeches, robinias
SEASON Late summer and autumn

Massaria Disease
DISEASE • *See p.155*

A strip of dead bark develops on the upper surface of larger branches, starting at the branch collar. Soft rotting of the wood of affected branches follows.

PLANTS AFFECTED London plane (*Platanus x hispanica*)
SEASON All year

Acute Oak Decline
DISEASE • *See p.106*

Cracks appear on the trunk and branches and develop weeping lesions. Affected trees may be killed within a few years.

PLANTS AFFECTED Mainly *Quercus robur, Q. petraea*
SEASON All year

Weedkiller Damage – Broadleaved
DISORDER
See p.196

Adventitious roots and other outgrowths form on stems (here of a brassica), causing them to become distorted. The foliage may not be affected.

PLANTS AFFECTED Plants without woody stems
SEASON Mainly late spring and summer

Leafy Gall
DISEASE • *See p.152*

Numerous small, bunched, distorted, thickened, or fasciated leaves grow, usually around the stem base (here of a pelargonium). Normal growth is also present.

PLANTS AFFECTED Annual or herbaceous plants
SEASON Mainly spring and summer

Broom Gall Mite
PEST • *See p.120*

Buds fail to grow into shoots and become very swollen. These galls gradually dry up and change from green to grey-brown. Heavily infested plants produce few flowers or new growth.

PLANTS AFFECTED Broom (*Cytisus*)
SEASON All year

Stem Problems CONTINUED

Cuckoo Spit
PEST • *See p.129*

Globules of frothy white liquid appear on stems and sometimes on foliage. Inside the froth is a yellowish-green, sap-feeding insect, which is the nymphal stage of a froghopper.

PLANTS AFFECTED Many plants
SEASON Late spring to early summer

Robin's Pin Cushion
PEST • *See p.177*

Roughly spherical swellings covered in reddish or yellowish-green, mossy leaves develop around the stems of certain roses and occasionally on foliage.

PLANTS AFFECTED Wild and species roses
SEASON Midsummer to early autumn

Mussel Scale
PEST • *See p.158*

Bark may become heavily encrusted with greyish-dark brown scales (here on box). These sap-feeding insects weaken the plant and cause shoots to die back.

PLANTS AFFECTED Apple, box, and some other shrubs
SEASON All year

Coral Spot
DISEASE • *See p.128*

Raised red, orange, or coral-coloured, hard pustules develop on dead or dying stems. Affected stems die back further and if the infection spreads to the crown, the whole plant may be killed.

PLANTS AFFECTED Woody plants
SEASON All year

Blossom Wilt and Spur Blight
DISEASE • *See p.117*

Raised, rounded, buff-coloured fungal pustules appear on twigs and stems (here of an apple tree), often associated with blossom wilt on adjacent flowers.

PLANTS AFFECTED Mainly apples, pears, and *Prunus*
SEASON Mainly late spring and summer

Forsythia Gall
DISEASE • *See p.137*

Rough-surfaced, near-spherical woody galls, usually 10–15mm in diameter, develop on stems. Several galls may be fused together. The growth and development of the stem is not affected.

PLANTS AFFECTED Forsythias
SEASON All year

Blind Shoots
DISORDER • *See p.116*

Shoots that should bear flower buds form, but no flower buds develop (here on a rose). Leaf and stem growth otherwise appear perfectly healthy, with no sign of dieback.

PLANTS AFFECTED Flowering shrubs
SEASON Mainly summer

Papery Bark
DISORDER • *See p.162*

Bark peels off as a thin, papery brown sheet. Small- to medium-sized shoots are most commonly affected, but any may show symptoms. Shoots can die back. Apple trees are frequently affected.

PLANTS AFFECTED Woody plants
SEASON All year (mainly spring and winter)

Stem Problems *CONTINUED*

Pineapple Gall Adelgid
PEST • *See p.168*

Shoot tips are swollen and resemble green pineapples by midsummer. The galls turn brown after the aphid-like insects emerge in late summer.

PLANTS AFFECTED Spruces (*Picea*)
SEASON All year

Physical Injury
See pp.78–79

Winds, heavy snow, browsing animals, or vandals may cause tearing of bark and injury to the wood tissues beneath. Jagged wounds are slow to heal and pathogens may enter more readily.

PLANTS AFFECTED Woody plants
SEASON All year

Raspberry Rust
DISEASE • *See p.175*

Bright orange spore masses develop on the stems, often appearing from ruptures in stems. Large fissures remain on the stems (here of a blackberry) when all the spores have been dispersed.

PLANTS AFFECTED Raspberries, blackberries
SEASON Late spring to autumn

Horse Chestnut Scale
PEST • *See p.146*

Brownish-black scales secrete a white, waxy substance that covers their eggs. Later the scales die and fall away, leaving the bark covered in white patches.

PLANTS AFFECTED Many trees and shrubs
SEASON All year; eggs in late spring

Asparagus Beetle
PEST • *See p.111*

Both grubs (*above left*) and adults (*above right*) eat the foliage and also gnaw bark from the stems, causing them to dry up and turn a yellowish brown colour.

PLANTS AFFECTED Asparagus
SEASON Late spring to early autumn

Fruit Tree Red Spider Mite Eggs
PEST • *See p.138*

The spherical winter eggs are less than 1mm in diameter, but there may be so many that they give the bark a reddish colour, especially on branch undersides.

PLANTS AFFECTED Apples, plums
SEASON Eggs early autumn to late spring

Stem Problems *CONTINUED*

Woolly Aphid
PEST • *See p.199*

Fluffy, white patches develop on branches, especially on pruning wounds and in cracks in the bark. Infestations on young shoots cause knobbly swellings where the aphids have been feeding.

PLANTS AFFECTED Apple, pyracantha, cotoneaster
SEASON Mid-spring to mid-autumn

Willow Bark Aphid
PEST • *See p.197*

Dense colonies of greyish-black aphids, up to 5mm long, develop on willow stems. The bark and ground under the tree becomes sticky with honeydew excreted by the aphids.

PLANTS AFFECTED Willows (*Salix*)
SEASON Midsummer to mid-autumn

Brown Scale
PEST • *See p.120*

Reddish-brown shells cover the bodies of sap-feeding insects or their eggs, especially on shrubs growing in sheltered places (here on a cotoneaster). Heavily infested plants lack vigour.

PLANTS AFFECTED Many woody plants
SEASON All year

Woolly Vine Scale
PEST • *See p.199*

In early summer this scale insect deposits its eggs under pads of white wax (here on a pyracantha), which can be drawn out in long threads. Heavily infested plants lack vigour.

PLANTS AFFECTED Vines, currants and others
SEASON All year (eggs in early summer)

Pine Adelgid
PEST • *See p.168*

Small, black, aphid-like insects suck sap from the young shoots. The adelgids are hidden under a fluffy, white wax that is secreted from their bodies. This can resemble a mould.

PLANTS AFFECTED Pines (*Pinus*)
SEASON Mid-spring to late summer

Mealybugs
PEST • *See p.155*

Pinkish-grey or white insects infest the stems and leaf axils. They are often covered with a fluffy, white, waxy substance. Honeydew excreted by the mealybugs allows sooty mould to grow.

PLANTS AFFECTED Greenhouse and house plants
SEASON All year

Small Ermine Moths
PEST • *See p.184*

Black-spotted, creamy-yellow caterpillars feed together under white silk webbing that they spin over their feeding area. If numerous, they can cause defoliation.

PLANTS AFFECTED Euonymus, apple, *Prunus padus*
SEASON Early to late summer

Stem Problems *CONTINUED*

Witches' Brooms
DISEASE • *See p.198*

Closely-packed, densely-branched clusters of twigs grow from normal stems. The stems do not die back, but bear numerous small leaves. Adjacent growth is perfectly normal.

PLANTS AFFECTED Trees
SEASON All year

Willow Scab
DISEASE • *See p.197*

Irregular, dark spots develop on the young stems and may cause girdling. Olive-brown pustules are found on the undersurface of leaves and on affected stems. Stems may die back.

PLANTS AFFECTED Willows (*Salix*)
SEASON All year (particularly spring)

Slugs
PEST • *See p.183*

The softer parts of leaves and leaf stalks are eaten (here of celery, causing elongate grooves between the veins). Damaged parts turn brown and may be infected by secondary rots.

PLANTS AFFECTED Soft tissue of many plants
SEASON Spring to mid-autumn

Lupin Aphid
PEST • *See p.154*

Large, whitish-grey aphids cluster on the flower stems and underside of leaves. Heavy infestations cause the plant to wilt. Honeydew and sooty mould soil the foliage.

PLANTS AFFECTED Lupins
SEASON Late spring to late summer

Deer
PEST • *See p.131*

Various species of deer come into gardens and eat foliage, flowers, and stems. Nibbled stems invariably have ragged cuts caused by a deer's lack of incisor teeth in its upper jaw.

PLANTS AFFECTED Roses and many others
SEASON All year

Wisteria Scale
PEST • *See p.198*

Large, blackish-brown, domed scales encrust the woody stems of suitable host plants. Heavy infestations cause plants to lack vigour, and they may die back.

PLANTS AFFECTED Wisteria, *Acer, Prunus, Parthenocissus*
SEASON Spring to early summer

Bud Problems

Blindness of Bulbs
DISORDER • *See p.117*

Flower buds fail to form, or do not develop fully (here on a daffodil) and open partially or not at all. Leaves are unaffected and appear healthy.

PLANTS AFFECTED Many bulbs
SEASON Spring

Big Bud Mites
PEST • *See p.115*

Infested buds (here on a blackcurrant stem) are abnormally swollen and rounded compared to normal buds. They usually dry up and fail to open in the spring.

PLANTS AFFECTED Blackcurrant, hazel, yew
SEASON Mid-winter to early summer

Rhododendron Bud Blast
DISEASE • *See p.176*

Buds form but turn dry and brown. Numerous tiny, black, bristle-like fungal growths develop on the surface. Affected buds remain on the plant.

PLANTS AFFECTED Rhododendrons
SEASON All year (evident late spring and summer)

Flowers

Most gardens are packed full of plants grown largely for their flowers, be they annuals, biennials, perennials, shrubs, climbers, or trees. In fruit and vegetable gardens, flowers are often equally important, as without them the edible fruits would never develop. Keeping flowers free from pests and diseases is therefore vital.

Flowers are usually short-lived, yet during their brief lifespan there may well be several pests or pathogens that could attack and spoil them. They may be discoloured (*p.60*) or disfigured by infestations of pests (*p.61*), deformed (*pp.62–64*) or damaged by pests (*p.65*). In many cases, however, even if a few flowers are spoilt, later flowers may bloom unscathed, having missed the attentions of the pest or the infection period of the pathogen.

Flowering in general is also greatly influenced by cultural and weather conditions, in particular soil moisture and fertility and the level of sunshine. To a large extent, growing a plant in a suitable spot ensures a good display of flowers.

Discoloured Flowers

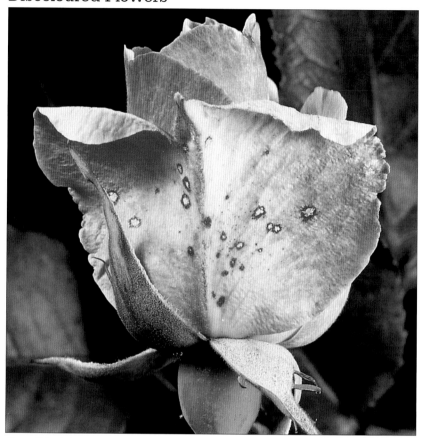

Grey Mould
DISEASE • See p.142

Spots of discoloration appear on the petals (here of a rose). Spots may be edged with a dark ring. No fungal growth is apparent. In extreme cases the whole flower deteriorates.

PLANTS AFFECTED Many flowers
SEASON Mainly spring and summer

Rhododendron Petal Blight
DISEASE • See p.177

Spots of discoloration develop on the petals. These enlarge and appear water-soaked. The flower collapses and rapidly discolours all over.

PLANTS AFFECTED Rhododendrons and azaleas
SEASON Spring

Viruses
DISEASE • See p.194

Petals show "colour breaking", with streaks of discoloration (here on a tulip). The petals do not deteriorate. Flowers may be reduced in number and size and show distortion.

PLANTS AFFECTED Many plants
SEASON Mainly spring and summer

Western Flower Thrips
PEST • See p.196

A pale flecking develops on petals (here of a gloxinia), and yellowish-brown insects are seen on flowers and foliage. Infested blooms have a shortened life.

PLANTS AFFECTED Mainly greenhouse plants
SEASON All year

Pests Visible on Flowers

Slugs and Snails
PESTS • *See pp.183, 184*

Damage caused by slugs and snails is usually indistinguishable. On daffodils, above, foliage is ignored but petals can be reduced to brown-edged fragments. A silvery slime trail is sometimes left behind.

PLANTS AFFECTED Wide range of plants
SEASON Mainly early to late spring

Apple Sucker
PEST • *See p.110*

Flattened, pale green insects up to 2mm long suck sap from the blossom trusses. Heavy attacks make the petals turn brown and can be confused with frost damage.

PLANTS AFFECTED Apple trees
SEASON Mid- to late spring

Cuckoo Spit
PEST • *See p.129*

A frothy, white liquid covers yellowish-green insects (here on *Potentilla*), which are the immature nymphs of froghoppers. They also suck sap from foliage and stems (*see p.56*).

PLANTS AFFECTED Many plants
SEASON Late spring to midsummer

Pollen Beetles
PEST • *See p.170*

Black or bronzy-green beetles, 2mm long, crawl around the centre of blooms, where they feed on pollen. No direct damage is caused, but they can be a nuisance on cut flowers in the house.

PLANTS AFFECTED Many flowers
SEASON Mid-spring to late summer

Melon Cotton Aphid
PEST • *See p.155*

Small black or dark green aphids infest the flowers (here chrysanthemum) and the undersides of leaves. Sooty mould grows on the aphids' excrement.

PLANTS AFFECTED Chrysanthemums and others
SEASON Mainly summer to autumn

Rose Aphids
PEST • *See p.178*

Green, yellowish-green, or pink aphids cluster on flower buds and stems. Heavy infestations reduce the quality of the blooms. White cast aphid skins litter the buds and foliage.

PLANTS AFFECTED Roses
SEASON Mid-spring to late summer

Deformed and Distorted Flowers

Glasshouse Red Spider Mite
PEST • See p.141

Although mainly a foliage pest (*see pp.36, 38*), red spider mites also infest flowers, especially chrysanthemums. Yellowish-black mites crawl among the petals.

PLANTS AFFECTED Many greenhouse plants
SEASON Mid-spring to mid-autumn

Peony Wilt
DISEASE • See p.166

Stem tissue is discoloured and shrinks inwards, commonly at the base or, as above, below flower buds. Fuzzy grey fungal growth develops and black sclerotia are found on or within stems.

PLANTS AFFECTED Peonies
SEASON Spring to autumn

Fasciation
DISORDER • See p.135

Stems become fused together, producing broad, ribbon-like stems (here on delphinium). The stems usually bear leaves and flowers, but become curled and distorted.

PLANTS AFFECTED Many plants
SEASON All year on perennial stems

Smuts
DISEASE • See p.184

The anthers of flowers become swollen and distorted and packed full of masses of dark brown or black spores. Anthers burst to release spores. Flower stems may be distorted.

PLANTS AFFECTED Several plants
SEASON Mainly summer

Drought
DISEASE • See p.133

Leaves and flowers wilt and soft stems may do so also. If conditions do not improve, wilting may be permanent (as here on a lavatera) and flowers become shrivelled and brown.

PLANTS AFFECTED Most plants
SEASON Mainly summer

Frost Damage (Buds and Flowers)
DISORDER • See p.137

Petals become discoloured (here of a camellia), usually brown and either soft or, less frequently, dry. Exposed flowers are most badly affected.

PLANTS AFFECTED Many flowers
SEASON Mainly spring and early winter

Deformed and Distorted Flowers *CONTINUED*

Michaelmas Daisy Mite
PEST • *See p.156*

The flowers are converted by the feeding activities of this tiny pest into rosettes of green leaves. Infested plants are often stunted, with scarring on the stems.

PLANTS AFFECTED *Symphyotrichum novi-belgii*
SEASON Early summer to mid-autumn

Chrysanthemum Petal Blight
DISEASE • *See p.126*

Small brown spots or larger water soaked lesions develop on the petals. Flowers rot and die off rapidly. Leaf growth is not affected.

PLANTS AFFECTED Mainly chrysanthemums
SEASON Mainly late summer

Wind Damage
DISORDER • *See p.198*

Petals (here of a camellia) become scorched and discoloured, usually brown. Damaged areas are usually dry, not soggy. Flowers on exposed areas of the plant are most susceptible.

PLANTS AFFECTED Many plants
SEASON Mainly spring

Hemerocallis Gall Midge
PEST • *See p.143*

Infested buds are squat and abnormally swollen; they fail to open, go brown, and dry up. Many small white maggots live in the buds at the base of the petals.

PLANTS AFFECTED Daylilies (*Hemerocallis*)
SEASON Late spring to midsummer

Fuzzy fungal growth begins to cover whole flowerhead

Flowerhead loses shape and dies off

Gladiolus Thrips
PEST • *See p.140*

Creamy-yellow or black, narrow-bodied insects, up to 2mm long, feed in the flower buds and on the foliage. Infested blooms have pale mottled petals, but buds may turn brown and fail to open.

PLANTS AFFECTED Gladioli
SEASON Mid- to late summer

Grey Mould
DISEASE • *See p.142*

At this advanced stage of grey mould, the flowers rapidly become covered in dense, grey fungal growth (here on a pelargonium). Dieback may spread. For earlier symptoms *see p.60*.

PLANTS AFFECTED Many flowering plants
SEASON All year

Deformed and Distorted Flowers *CONTINUED*

Aquilegia Gall Midge
PEST • *See p.110*

Flower buds fail to open properly and the petals turn brown. Small, pale orange maggots may be found in affected buds.

PLANTS AFFECTED Columbine (*Aquilegia*)
SEASON Late spring to early summer

Rose Balling
DISORDER • *See p.178*

Outer petals on an unopened or only partially opened bud are pale brown and dry. Initially, inner petals are unaffected, but they may die off when secondary organisms invade.

PLANTS AFFECTED Roses
SEASON Mainly summer

Proliferation
DISORDER • *See p.173*

Flower buds form within the centre of the flower (here a rose). These remain as buds or develop into stems bearing flowers and occasionally leaves. Only the flowers or buds are affected.

PLANTS AFFECTED Many roses, pears, and apples
SEASON Late spring and summer

Bolting
DISORDER • *See p.117*

Plants grown for their leaf crop become elongated and produce flowers unusually early (here on a brassica). This limits the cropping period. Early cultivars are particularly susceptible.

PLANTS AFFECTED Beetroot, brassicas, lettuces
SEASON Mainly summer

Agapanthus Gall Midge
PEST • *See p.106*

Flower buds are distorted, turn brown, and fail to open. Small, pale yellow maggots may be found inside affected buds.

PLANTS AFFECTED African lily (*Agapanthus*)
SEASON Midsummer

Damaged Flowers

Nectar Robbing
PEST • *See p.159*

Some short-tongued bumblebees cannot reach the nectar from the front of some flowers and so bite a hole at the base of the flower. Bees taking nectar in this way do not assist with pollination.

PLANTS AFFECTED Often runner and broad beans
SEASON Summer

Sparrows
PEST • *See p.185*

House sparrows peck the petals of some flowers (here a crocus) to shreds, leaving fragments scattered nearby on the soil. They sometimes select certain colours, but are not consistent.

PLANTS AFFECTED Crocuses, primroses, and others
SEASON Early spring to summer

Bullfinches
PEST • *See p.121*

The dormant flower buds (here of a plum) are pecked off and eaten during the winter. The outer bud scales are discarded and dropped to the ground. Bullfinches also eat seeds.

PLANTS AFFECTED Fruits and some ornamentals
SUMMER Late autumn to mid-spring

Apple Blossom Weevil
PEST • *See p.108*

Infested flowers fail to open fully and the petals turn brown. A white grub or the pupa may be found inside these "capped" blossoms.

PLANTS AFFECTED Apples
SEASON Mid- to late spring

Ragged holes are eaten in petals

Male earwigs (top) have curved pincers, while females (bottom) have straight ones

Capsid Bugs
PEST • *See p.123*

Flower buds (here of fuchsias) abort so that infested plants (*above left*) are largely flowerless. Foliage is distorted and has many small holes, due to capsid bugs damaging leaves at the shoot tips.

PLANTS AFFECTED Many plants
SEASON Late spring to late summer

Earwigs
PEST • *See p.133*

Earwigs emerge after dark, usually in mild weather, to feed on petals (here of a cosmos) and on foliage at shoot tips. This results in holes developing in the blooms and the expanded leaves.

PLANTS AFFECTED Clematis, dahlias, and others
SEASON Late spring to early autumn

Lawns

Lawns are very popular and often one of the most widely used areas of a garden. Even if you do not want a perfect lawn, but prefer an area that is practical, serviceable, and a good shade of green, there will still be times when problems need to be dealt with.

There are several fungal infections and pests which are common. The majority of problems, whether caused by a pest, a pathogen, or poor cultivation, result in areas of grass turning yellow or brown. At first sight, differentiating between the various causes may not be easy. Most fungal problems do involve the development of visible (if small) fungal growths and a few, such as fairy rings or other toadstools, are very distinct. Lawn pest problems are generally easier to identify and culprits more readily visible.

In all cases, prompt action is advisable, as once turf grasses weaken or their growth becomes thinner, coarse grasses, mosses and weeds colonize rapidly. Among these, dog lichens (see facing page) are commonly mistaken for a disease, but in fact are part fungus and part alga.

Lawn Problems

Leatherjackets PEST • See p.152

These daddy-long-legs larvae generally cause damage when they are becoming fully grown. Patches of turf turn yellow-brown and holes may be seen where birds have been searching for the grubs.

PLANTS AFFECTED Turf, roots of many plants
SEASON Midsummer

Slime Mould DISEASE • See p.183

Numerous tiny grey, yellow, or off-white spheres are clustered together on individual blades of grass. These may appear to "dissolve" when wet. Turf growth and health are not affected.

PLANTS AFFECTED Many plants, including turf
SEASON Mainly summer and autumn

Burrowing Bees PEST • See p.121

Small, conical heaps of powdery soil form on the lawn. Each has an entrance hole at the top where a solitary bee leaves and enters. No significant damage is caused to the lawn.

PLANTS AFFECTED Turf
SEASON Mid-spring to midsummer

Chafer Grubs PEST • See p.125

Plump, creamy-white grubs with brown heads sever roots, making it easy for foxes, badgers, and birds to rip up turf to feed on them. The grubs are curved, with three pairs of legs (see p.19).

PLANTS AFFECTED Turf, other plant roots
SEASON Early autumn to early spring

Turf Thatch Fungal Mycelium DISEASE • See p.191

Bleached yellow or reddish patches of grass develop. Dense white or off-white fungal growth appears around the grass roots.

PLANTS AFFECTED Turf
SEASON All year

Turf Dollar Spot DISEASE • See p.191

Patches of grass die off and turn straw-yellow. The individual patches are usually 10cm or less in diameter. The fungus causing this disease is most prevalent during warm, damp weather.

PLANTS AFFECTED Turf
SEASON Mainly early autumn

Earthworms PEST • See p.133

Some earthworms deposit heaps of muddy excrement on the surface of lawns. These worm casts become smeared by feet or mowers and spoil the lawn's appearance.

PLANTS AFFECTED Turf
SEASON Mainly autumn to spring

Lawn Problems *CONTINUED*

Ants
PEST • *See p.107*

Heaps of fine soil appear on the lawn's surface above ant nests; if they are disturbed, black or reddish-yellow ants will be seen. Ant heaps get in the way of mowers and make the lawn uneven.

PLANTS AFFECTED Turf
SEASON Late spring to early autumn

Turf Red Thread
DISEASE • *See p.191*

Patches of grass begin to deteriorate. Numerous pale red or pink, gelatinous, horn-like fungal growths are attached to individual grass blades. Later, fluffy, pale pink fungal growth develops.

PLANTS AFFECTED Turf
SEASON Mainly late spring to early autumn

Dog Lichens
See facing page

Small, unevenly shaped, grey, leaf-like growths, off-white beneath, develop among grass, particularly when turf is sparse and weak. Lichen is not a pest or disease; it should be treated as a weed.

PLANTS AFFECTED Turf
SEASON All year

Turf Snow Mould
DISEASE • *See p.191*

Patches of dying grass develop and then enlarge and turn brown. Deteriorating grass may be covered in pale pink, fluffy fungal growth.

PLANTS AFFECTED Turf
SEASON Mainly autumn and winter

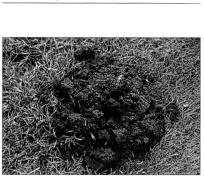

Moles
PEST • *See p.157*

Moles create a network of tunnels in the soil in which they live, and leave the excavated soil in heaps on the surface. Collapsed mole tunnels give the lawn an uneven surface.

PLANTS AFFECTED Turf
SEASON All year

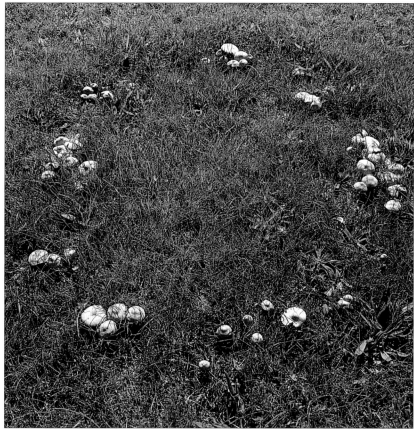

Turf Fairy Rings
DISEASE • *See p.191*

Rings of dead or dying grass, or sometimes other patterns of discoloration, develop on a lawn. One or more rings of lush grass grow adjacent to these. Brown toadstools grow out of the ring.

PLANTS AFFECTED Turf
SEASON Mainly late summer and autumn

Beneficial Garden Creatures

Gardens are inhabited by a wide range of insects and other animals, but only a small proportion are plant pests. Most are neither particularly helpful nor harmful to the gardener. There are, however, some that are very beneficial, as they feed on pests, pollinate flowers, or improve soil.

Predatory and parasitic insects may not be sufficient to prevent pests causing damage, but they do stop pests from having things all their own way. Pollinating insects are essential for good crops of most tree, bush, and cane fruits, and some vegetables. Any insects that move from one flower to another can effect the transfer of pollen.

Most insecticides do not discriminate between pests and beneficial insects. It is therefore important to avoid the unnecessary use of pesticides, saving them for plants that would be seriously damaged by pests or diseases if control measures were not taken.

Ladybirds
Most ladybirds (*see p.150*) prey on aphids, but some feed on scale insects, spider mites, or mealybugs, and others on powdery mildew fungal spores. Not all are red with black spots; some are yellow with black spots, others are orange or brown with white spots.

Flower Bugs
Flower bugs, also known as anthocorid bugs, are 3–4mm long and are general predators of aphids, thrips, other small insects, mites, and their eggs. They have needle-like mouthparts with which they pierce their prey and suck out the contents.

Parasitised Aphids
Several species of tiny parasitic wasp lay eggs in young aphid nymphs. The wasp grub develops as an internal parasite that kills the aphid, causing it to become very swollen and straw brown in colour. The adult wasp emerges through a hole it makes in the dead aphid's abdomen.

Ladybird Larvae
Although less familiar than the adult beetles, ladybird larvae are equally voracious predators of aphids and other pests. The larvae, up to 11mm long, are black with orange or white markings. A larva will eat 200–400 aphids before it is fully fed and ready to pupate.

Hoverfly Larvae

Only the larval stages (*see p.146*) of many hoverflies feed on aphids. The larvae have flattened, semi-transparent bodies, up to 16mm long. The adult flies have yellow and black striped abdomens, giving them a wasp-like appearance, although their bodies usually look flatter.

Tachinid Flies

There are many species of tachinid flies, between 4 and 18mm long. Most have larvae that are parasites of insects, especially caterpillars. The maggots feed inside the body cavity of their host insect, causing its death by the time the tachinid larvae are ready to pupate.

Earthworms

Earthworms (*see p.133*), up to 20cm long, burrow tunnels in the soil through which water can drain and air can reach plant roots. Worms feed on dead plant material and their presence in compost heaps speeds up the conversion of debris to compost.

Brown Centipedes

These centipedes, up to 30mm long, live in the surface layers of the soil and also shelter under stones, logs, and pots. They are active at night, preying on a wide range of small soil animals and their eggs. The front pair of legs curve forward and are modified to act as jaws.

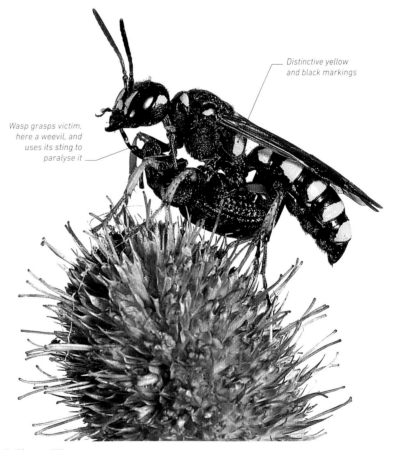

Distinctive yellow and black markings

Wasp grasps victim, here a weevil, and uses its sting to paralyse it

Snake Centipedes

Snake centipedes live below soil level and are often found when soil is dug. They have slender, yellowish-brown bodies, up to 60mm long, adapted for moving through the soil where they hunt for small insects. If disturbed they writhe in a snake-like manner.

Solitary Wasps

Unlike social wasps, these tunnel out individual nests in rotten wood and sandy soil, or make use of beetle tunnels in dead wood. They provision their nests with aphids, flies, weevils, or other insects to feed their larvae. There are many species of various sizes.

Bumblebees

Bumblebees make their nests in the ground, often down abandoned mouse holes. They are valuable pollinating insects as they will visit flowers in poor weather when honeybees stay in their hives. At full strength in midsummer a bumblebee nest houses 100–500 bees.

Honeybees

Many plants, especially most fruits, require insects to carry pollen from one flower to another to fertilize the flowers and allow seeds and fruits to set. The hairy bodies of honeybees, 12mm long, become heavily dusted with pollen when they visit flowers to collect nectar.

Wolf Spiders

Most spiders spin sticky webs to capture flying or crawling insects, but wolf spiders, 5–8mm long, run rapidly over the soil or vegetation to hunt down their prey. The females are frequently seen in early summer carrying their silken egg sacs under their abdomens.

Parasitic Wasps

Many species of parasitic wasps lay their eggs in the larvae, pupae, or eggs of other insects. The larvae develop inside the host insect, which is killed when the parasite is ready to pupate. Some wasps have long egg-laying organs to enable them to reach concealed insects.

Shrews

These small, mouse-like predators, which have elongate snouts, are rarely seen but may be numerous in hedgerows and rough grass. They are active during the day and at night, when they feed on a wide range of insects, slugs, woodlice, and earthworms.

Devil's Coach Horse

These beetles, 25mm long, are often found sheltering under stones, logs, or pots. They are one of the staphylinid or rove beetles, which are characterized by short wing cases that leave most of the abdomen uncovered. They eat slugs and soil pests such as cutworms and leatherjackets.

Song Thrush

The song thrush specializes in feeding on snails. It breaks the snail's shell by smashing it against a stone. Song thrushes often make regular use of certain stones and these "anvils" become littered with pieces of snail shell. Their diet also includes worms, insects, and berries

Green Lacewing

Lacewings, 12–15mm long, derive their name from the many veins that give the wings a lace-like appearance. Other types of adult lacewings (*see p.150*) have black or brown bodies. The larval stages are voracious predators of aphids and sometimes other small insects.

Carabid Beetles

Also known as ground beetles (*see p.143*), these insects live mainly on the soil surface, but some climb up plants in search of insects to eat. They are from 2 to 35mm long and mostly black, brown, or metallic green. The larvae prey on a range of insects, slugs, and pest eggs.

Common Toad

Toads, like frogs, need ponds in which to breed in the spring, but for the rest of the summer they may be found well away from water. They are mainly active after dark, when they feed on a variety of insects, woodlice, worms, and spiders.

Beetle's jaws are strong for killing and eating insects

Long legs to help chase prey

Hedgehogs

Although sometimes active in daylight, hedgehogs are largely nocturnal. Their diet includes slugs, worms, and insects, particularly beetles. Hedgehogs need somewhere sheltered where they can hibernate undisturbed during the winter months.

Tiger Beetle

This colourful predator, 11–13mm long, occurs mainly in sandy areas. It is a type of ground, or carabid, beetle. Its larvae live in pits in the soil from which they emerge to seize passing insects. The adults run over the soil surface and readily fly up in sunny weather.

Garden Health and Problems

Garden plants co-exist with many other living
things – some beneficial, many benign, but
some harmful. Favourable growing conditions
and good gardening practice make plants
strong and more able to tolerate potentially
harmful organisms, but when a serious attack
by pests or diseases occurs, the gardener must
choose an appropriate method with which to
bring the problem under control.

The Balance of Life

Every garden forms a miniature ecosystem that is unique, with its own combination of plants, soil conditions, topography, and climate. These factors, together with the interaction of the various creatures – including predators, parasites, pathogens, saprophytes, and herbivores – that live in the garden, determine what thrives, what survives and what deteriorates or dies. The aim of the gardener is to encourage desirable growth – usually that of ornamental and edible plants – but this must be achieved within a healthy ecological balance that involves other life forms. A garden purged of insects or fungi, for example, would soon become unproductive. The activities and life processes of other plants and animal life, together with their death and decay, are crucial to the growth and health of garden plants. Since all the organisms in a garden influence the overall "working" and functioning of the cultivated plants and the garden as a whole, any action taken by you may have an effect on plants or animals other than those being treated. When considering pest or disease control it is important to consider such wider impacts.

Interaction in the Garden

Even in the smallest corner of the garden, plants, creatures, and environmental factors interact to form a rich and complex ecological system. Any action taken by the gardener may affect numerous links in the varied and interlocking cycles and webs of life.

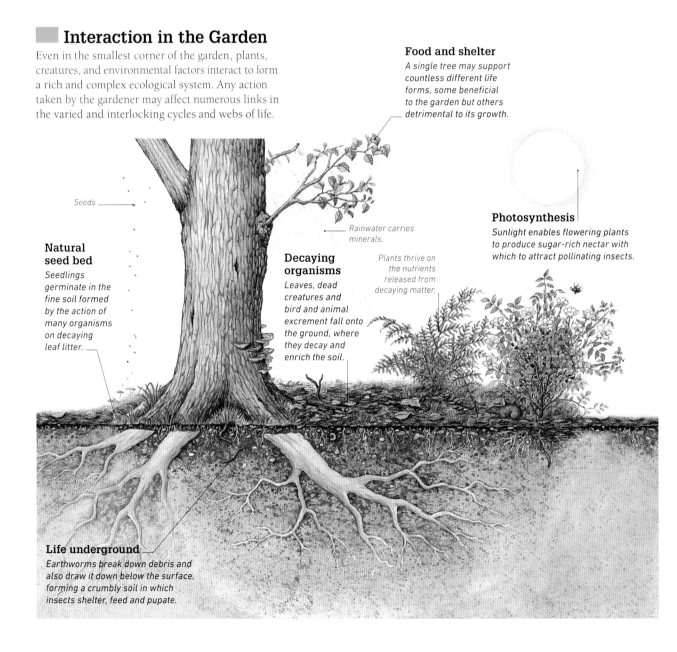

Food and shelter
A single tree may support countless different life forms, some beneficial to the garden but others detrimental to its growth.

Seeds

Rainwater carries minerals.

Photosynthesis
Sunlight enables flowering plants to produce sugar-rich nectar with which to attract pollinating insects.

Natural seed bed
Seedlings germinate in the fine soil formed by the action of many organisms on decaying leaf litter.

Decaying organisms
Leaves, dead creatures and bird and animal excrement fall onto the ground, where they decay and enrich the soil.

Plants thrive on the nutrients released from decaying matter.

Life underground
Earthworms break down debris and also draw it down below the surface, forming a crumbly soil in which insects shelter, feed and pupate.

The Energy for Growth

Plants manufacture food for energy by the process of photosynthesis, in which the action of sunlight on green leaf pigments enables plants to combine carbon dioxide from the air with water to make sugars. As well as providing energy for growth, sugars may be used by plants for other purposes, such as aiding pollination and seed dispersal. Many plants are insect-pollinated, and the secretion of sugar-rich nectar by flowers is an effective means of encouraging bees and other insects to visit blooms. While doing so insects pick up pollen when they come into contact with the anthers, and this is transferred to the stigmas of flowers that are subsequently visited.

Sugars also make fruits and berries sweet and attractive to birds and mammals. The seeds are dispersed away from the parent plant when they are spat out or have passed through the animal's gut.

Essential Nutrients

In addition to sugars, plants need essential nutrients (*see pp.76–77*), which are absorbed by the roots from the soil or, in the case of parasitic plants, from their hosts. For the cycle of life to continue, these nutrients must be returned to the soil. Dead or waste plant or animal matter is broken down by the actions of fungi, bacteria, earthworms, and many other organisms that feed on decaying material. Certain larger fungi and many micro-fungi, described as saprophytic, live on dead organic material and, by breaking it down, release nutrients that are then used by other organisms. Their primary function, however, is one of breaking down debris, which would otherwise accumulate. In a similar way, numerous garden creatures live on debris and are to be encouraged because of their recycling function.

Recycling Activity
Woodlice occasionally cause plant problems but have a valuable role in processing garden debris.

As organic matter deteriorates it releases and uses certain nutrients and so alters the range and quantity of materials available to plants and other organisms in the garden. Although decaying plant material may harbour plant pathogens, it may also provide a useful overwintering or breeding site for beneficial animals, so it is important to reach a compromise between extremes of garden tidiness.

Food and Feeders

Plants form the basis of all terrestrial food webs. There is a huge range of invertebrate animals, including nematodes, millipedes, insects, mites, woodlice, slugs, and snails, that

Plant Associations
Gardeners' time frames often differ from those of nature. While mistletoe on tree branches (usually, of apple) will weaken the host over many years, most gardeners prefer the short-term attraction and novelty value of its presence.

feed directly on plants. Many of them specialize in feeding on a particular part of a plant; some may feed on a wide range of plants, while others may be very specific in their choice of food plant. Most garden pests and disease-causing organisms have a relatively restricted host range, attacking only one type of plant, or a few closely related ones. An outbreak of one pest or disease could therefore remove the main food source of another potential pest or pathogen, so indirectly minimizing the damage it causes. Plant pests in turn are under attack from predators and parasites, which help to prevent pests becoming so abundant that plants are killed.

Other Beneficial Associations

Fungi, bacteria, and viruses are equally active in attacking plants, but there are also some that have a beneficial effect. Many plant roots have developed a mutually beneficial relationship with certain fungi or bacteria that help host plants to absorb nutrients from the soil (*see overleaf*).

Pest Predator
Toads eat a wide range of pests, including slugs, snails, and insects, making them valuable garden allies.

Soil and Nutrients

The structure and texture of soil, and the nutrients it contains, has a great impact on plant growth. Texture – whether a soil is light and sandy, heavy and clayey, or something in between – influences plants' ability to develop strong root systems through which they take up water and nourishment, and affects the soil's ability to retain moisture and nutrients. These nutrients comprise a range of elements and minerals. Although some are needed in only small quantities, deficiencies due to natural factors or caused by heavy demands by plants that go unreplenished, will cause symptoms in plants that are often mistaken for pest or disease damage (*see table, facing page*). Plants growing in containers are more prone to mineral deficiencies than those growing in open soil, because potting compost soon loses its nutrient content and the plants' restricted roots are unable to seek out nutrients elsewhere. Sometimes, pest and disease problems follow on from deficiency, as the plant is weakened and more vulnerable to attack.

Soil pH also influences the range of nutrients that are in a form available to plants. For example, soils of a high pH (alkaline) may "lock up" iron and manganese so that plants become chlorotic. In an alkaline soil most trace elements are less readily available than in an acidic or neutral soil.

Cross-section, Garden Soil
Garden soil has a distinct structure, with three layers – the topsoil, the subsoil and the bedrock (not shown here). The topsoil is the uppermost layer and is usually a dark brown colour. It contains a high level of organic matter and soil organisms and is the layer in which you dig and plant. The subsoil has a much lower level of organic matter, nutrients and organisms.

A low-pH (acidic) soil is more likely to produce plants that suffer a deficiency of magnesium. An unusually high level of one nutrient will influence the levels or availability of another. Commonly, tomatoes fed with a high-potash feed may also show yellowing between the veins and around the edges of older leaves due to magnesium deficiency.

The Nitrogen Cycle

Plants need nitrogen to grow but they can only absorb and process it in the form of nitrates. Bacteria in the soil play a vital role in converting atmospheric nitrogen and soil-borne nitrogen compounds into usable nitrates.

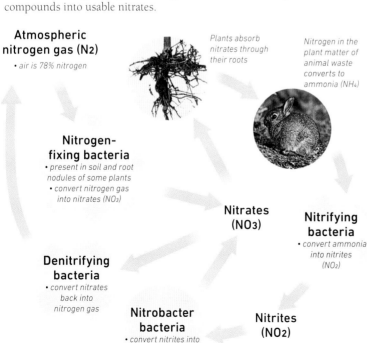

Atmospheric nitrogen gas (N2)
• air is 78% nitrogen

Plants absorb nitrates through their roots

Nitrogen in the plant matter of animal waste converts to ammonia (NH4)

Nitrogen-fixing bacteria
• present in soil and root nodules of some plants
• convert nitrogen gas into nitrates (NO3)

Denitrifying bacteria
• convert nitrates back into nitrogen gas

Nitrobacter bacteria
• convert nitrites into nitrates (NO3)

Nitrates (NO3)

Nitrifying bacteria
• convert ammonia into nitrites (NO2)

Nitrites (NO2)

Balancing Nutrient Levels

Checking the nutrient status and soil pH is a worthwhile job, especially when taking on a new garden with unfamiliar soil. Various kits are available for carrying out your own tests (*see facing page*), although laboratory testing is now easily accessible through online companies and will provide a more accurate analysis; testing every five years is sufficient. Tests will enable you to determine the soil pH and give an outline of the levels of major nutrients. Serious deficiencies are quite rare in gardens and feeding with a general fertilizer or regular use of plenty of organic material and good husbandry is usually enough to keep most plants adequately fed. Occasional use of specific fertilizers to encourage flowering and fruiting or improve growth, may be beneficial. It is worth bearing in mind, however, that overfeeding can be damaging both to plants and soil organisms.

Other Soil Life

Many garden creatures, including a number of pests, spend their lives, or a stage in their lives, within the soil. However, gardeners

Using a Kit to Test Soil Nutrients

1 Preparing This test for phosphorus is part of a kit to test the pH and major nutrient levels of soil. Mix 1 part soil and 5 parts water and allow to stand until the liquid is fairly clear. Use the pipette to transfer liquid to the testing chambers.

2 Testing Make sure the water extract reaches the level marks on the colour chart. Open the correct capsule (blue in this test) and pour the powder into the chamber. Fit the cap onto the tester and shake until the powder has dissolved.

3 Results Allow the colour to develop in the test chamber for a few minutes. Then compare this with the colour chart against a white background and in natural daylight. Read off the results and take further action if required.

are far better occupied in improving soil texture and composition than in attempting to control its inhabitants. Although it is easy to become preoccupied with potentially damaging garden organisms, the vast majority of creatures in soil are harmless and indeed some are beneficial (*see pp.68–71*). The activity of each type of organism affects the population levels of others and so any action you take may well have a knock-on effect on something else.

Some animals and fungi can enhance the productivity or the performance of plants; examples include pest predators and parasites, and mycorrhizae – fungal associations formed on the roots of plants, which allow them to absorb nutrients more efficiently. Mycorrhizae are sometimes visible in the soil as tiny fungal strands attached to the roots – these are often presumed to be harmful. Some also produce fungal fruiting bodies or toadstools around the base of the plant; again it may be presumed, incorrectly, that all toadstools around plants are harmful.

Nitrogen Fixers

Certain bacteria that "fix" nitrogen, making it available to the plant, inhabit nodules on the roots of some plants. If these roots are left in the soil once the crop has been harvested, they will improve the nutrient status of the soil for any plants grown there subsequently. Nitrogen-fixing bacteria are invisible to the naked eye but the nodules are seen as distinct swellings on the roots (*see facing page*). The presence of root nodules on peas and beans can be used to some benefit when following a rotation system for growing vegetables (*see p.159*).

NUTRIENT REQUIREMENTS AND ASSOCIATED PROBLEMS

Element	Specifically aids	Symptoms of deficiency	Possible causes	Remedies
Nitrogen (N)	Leaf and shoot growth.	Pale foliage, poor growth, and may develop reddish or yellow discoloration. Older leaves affected first, as the nitrogen is moved from these into the newer growth when in short supply.	Most common on soils with low organic matter or where undecomposed organic matter is added to soil in large quantities.	See p.159
Phosphorus (P)	Root growth.	The plant's overall growth, including flowering and fruiting and the growth and development of seeds, is affected.	Most common on acidic soils or following heavy watering or rain.	See p.167
Potassium (K)	Flowering and fruiting; wood ripening; tissue strengthening.	Leaves may show scorching of the tips. Fruiting and flowering may be reduced and on tomatoes it causes blotchy ripening.	Most common on sandy soils and on chalk.	See p.170
Magnesium (Mg)	Chlorophyll formation (the green pigment in plants).	Leaves, particularly the older ones, show yellowing between the veins and around the edges. When it is in short supply, magnesium is moved from the older tissues into the new ones, so young foliage rarely shows any symptoms.	Most common on sandy soils and following heavy watering or rain, or heavy applications of potassium.	See p.154
Calcium (Ca)	Formation of plant cell walls.	The most commonly seen symptoms are bitter pit on apples and blossom end rot of tomatoes. Foliage may also be reduced in size and distorted.	Most common on acidic soils, and on plants growing in any soil that is not adequately watered to allow the plant to take up calcium.	See p.122
Iron (Fe)	Chlorophyll formation.	Leaf yellowing, mainly between the veins and most apparent on the young leaf growth.	Most common on alkaline soils.	See p.148
Manganese (Mn)	Chlorophyll formation.	Yellowing between the veins on young leaves is the most common symptom.	Most common on alkaline soils and waterlogged or poorly draining soils.	See p.155

Natural Upsets

In any natural environment there are factors that are not ideal for plant growth. These naturally occurring problems are often the result of adverse weather conditions that either cause physical wounds through which infections may enter, or, more commonly, simply weaken the entire plant so that it is more likely to succumb to pests and diseases.

Often gardeners can compound problems, either through "contributory negligence" – leaving drought-stricken plants unwatered, for example – or by unintentionally inflicting damage such as that caused by poor pruning or by overtight stem or tree ties. By far the most common contributory factor, however, is the poor choice of plants for a given site. If a plant is not suited to the prevailing conditions and climate, or microclimate, it will rarely thrive and hence will be more prone to problems. A less than fully hardy plant may be so set back in a cold, exposed site each year that it never makes a good specimen. A moisture-loving plant will fare badly on dry soil; a Mediterranean plant may rot in waterlogged ground. Always check before buying that the plant stands a good chance of survival in the conditions your garden offers.

Water and Drought
Water is a fundamental plant requirement, used in photosynthesis (*see pp.74–75*), to maintain turgidity (the ability of stems to stand upright) and to transport nutrients into and around the plant tissues.

▉ Damage Caused by Environmental Factors

Frost-damaged Pieris japonica leaves

Frost Damage
A scorched appearance is the most common symptom of frost damage. Soft new growth is most vulnerable to damage, with blackening and withering of shoot tips; pruning should not be left so late in the season that regrowth cannot ripen.

Lightning Damage
The larger and more ragged wounds are, the more slowly they will heal. Until a wound has completely callused over with scar tissue, it is vulnerable to invasion by harmful pathogens.

Animal Vandalism
Few plants can withstand repeated injury caused by browsing deer and rabbits, which may cause considerable problems in gardens in rural areas or close to parkland.

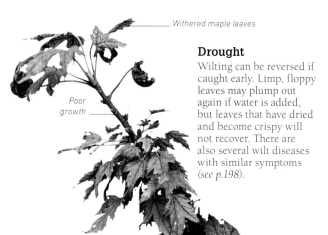
Withered maple leaves

Poor growth

Drought
Wilting can be reversed if caught early. Limp, floppy leaves may plump out again if water is added, but leaves that have dried and become crispy will not recover. There are also several wilt diseases with similar symptoms (*see p.198*).

Plants vary in their requirement for water and some are better than others at surviving dry conditions. Wilting is the most common symptom of drought; if soil moisture levels are restored fairly rapidly it can be reversed. Prolonged drought can, however, cause parts or all of the plant to die. If water levels are consistently low, but never in drastically short supply, the plant may survive, but be small and fail to perform well. Poor flowering, buds that fail to develop, and small fruits with tough skins are common symptoms. Dry soil is most obviously associated with dry weather, but light soil texture and drying winds may also be factors.

Waterlogging and Irregular Watering

Excess water can be just as harmful as drought. If soil is waterlogged, plant roots may deteriorate rapidly, largely due to the lack of oxygen. Waterlogged soil is also more likely to contain harmful pathogens, such as damping-off or foot- and root-rotting fungi. In the early stages, however, chlorosis or yellowing, often associated with the veins of leaves, is a common symptom.

Plants also find it difficult to adapt to an erratic water supply. Plants growing in potting compost in containers are very susceptible, as they tend to dry out readily and may then be overwatered by

Poor Pruning
Stubs left after pruning are readily colonized, as here, by rotting fungi that may spread.

an over-enthusiastic gardener. Symptoms include poor or slow growth, bud drop, and cracked or split fruits.

Extremes of Temperature

In addition to drought symptoms, high temperatures often cause scorching of leaves, particularly on plants growing under glass. Low temperature injury is most commonly seen as frost damage (*see facing page*), causing withering of shoots, foliage, or flowers or even death of the whole plant. Plants growing in very cold areas or in containers may also suffer root injury due to freezing of the root ball. This may then be followed by poor growth, dieback, or even death. Fruit set is frequently affected by low temperatures as the activity of pollinating insects may be restricted, and blossom damaged.

Hail may cause considerable damage, particularly when the hail storm is heavy and plant growth is still fairly young and soft. Symptoms can include leaf spotting, and in extreme cases petals or young leaves may be ripped or display numerous tiny holes. Snow, too, can injure leaves or weigh down and break branches. This problem is particularly common when a heavy fall of snow has

Water-filled flesh

Cross-section

Natural Protection

Cacti (*above and right*) are very well adapted to drought conditions. Evaporation from the plant's surface is minimized by the thick, tough outer skin, and the spines (reduced leaves) protect the water-rich, fleshy core from browsing animals.

partially melted and re-frozen. In some cases, however, snow may act as an insulating "blanket" and prevent cold wind or frost injury of the plants it covers.

Strong Winds

Wind may cause physical injury and scorching of foliage and a prevailing strong wind may also prevent a plant growing to its full potential size, or even alter its shape. Wind blows moisture away from transpiring leaves and so will, in turn, increase the plant's water requirements, making it more likely to suffer from drought if a plentiful supply of water is not available. In very windy seasons or sites, pollination may also be poor due to pollinating insect activity being restricted.

Healthy Primula obconica *flowers*

Leaves turn yellow and flop over

Flowers dying off

Waterlogging

The water in saturated soil displaces air between soil particles, creating an anaerobic environment in which plants cannot take up water. The symptoms therefore resemble those of drought, and plants may mistakenly be given yet more water.

Types of Plant Disease

The disease-causing organisms or pathogens most prevalent in gardens are fungi and viruses. In addition, several bacterial infections and a few diseases caused by phytoplasmas occur.

Fungal Infections

This major group of infections may be caused by microfungi that are extremely small but usually still visible to the naked eye. Examples include powdery mildews, fungal leaf spots, smuts, and rusts. A few fungal infections are systemic (found throughout the plant), but the majority are restricted to certain plant parts. The larger fungi – those that produce easily visible fungal structures such as fungal brackets or toadstools – are less common but may prove extremely damaging.

Virus Infections

Virus particles are invisible even with the aid of a normal microscope, and can only be seen with extremely high-power microscopes. The symptoms they cause are, however, generally easily visible, most commonly as yellow leaf markings, distortion, and stunting. Virus particles are systemic so are found throughout an infected plant, but their symptoms may only be apparent on one area, or their appearance may vary from one part of the plant to another. Virus infections may occur alone or in combination with other virus infections. It is impossible to identify precisely which are involved from the symptoms alone.

Oomycete Infections

Infections may also be caused by types of organisms known as oomycetes, which include downy mildews and *Phytophthora*. These are often referred to as "fungus-like organisms" but are actually more closely related to algae. They produce zoospores, which are single-celled structures that spread by propelling themselves with a swimming motion through moisture, for example on a leaf surface or through water in soil. Oomycetes therefore tend to cause more serious problems in damp or wet conditions.

Bacterial Infections

Problems caused by bacterial infections are usually localized, but as many cause rapid deterioration of plant tissues, their effect may be quite dramatic. Sometimes the bacteria are apparent as bacterial ooze, for example on a tree infected with bacterial canker or fireblight. Many bacterial infections that result in softening of plant tissues are accompanied by a distinct, and often unpleasant, smell. Phytoplasmas are similar to bacteria (except they do not have a cell wall) but much less common.

How Diseases Are Spread

Most fungal infections are spread from plant to plant or around a plant by spores. Their type, size, and quantity depends largely on the specific fungus. Most spores are spread either on air currents or by rain splash. Bacteria are spread by similar means. Some infections may also be

▧ Microscopic and Sub-microscopic Organisms

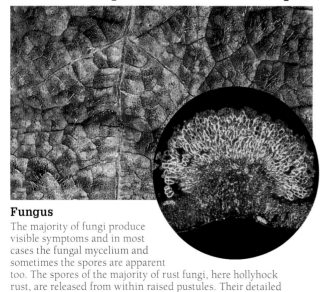

Fungus
The majority of fungi produce visible symptoms and in most cases the fungal mycelium and sometimes the spores are apparent too. The spores of the majority of rust fungi, here hollyhock rust, are released from within raised pustules. Their detailed structure is only visible (*inset*) with the aid of a microscope.

Virus
Infection by virus particles (*inset*) usually results in visible symptoms, such as leaf distortion (here from cucumber mosaic virus) and yellow patterning, often accompanied by general poor growth and stunting. Virus-infected plants may also be symptomless, or the extent of the symptoms may vary greatly.

Spreading Disease
Using cutting tools such as secateurs on an infected plant and then on a healthy one without disinfecting them can spread disease.

Colour Breaking
A characteristic symptom of virus attack is slashing or striping of flower colour. Usually unwanted, it has in some cases, particularly in tulips, been developed through breeding to produce novel petal patterns in otherwise healthy plants.

carried by animals, including insects. Soil-borne infections are usually spread either in association with soil particles or in soil moisture. The zoospores of fungus-like infections, such as *Phytophthora*, spread by propelling themselves through moisture. Some soil-borne infections, such as honey fungus, may be spread using specialised fungal structures called rhizomorphs (or "bootlaces"). Weather conditions may affect disease spread and development, and in most cases moisture on the plant surface encourages infection and spread. Humidity and the resultant film of moisture on plants may also cause a marked increase in disease spread and development.

Viruses are most commonly spread when sap from an infected plant is transferred to another potential host. Pruning tools, physical injury, grafting, and the taking of cuttings are all common routes. Even handling – for example, dividing – an infected plant may be sufficient to transfer virus particles to a healthy host. Some viruses are also transmitted by garden creatures which, because of this

role, are known as vectors. The most common are sap-sucking pests such as aphids, leafhoppers, and nematodes. Viruses may also be seed-borne.

Increasing Vulnerability
A plant which has suffered a severe attack of, say, powdery mildew, with consequent loss of leaves, may be severely stressed, particularly if infection occurs in several consecutive years. The plant may well then be more prone to attack by another pathogen, such as a leaf spot or honey fungus. In a similar way an apple that has been attacked by codling moth is very prone to attack by brown rot fungus, which gains entry via the wounded skin of the fruit.

Stresses brought about by unsuitable growing conditions or extremes of weather (*see pp.78–79*) and problems associated with nutrient deficiencies and phytotoxicity (damage caused by unsuitable nutrient levels or use of unsuitable chemicals) are often responsible for lowering a plant's ability to fight off infection and disease development.

MOST COMMON PLANT DISEASES

Disease	Type	Method of spread	Conditions making plant more susceptible	Season
Powdery Mildews	Fungi	Water/rain splash; air currents.	Dry soil/root conditions; moist or humid environment around leaves.	Spring, if dry, and summer to autumn.
Downy Mildews	Fungus-like organisms	Water/rain splash; air currents.	Humid or wet air around foliage; damp spring and autumn weather; stagnant conditions. Plant debris in vicinity; dead/damaged plant parts.	All year.
Rusts	Fungi	Water/rain splash; animals; wind.	Damp, warm, stagnant conditions; moisture film on leaves.	Mainly damp summers and autumn.
Fungal Cankers	Fungi	Water/rain splash; air currents.	Damage to bark from pruning wounds; pest attack; frost crack.	All year.
Fungal Wilts	Fungi	Pruning tools; infected root material; infested soil.	Damage creating open wounds when spores in vicinity.	All year.
Bacterial Soft Rots	Bacteria	Water splash, water films; pruning tools; insects.	Damage to the plant's surface allowing entry of bacteria; poor storage conditions.	All year.
Fungal Leaf Spots	Fungi	Mainly water/rain splash; some air currents.	Poor growing conditions; weather extremes; other diseases.	Mainly spring and summer.
Honey Fungus	Fungus	Underground rhizomorphs (fungal strands); root contact.	Any other predisposing stresses such as other diseases, poor growing conditions, weather extremes or excessive pruning.	All year.
Cucumber Mosaic Virus	Virus	Sap-sucking pests; handling; pruning and taking cuttings; occasionally seed-borne.	Presence of any factors which encourage spread (*see left*).	Mainly late spring and summer.
Phytophthora	Fungus-like organisms	Soil- and water-borne.	Injury to plant; over-wet soil conditions.	All year.

Types of Garden Pest

Animals become pests when their activities have an adverse effect on plants. They may do this by reducing plant vigour so that yields of fruits and vegetables, or the appearance of ornamental plants, are impaired. In addition to the direct damage caused by their feeding activities, pests can have indirect effects on plants. Those feeding on the edible parts of fruits and vegetables may not be numerous enough to reduce yields, but nevertheless make them unpalatable due to the presence of insects and their excrement. Many sap-sucking insects soil their host plants with a sugary excrement called honeydew that encourages the growth of sooty moulds.

Pests can also transmit some virus and fungal diseases of plants on their mouthparts or bodies. Some plant diseases are dependent on the presence of pest-damaged tissues and cannot become established without this initial help.

While some garden pest problems are local, others are the result of migrations from elsewhere. Pests may arrive on purchased plants or gifts. Pollen beetles and flea beetles can fly in from agricultural crops, such as oilseed rape, while cabbage white butterflies, which are capable of longer flights, may come to Britain from elsewhere in Europe.

Snail uses its rough "tongue" or radula to break up plant tissues

Solitary Feeders
While snails can occur in great numbers in gardens, they are essentially lone operators when browsing on plants. Each individual is, however, capable of causing significant damage.

Pests affect plants in a variety of ways. Some attack specific parts of the plant, while others feed more generally.

Sap-suckers
Sap-sucking pests such as aphids and scale insects have needle-like mouthparts that are inserted into plants in order to suck sap. No visible holes are made in the leaves, but many sap-feeders have saliva that is toxic to plants and this results in leaf distortion and/or discoloration. Some sap-sucking pests – including aphids, leafhoppers, and thrips – can transmit virus diseases. Most sap-suckers infest the foliage, stems, and flowers, but there are some aphids and mealybugs that attack roots.

Flower Pests
Relatively few pests specialize in feeding on flowers or flower buds, but these can be attacked by many general plant pests such as slugs, earwigs and caterpillars. Pests specifically associated with flowers include pollen beetles, hemerocallis gall midge, apple blossom weevil, and Michaelmas daisy mite. Some birds, such as bullfinches and house sparrows, also destroy flowers and flower buds.

Gall-formers
Some pests secrete chemicals into their host plants when they deposit eggs or as they feed. This induces a variety of abnormal growths known as galls that enclose the causal animals. The gall-forming habit is found among many types of insects, mites, and nematodes. Plant galls are also caused by some fungi and bacteria.

Root-feeders
The larvae of a wide range of insects bore into or eat roots, corms, bulbs, and tubers. Their activities are often slow to be discovered, and can be difficult to control. This may require disposing not only of the affected plant

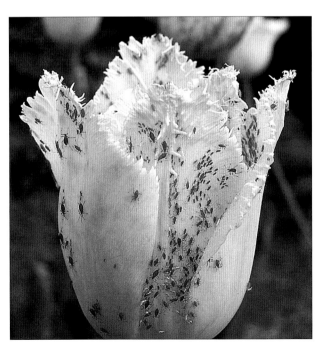

Pests En Masse
A single aphid or even a small colony presents no threat to a healthy plant. The speed at which they multiply, however, quickly results in heavy infestations that can cause damage.

Feeding Damage
Feeding by pests can spoil the appearance of a plant. It can also create open wounds that may act as entry points for secondary infections.

but of the soil in the vicinity of the rootball. Many root pests arrive as eggs laid by winged insects; others may be brought into gardens via container-grown plants. Plants purchased or received as gifts should be tipped out of their pots and the rootball examined for signs of vine weevil grubs (*see p.19*) before planting.

Fruit Pests
Fruits, berries, and seeds are attacked by the larvae of many insects, including a variety of moths, sawflies, and beetles. Wasps, birds, and squirrels can also be a problem. Some pests, such as capsid bugs, reduce the quality of fruits by causing skin blemishes; others render them inedible. Damage to ripening fruits exposes them to infection by fungal rots, so they cannot be stored.

Stem Pests
Most pests attacking plant stems feed from the outside and include most of the sap-sucking pests. Bark beetles feed as larvae under the bark of woody plants while the larvae of leopard and clearwing moths bore into the centre of stems. Mammals, such as rabbits, deer, voles, and squirrels, can damage bark, especially during the winter and spring.

Leaf Pests
Plant foliage has a greater variety of pests than any other part of a plant. In addition to the sap-sucking pests listed above, there are pests that feed on the leaf tissue itself, such as slugs, snails, earwigs, caterpillars of sawflies, moths and butterflies, leaf beetles and weevils. Some insects, such as the larvae of certain flies, moths, sawflies, and weevils, feed within the leaves as leaf miners.

How Pests Attack

Whiteflies

Sap-suckers
Adelgids, aphids, capsid bugs, froghoppers, leafhoppers, mealybugs, scale insects, spider mites, suckers, thrips, whiteflies.

Earwig

Flower Pests
Aphids, apple blossom weevil, bullfinches, caterpillars, earwigs, hemerocallis gall midge, house sparrows, Michaelmas daisy mite, pollen beetles, slugs, snails, thrips.

Vine Weevil Grubs

Root-feeders
Cabbage root fly larvae, carrot fly larvae, chafer grubs, cutworms, eelworms, leatherjackets, onion fly larvae, root aphids, root mealybugs, slugs, swift moth caterpillars, vine weevil grubs, wireworms.

Gall of a Gall Wasp

Gall-formers
Adelgids, aphids, eelworms, gall midges, gall mites, gall wasp larvae, sawflies, suckers.

Wasps

Fruit and Seed Pests
Apple sawfly, bean seed beetles, birds, codling moth larvae, nut weevil, pea moth larvae, plum moth larvae, plum sawfly, raspberry beetle, squirrels, strawberry seed beetle, wasps.

Scale Insects

Stem Pests
Many sap-sucking pests (see above), plus bark beetles, clearwing moth larvae, deer, leopard moth larvae, rabbits, squirrels, voles.

Caterpillar

Leaf Pests
Many sap-sucking pests (see above), plus adult vine weevil, earwigs, eelworms, flea beetles, lily beetle, moth and butterfly caterpillars, sawfly larvae, slugs, snails.

Invertebrate Life Cycles

The life cycles of invertebrate pests are as diverse as the pests themselves, but they can be divided into two broad categories: complete and incomplete metamorphosis. The latter involves little more than a gradual increase in size as the pest reaches maturity, with the immature stages usually resembling the adult animal. This form of growth is shown by millipedes, woodlice, mites, and some insects, including earwigs, aphids, leafhoppers, froghoppers, whiteflies, mealybugs, scale insects, capsid bugs, and thrips. The feeding habits of the immature stages are often the same as the adult and so all active stages in the life cycle are capable of causing damage.

Complete Metamorphosis

Insects such as butterflies, moths, beetles, flies, ants, bees, sawflies, and wasps, undergo complete metamorphosis. Here, the larval stages are very different in appearance from the adult and there is a pupal stage in which the larva, usually enclosed within a protective casing, changes into the adult insect. With these insects the feeding habits of the larvae are often different to those of the adult and in pest species it is usually the larvae that cause damage to plants.

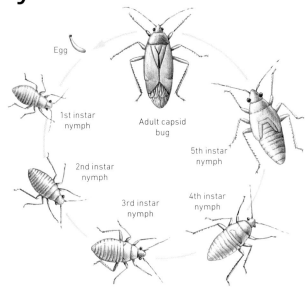

Incomplete Metamorphosis, Capsid Bug
Invertebrates that undergo incomplete metamorphosis look broadly similar at each stage of their lives and gradually acquire adult features.

For example, caterpillars of butterflies and moths feed on leaves while the adults feed on nectar. Some beetle pests, however, cause damage at both the larval and adult stages.

How Invertebrates Develop
Invertebrates generally start life as eggs, although many aphids give birth to live young for most of the year, only resorting to producing overwintering eggs in autumn. Invertebrate animals, such as insects, mites, woodlice, and millipedes, do not have any bones or internal skeletons. Instead they have a hard outer covering to their bodies. Like a suit of armour, this restricts growth, so must be shed from time to time to allow the animal to increase in size. This process is known as ecdysis. After the old outer skin has been cast off, the soft new skin underneath is stretched by the animal before it hardens, thus creating more space for growth. The number of stages, or instars, between the newly-hatched nymph or larva and the adult animal is usually between four and twelve, depending on the species.

How Invertebrates Reproduce
Most adult invertebrates are either male or female and mate in a conventional manner. There are, however, exceptions. Slugs and snails are hermaphrodites with both male and female genitalia. They mate with others of their species and both partners will subsequently lay eggs. Some insects are capable of breeding without

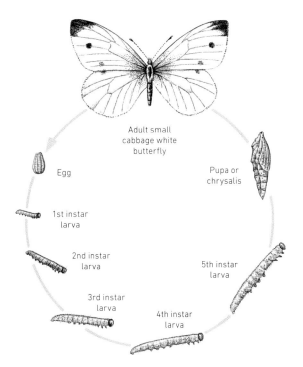

Complete Metamorphosis, Cabbage White Butterfly
Insects such as butterflies, moths, beetles, and flies undergo a complete transformation in appearance and often differ in feeding habits during their lives.

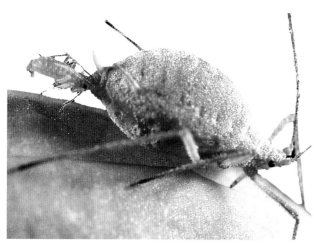

Aphid Giving Birth
Aphids give birth to live young in spring and summer, but as day length shortens and temperatures fall in autumn, they lay eggs instead, which will overwinter.

mating (parthenogenesis). Some, such as vine weevil and aruncus sawfly, only occur as females when adult. Aphids occur as females only during spring and summer; males are produced in late summer when mating is necessary to produce overwintering eggs. Some gall wasps have alternating asexual all-female generations and sexual generations with males and females.

Invertebrate Lifespans

The length of time required to complete the life cycle depends on factors such as temperature, the availability of food, and the nature of the animal. Many invertebrates are able to complete their development within a year, though some require much less time, and others take longer. Wireworms, for example, feed on roots in the soil for up to five years before they pupate and become adult click beetles. Aphids, in part by cutting out the egg stage, are able to go from young nymph to adult insect in as little as seven days. Invertebrates that have more than one generation in a year complete their development more rapidly in warmer weather. Glasshouse red spider mite takes 55 days to complete its life cycle at 10°C (55°F), but only 12 days at 21°C (70°F). Most outdoor pests have one to three generations a year. Glasshouse pests like whitefly and red spider mite can go through as many as 12 generations a year due to the warm, sheltered conditions in which they live.

Seasonal Changes

Pests have evolved alongside the plants they feed on and so they begin to attack their hosts at the time of year when the appropriate plant tissues are available. This is strongly related to temperature and day length, which

not only control plant growth processes such as bud burst, flowering, fruiting, and aging, but also bring pests out of winter dormancy and allow egg hatching. Similarly, the life cycles of predatory insects, such as ladybirds, are related to that of their prey, aphids.

Surviving the Winter

The means by which invertebrates overwinter varies, but each species will usually have one particular stage in its life cycle in which it survives the colder months of the year. This may be the egg, one of the nymphal or larval stages, the pupa or adult. The overwintering stage generally occurs in some sheltered place, such as in the soil, in the crevices of bark, or inside dense shrubs such as conifers. Many insects and mites go into a form of hibernation called diapause from which they will not emerge until they have experienced a sufficient period of cold. This prevents them emerging prematurely whenever the winter weather turns mild. Some glasshouse pests, such as whitefly, continue feeding and breeding throughout the year, albeit more slowly in winter than in summer.

Annual Populations

While some pests are a problem every year, others are only occasional problems. Serious outbreaks may be due to a combination of favourable weather conditions and a relative absence of predators, parasites, and diseases of the pest. Minor pests often need a run of several good reproductive cycles before they are numerous enough to become a significant problem. Hot, dry summers favour certain pests such as spider mites and thrips, while slugs and snails thrive under damp conditions. Cold winters do not kill pests as all those native to northern Europe have a cold-tolerant overwintering stage. Mild winters can allow some pests, such as aphids, that would normally go dormant, to continue feeding and breeding, resulting in higher populations in spring.

LADYBIRD LARVA EATING APHID   ADULT LADYBIRD EATING APHID

The Beneficial Ladybird
Both larval and adult stages of some ladybirds feed avidly on aphids, making them doubly valuable to gardeners. Some are also predators of scale insects, thrips, mealybugs, or mites.

Approaches to Control

When something goes wrong in a garden and pests or diseases are to blame, the first consideration is whether control measures are necessary, and if so what action should be taken. Not all infections or infestations merit control. Powdery mildew infections at the end of the season on herbaceous plants may look unsightly, but as the leaves are going to die back shortly anyway, the infection has little detrimental effect and may not be worth treating. Similarly, it may be easier and more sensible to pick off one or two caterpillars on a cabbage than attempt to control them with an insecticide.

Some problems cause rapid and severe damage, while others cause limited damage and are probably more worrying to the gardener than they are to the plant. It is important to balance the need to treat with the damage

actually caused. Farmers and growers have an "economic threshold" – the point at which a pest or disease is causing or is likely to cause sufficient loss to merit the time and money spent spraying. This is something the gardener should consider, modifying the concept to include the extent to which a problem spoils a plant's appearance, or ability to crop before reaching for the sprayer.

Using Garden Pesticides and Fertilizers

Pesticides can offer rapid, safe, and usually effective solutions to some pest or disease problems when used correctly (*see pp.94–95*). Garden products available today have been carefully formulated and tested so that adverse side-effects should be minimal (*see pp.96–97*). Most gardeners who do not follow organic gardening

▉ Integrated Control in the Garden

The integrated approach to controlling problems combines appropriate plant choice and good gardening practice, preventive measures to avoid problems, and sensible use of pesticides only when warranted by serious pest or disease attack.

Pesticide sprays
Spraying can combat serious crop pests and diseases.

Avoiding problems
Traps and barriers (see pp.86–89) prevent many pest problems from occurring.

Special care
Roses rarely escape problems and should be inspected regularly.

Grease band

Compost heap
Composting healthy plant material returns nutrients to the soil.

Tidy flower beds
Tidy beds with well-spaced plants minimize conditions that encourage disease.

Barriers thwart flying pests

Crop rotation prevents a build up of problems

Using a trellis
Climber held away from wall by trellis allows good air circulation.

Out of reach
Training keeps crops off the ground, where they may be eaten by insects, slugs, and mice.

Lawn problems
The perfection or otherwise of a lawn, and thus the amount of chemicals that need be used, is generally entirely a matter of personal choice.

Companion planting
Flowering plants can attract beneficial insects that may help control vegetable pests.

Edible crops
Chemicals formulated for garden crops are safe to use on listed fruits and vegetables, provided that harvest intervals are observed.

methods are also happy to use non-organic fertilizers. The non-organic approach to gardening is still more widespread than the organic, although many people who would not call themselves organic gardeners do also use organic methods.

Organic Gardening

Organic methods of gardening have become increasingly popular. Many gardeners have started to question the use of too many garden pesticides and fertilizers, and some use purely organic methods. Organic gardening avoids the use of manufactured chemicals, but organic gardeners can use pesticides, such as pyrethrum and plant oils, as these are made from natural ingredients.

The organic approach also advocates the use of only naturally occurring materials, such as manure, for improving soil fertility, and materials such as leaf mould for mulching or improving soil texture. It embraces other environmental concerns, too, such as the use of coir or bark-based growing media in preference to peat-based products, and of recycled materials wherever possible.

Organic gardening also makes full use of what are, essentially, good gardening techniques, such as pruning out infected material and good garden hygiene (*see pp.100–104*). Resistant or tolerant plant cultivars and trap and barrier methods (*see pp.88–91*) are also important to organic gardeners.

To succeed at organic gardening it is also important to encourage populations of beneficial creatures, such as naturally occurring pest predators and parasites (*see pp.92–93*). The avoidance of pesticide use makes this easier, but it should be remembered that some of the products accepted by many organic gardeners (for example, pyrethrum) are not selective in their activity and so may harm the beneficial organisms that organic gardeners try so hard to encourage.

Integrated Control in the Greenhouse

The enclosed environment of a greenhouse can increase problems, but may also assist control. While many pests flourish in the warm, sheltered atmosphere, they may also be targeted much more successfully by traps, sprays, or the release of the predators or parasites known as biological controls.

Irrigation system
A simple automatic system ensures that plants receive an even and adequate water supply, avoiding drought stress. Watering plants from below, rather than above, also minimizes disease problems by preventing a film of water from resting on leaves and stems.

Shading

Sticky traps can be successful within enclosed spaces

Temperature control
A greenhouse heater, shading, and a thermometer are aids in preventing high- and low-temperature injury.

Healthy plants
Plants will be much more resistant to problems if they are well cared for.

Heater with automatic switch

Out of reach
Gap behind staging foils pests crawling up the walls to reach plants.

Special conditions
Plants can be given more favourable conditions and special growing mediums in a controlled environment.

Rodent control
Traps to protect crops enable mice to be evicted from the greenhouse.

Beneficial creatures
Frogs and toads, which are insectivorous, are welcomed in the greenhouse by most gardeners.

Floor kept well swept, clean and dry

Saucers of water prevent insects from crawling up legs of staging

Pesticide-free Controls

Relying solely on pesticides to control pests and diseases can bring long-term problems in the form of pesticide resistance, harm to non-pest animals, and consequent damage to the environment. Integrated control is a more environmentally friendly approach to tackling pests and diseases. It includes recognizing the role that natural enemies play in reducing pest numbers and using complementary control measures that help pest predators and parasites to survive. This may involve more careful monitoring of pest and disease infestations to determine whether pesticide treatments are needed and, if so, when they should be applied to achieve the best effect. This approach can significantly reduce the number of spray applications.

Preventive measures (*see pp.100–104*), such as removing badly infested or diseased plants, ensuring plants have adequate water and nutrients, growing plants suitable for the local soil and climate conditions, and selecting cultivars with pest or disease resistance where available, are all part of an integrated control approach to limiting pest and disease problems, as is the use of biological controls (*see pp.92–93*). Picking off pests by hand at an early stage can prevent many problems from escalating.

A Beer Trap for Wasps
Half-fill a jam jar with beer and seal it with some paper and an elastic band. Make a small hole in the paper and, in their attempts to reach the beer, the wasps will drown.

Pests that are more active at night, such as slugs and snails, are best gathered up with the aid of a torch on damp evenings. On plants grown in or around garden ponds, you should always pick off any pests by hand, since it is dangerous to fish and other pond life to apply pesticides in or near water.

There is also a number of traps and barriers that either prevent pests reaching their intended goal, or draw them to a more tempting destination. Some, such as beer traps for slugs and wasps, kill the pests; others are monitoring devices that indicate when controls will be most effective.

Pheromone Traps

Pheromones are volatile chemicals produced by some insects as a means of communicating with others of the same species. They play an important role in maintaining the organization of social insects such as ants, bees, and wasps. Pheromones are also used by some insects, usually females, as a means of attracting males for mating purposes.

The pheromones of some pest species, such as codling moth, plum moth, box tree moth, leek moth, and carnation tortrix, have been identified and the chemicals synthesized for use in pheromone traps, which are sold in kit form (*see above right*). Once the trap is in position male moths fly into it, expecting to find a virgin female, but instead get caught on the sticky sheet. The main function

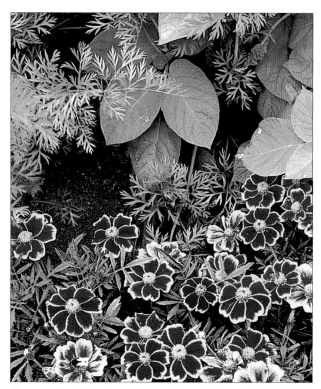

Companion Planting
Companion plants were once thought to deter pests from crops. It is more likely that they attract pest predators and initiate a form of biological control (*see p.92*), though the effect is not certain.

How to Assemble a Pheromone Trap

1 Kit contents A self-assembly pheromone trap kit will include a fold-together, open-sided box with hanging wire, some sticky sheets and some synthetic pheromone lures.

2 Assembling Make up the box and place a sticky sheet in the bottom of it. Put a pheromone lure on top of this sheet. The sheet and pheromone should be replaced after six weeks.

3 Hanging Hang the trap by its wire on the tree, at about head height and within the central canopy. It should be put in position at the expected beginning of the moth's flight period.

of these traps is to record the moths' flight and egg-laying period accurately so that, if necessary, insecticides can be applied at the right time to control the newly-hatched larvae. On isolated host plants pheromone traps may catch sufficient male moths to reduce the mating success of females, and so limit the number of fertile eggs that are laid. Traps baited with the scent of raspberry plants are also available to capture adult raspberry beetles.

Grease Bands and Sticky Traps

Some insects cannot fly and it is possible to prevent them gaining access to their host plants by placing a sticky barrier in their path. Grease bands are mainly applied to the trunks of fruit trees to trap winter moth and other related species that have wingless females. These emerge from pupae in the soil during autumn to spring and crawl up the trunk to lay eggs on the twigs. Grease bands applied to the trunk and tree stake in mid- to late autumn will reduce the number of eggs laid.

Other crawling insects that can be deterred by sticky or greasy barriers are earwigs, ants, and adult vine weevils.

A band around the base or the rim of containers can be very effective. Smearing grease around stems stops earwigs climbing up into flowers such as chrysanthemums and dahlias. Flea beetles can be made to jump on to a sticky trap: grease a piece of wood or card, and draw it, grease side down, along the tops of plants, being careful not to touch them with it, but agitating the foliage with your hand or a stick. Flea beetles jump when disturbed, and so will be caught on the trap.

Sticky traps are also useful in greenhouses for catching winged insects, such as whiteflies, thrips, and fungus gnats. These traps are yellow plastic strips covered in non-drying glue. They are suspended above the plants and when insects alight they are unable to escape. Such traps may not give adequate control on their own, but they give an early warning of the presence of pests. It is often easier to see whiteflies on a yellow trap than to find them by searching the underside of leaves. Early detection of an infestation allows the pest to be controlled, either by introducing biological controls or by using pesticides, before the plants are damaged.

Type of control	Name of active ingredient	Problems used against
ORGANIC CONTROLS CURRENTLY APPROVED		
Contact insecticides	Fatty acids	Aphids, mealybugs, spider mites, scale insects, thrips, and whiteflies.
	Plant oils, inc. winter wash	Aphids, mealybugs, spider mites, scale insects, thrips, and whiteflies.
	Pyrethrum	Aphids, leafhoppers, mealybugs, red spider mites, scale insects, thrips, and whiteflies. Also small caterpillars and some beetle pests.
Slug bait	Ferric phosphate	Slugs and snails

HANGING A STICKY TRAP

Other Traps

Other traps can be used to protect flowers from earwigs, and roots from wireworms. Grapefruit skin and beer traps can be used against slugs and snails. The earwig trap uses an inverted flower pot filled with dry grass and placed over a cane among susceptible plants. Earwigs feed at night and will crawl into the trap to shelter during the day. You can then remove the pot and pick out the earwigs. To limit wireworm damage, fix a potato on a stick and bury it: the wireworms will move in, preferring it to plant roots. The potato with its pest inhabitants can be regularly removed and renewed.

Hollowed-out grapefruit halves can be propped up on stones on the ground in flower beds; slugs and snails are attracted to the smell, crawl in, and remain there. The grapefruit skin can be picked up in the morning, complete with the slugs, which can then be disposed of. A jar half-full of beer can be almost buried in the ground, where the smell will attract slugs and snails, which fall in and drown. The addition of some porridge oats makes it an even more effective bait. (For how to make a beer trap for wasps, *see p.88*).

Barriers

There are various barriers that can be used in gardens to deter pests and provide an alternative to pesticide use. They range in size from fruit cages (*see below*) and large glass cloches to individual plastic cloches (*see facing page*).

To protect smaller fruit bushes, such as strawberries or currants, against bird and animal damage, use mesh netting to construct a low frame, making sure that the base is secured with bricks to prevent pests getting in underneath. Apple and pear trusses can be individually enclosed in bags made from muslin or old tights. Although time-consuming, this can be effective and should be done when the fruit is almost ripe – the time it is most likely to be attacked by wasps or birds.

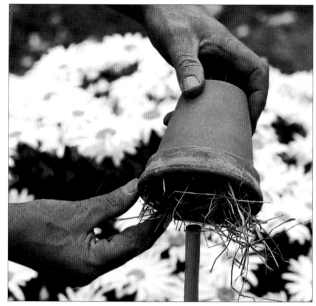

Earwig Trap
Earwigs are a particular nuisance on chrysanthemums because they crawl to the top of stems and eat the buds and petals. This earwig trap is an easy and effective control.

A variety of traps can be used in a greenhouse (*see p.87*); these include putting legs of tables or any other staging in saucers or pots of water. This will prevent insects climbing up onto the plants.

Some animals, such as hedgehogs and frogs, are highly beneficial in a garden (*see pp.68–71*) and should be encouraged. However, others, such as deer, rabbits, and moles, can cause extensive damage. Fences can be used against rabbits and deer; wire netting buried at least 15cm below ground with 75cm above should deter rabbits. Deer will need a much taller barrier (*see p.131*). Mice (*see p.156*) and moles (*see p.157*) need traps.

FRUIT CAGE

PROTECTIVE FLEECE

Protective Covers
The fruit cage (*far left*) is made from wire or plastic netting supported by metal posts. This type of cage can protect small fruit trees or bushes from bird and animal attack. The netting (*left*) protects and insulates plants and seedlings, while allowing light and air to reach them. This tunnel is constructed by using strong wire hoops to support the insect barrier fabric.

Insect Barrier Netting

In recent years several types of fine mesh netting have become available. These are lightweight materials, usually white or green in colour and made of polypropylene or other plastics, that can be laid over developing vegetable crops. These nets give the young plants valuable protection against wind and cold, and also exclude pests such as pigeons, cabbage butterflies, carrot fly, cabbage root fly, and flea beetles.

When using netting to give protection against insect pests, it is also essential to adopt a crop rotation system (*see p.104*). If, for example, carrots are sown where carrots or parsnips were grown the previous year and then covered with netting, adult carrot flies may emerge from pupae buried in the soil and will then be trapped under the covering with the crop.

Brassica Collars

Transplants of brassicas and other related vegetables, such as swedes, are prone to attack by cabbage root fly. A ready-made collar, or one made using a piece of carpet underlay, or cardboard, is placed around the stem of the young plant. This prevents the female flies from laying their eggs in the soil; instead, they will lay them on the collar, where they will dry up before hatching.

Hot Water Treatment

This form of pest control is mainly carried out by commercial growers, but can be attempted by gardeners. It is used against certain bulb pests, such as narcissus

An Individual Cloche
A cut-off plastic bottle placed over young plants in early spring makes an ideal slug deterrent and also acts as a mini-cloche, enabling soil around the roots to warm up more quickly.

eelworm and bulb fly, and also against eelworms on some herbaceous plants such as chrysanthemum. The technique involves immersing dormant bulbs or plant stools in water held at a temperature that is high enough to kill the pest without harming the plant. For details of hot water treatments, see chrysanthemum eelworm (*p.126*) and narcissus eelworm (*p.158*).

Care must be taken to ensure that the temperature is kept constant throughout the period of treatment, since too much heat damages the plants while insufficient heat allows the pest to survive. After treatment, bulbs are planted out in fresh soil, while herbaceous plants are potted up to produce cuttings.

Repellents

Repellents are used principally against birds and mammals, where the aim is to deter feeding or encourage them to move elsewhere, rather than to cause them any harm.

Humming tapes are used to protect plants against pigeons and other birds, while devices that produce ultrasonic sound are available for frightening away moles, deer, and cats. Other means of keeping pets and pest animals away from plants rely on using irritant substances, such as pepper dust, or compounds that give plants unpleasant tastes or smells. Repellents may need frequent replenishment to maintain any effect. It is also advisable to change the type of repellent in use to avoid the target animals becoming familiar with the sound, smell, or taste and losing their fear of it.

Brassica Collar
Prevent female cabbage root flies from laying their eggs in the soil around the stems of vulnerable young brassicas by placing a disc or collar around the stem base.

Biological Controls

This approach to control involves enlisting the help of pests' natural enemies – predators, parasites, and fungal, bacterial, or viral diseases. These natural enemies limit pest populations, but are not always successful in stopping them from damaging plants. This is often because pests breed more rapidly in the spring and early summer so that heavy infestations are already present by the time beneficial insects become numerous.

Biological control occurs naturally in gardens: ladybirds eat aphids, for example, and thrushes attack snails. Today, however, the term "biological control" is more commonly taken to refer to the use of natural enemies that are introduced under suitable conditions to give effective control. This approach has more success in greenhouses where specific predators and parasites can be purchased for use against most of the major pests. There are now several biopesticides for the control of plant diseases on the professional market, with varying success rates, but these are not currently available to home gardeners due to the need for protective gear so that live microorganisms are not inhaled.

Benefits of Biological Control
Biological controls may be more suitable to use than pesticides. When used on food plants, such as tomato

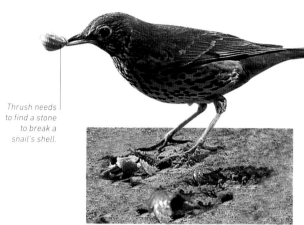

Thrush needs to find a stone to break a snail's shell.

A Useful Bird
While a few birds are a nuisance in the garden, song thrushes have a useful role, feeding on snails. Conveniently placed flat stones in or near beds and borders provide them with anvils on which to smash the shells.

and cucumber, for example, there is no problem with pesticide residues on the edible parts. Similarly, the risk of pesticides causing spray damage is avoided.

Predators and parasites can seek out pests living in places that are difficult to reach with pesticides. Once released, natural enemies can be left to get on with controlling the pests; pesticides often require repeat applications, particularly during the summer when high temperatures allow pests to breed rapidly. Biological controls are usually selective, controlling particular pests with no harmful effects on non-target species. Some pests, such as glasshouse whitefly and red spider mite, can become resistant to pesticides, making biological control the only effective treatment.

Encouraging Pest Predators
The mixture of flowers and other plants found in a typical garden is helpful in providing, directly and indirectly, food and shelter for a wide range of beneficial insects, birds, and pest-eating mammals such as bats, shrews, and hedgehogs. Installing a pond also helps as it provides a breeding site for frogs and toads. In the autumn many of these pest predators will seek out sheltered places where they can survive the winter. A thorough tidying-up of the garden at the end of the summer may be pleasing to the eye, but could result in the loss of many potential allies in the battle against pests.

The type of pesticide used on garden plants can also have a significant effect on predators. Some pesticides are more selective than others for pests and their use limits

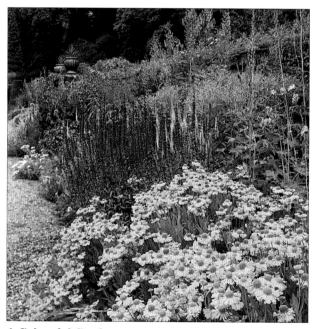

A Colourful Garden
Bright flowers, particularly in shades of purple and yellow, will attract many types of beneficial insects, such as hoverflies, whose larvae feed on greenfly and other aphids.

the harm caused to non-pest animals. Organic sprays, for example, can give good control of aphids, but, because of their short persistence, have limited impact on most beneficial insects and mites.

Buying and Using Biological Controls
The predators, parasites, and pathogenic agents that are commercially available are intended for use against specific pests. It is therefore essential that pests are accurately identified. The various biological controls are compatible with each other and can be used in combination; however, they must not be used together with pesticides, or immediately afterwards. Predators and parasites are very susceptible to most insecticides and must be regarded as an alternative to pesticides rather than an additional treatment. Synthetic pyrethroids, such as deltamethrin, cypermethrin, and lambda-cyhalothrin can be harmful to predators and parasites up to eight weeks after they are applied.

   All of the biological controls are living animals, which should be released onto the infested plants, or the soil, or potting compost, as soon as they have been obtained. They can sometimes be bought from garden centres, but more usually are sold by mail-order firms. A list of biological controls and suppliers is given on the RHS website: rhs.org.uk.

Applying Biological Controls
Pathogenic nematodes are watered into the soil or potting compost. The nematodes must have moist, well-drained soil to enable them to survive and move between the soil particles. The vine weevil nematodes and the slug nematode can work in a soil temperature of 5°C (41°F) upwards, but nematodes used against chafer grubs and

Predatory Midge
Aphids can be among the most damaging and common of pests. The larvae of a tiny fly, *Aphidoletes aphidimyza*, though only up to 3mm long, will kill and eat up to 80 aphids each.

leatherjackets need soil temperatures of 14–21°C (57–70°F). It is best to apply nematodes in the evening to avoid rapid drying out of the soil surface. The other predators and parasites that are available to gardeners (*see the table below*) need time to breed before they are numerous enough to overcome pests. It is therefore necessary to introduce these biological controls before pest infestations become heavy. Although they can survive lower temperatures, these predators and parasites need warm daytime temperatures, usually at least 21°C (70°F), and good light, if they are to be effective. Because of this their period of use is likely to be between early spring and mid-autumn. The bacterium *Bacillus thuringiensis* is no longer available to amateur gardeners for caterpillar control.

TYPES OF BIOLOGICAL CONTROL

Pest	Name of biological control	Type of predator or parasite	How it works
Aphids	*Aphidoletes aphidimyza*	Predatory midge	Larvae eat aphids.
	Aphidius species	Parasitic wasps	Larvae develop inside the aphids' bodies.
Chafer Grubs	*Heterorhabditis bacteriophora*	Pathogenic nematode	Infects chafer grubs with a fatal disease.
Fungus Gnats (Sciarid Flies)	*Hypoaspis miles*	Predatory mite	Feeds on the flies' larvae.
Glasshouse Red Spider Mite	*Phytoseiulus persimilis*	Predatory mite	Eats the eggs, nymphs, and adult spider mites.
Glasshouse Whitefly	*Encarsia formosa*	Parasitic wasp	Attacks the pest's scale-like nymphal stages.
Leatherjackets	*Steinernema feltiae*	Pathogenic nematode	Infects leatherjackets with a fatal disease.
Mealybugs	*Cryptolaemus montrouzieri*	Predatory ladybird beetle	Both the adult and larval ladybirds eat the eggs and active stages in the mealybug's life-cycle.
Slugs	*Phasmarhabditis hermaphrodita*	Pathogenic nematode	Infects slugs with a fatal disease.
Vine Weevil	*Steinernema kraussei*	Pathogenic nematode	Infects vine weevil grubs with a fatal disease.
Thrips	*Amblyseius* species	Predatory mites	Eat the immature thrip nymphs.

Pesticide Controls

Products for keeping plants in good health were first developed in the nineteenth century, when they included some controls based on arsenic, cyanide, mercury, and other chemicals far too dangerous to be used today. Only the plant-derived chemical pyrethrum survives as an insecticide from that time and is still in use in gardens.

Safety Standards

Safety testing and regulation of pesticides has become progressively more demanding as the concern about potential adverse environmental effects has grown. Before any new product can be marketed it has to be thoroughly researched to test factors such as its effectiveness, its persistence, effects on non-target plants and animals, and its long- and short-term effects on humans. These tests start in the laboratory and build up to field trials, with applications to market a new product taking about ten to twelve years to compile. No other chemicals, apart from pharmaceuticals, are as thoroughly tested before they can be sold to the public.

Pesticides are developed for the agricultural and professional horticulture market and only a small proportion of these pesticides available to professional growers are subsequently marketed for garden use. These pesticides are considered to be at the safer end of the toxicity spectrum and are supplied as formulations that can be applied without the need for special protective clothing.

Pesticides on Edible Plants

Not all garden pesticides are approved for use on fruits, vegetables, and culinary herbs. Organic pesticides, such as pyrethrum, fatty acids, and plant oils, have very little persistence and low toxicity. For these reasons, they can be used on edible plants without any limitations on the type of plant treated; for a list of approved organic controls, *see table p.89*. With synthetic pesticides the situation is very different. Only those edible plants named on the product instructions can be treated (*see table, opposite*). The instructions may also indicate the maximum number of applications permissible during the growing season, plus the time intervals between applications and between the last treatment and harvesting the crop.

CONTROLS CURRENTLY APPROVED		
Type of control	**Name of active ingredient**	**Problems used against**
Systemic insecticide	Acetamiprid	Sap-sucking insects, including aphids, capsid bugs, leafhoppers, mealybugs, scale insects, thrips, and whiteflies. Also leaf miners and leaf beetles, such as lily beetle and viburnum beetle. A compost drench formulation is used against vine weevil grubs.
Contact insecticides	Cypermethrin	A wide range of insects, including aphids, beetle pests, caterpillars, and whiteflies.
	Deltamethrin	Sap-sucking insects, including aphids, apple and pear suckers, capsid bugs, leafhoppers, mealybugs, scale insects, thrips, and whiteflies. Also caterpillars, including codling and plum moth, flea beetles, sawfly larvae, weevils, and other beetle pests.
	Lambda-cyhalothrin	Sap-sucking insects, including aphids, apple and pear suckers, capsid bugs, leafhoppers, mealybugs, scale insects, thrips, and whiteflies. Also caterpillars, including codling and plum moths, flea beetles, sawfly larvae, weevils, and other beetle pests.
	Permethrin	Dust and aerosol products for controlling ants, wasp nests, woodlice, and other crawling insects.
Lawn drench	Cypermethrin	Ant nests in lawns.
Glasshouse fumigant	Permethrin	An end-of-season clean-up of greenhouse pests, such as aphids and whitefly.
Slug bait	Metaldehyde	Slugs and snails.
Rat poisons	Brodifacoum / bromadiolone / difenacoum	Rats and mice.
Systemic fungicides	Myclobutanil	Powdery mildew, rose black spot, and rust.
	Tebuconazole	Varies with product but may include: box blight, powdery mildew, rose black spot, and rust.
	Trifloxystrobin	Lawn diseases such as turf red thread and turf snow mould.
	Trifloxystrobin + tebuconazole	Varies with product but may include: box blight, leaf spots, powdery mildew, rose black spot, rust, and white blister.
	Triticonazole	Powdery mildew, rose black spot, and rust.
Surfactant and nutrient mixes	Various plant or algal extracts and nutrients	Small pests, such as aphids, mealybugs, red spider mites, thrips, and whiteflies. Also powdery mildew.
Disinfectants	Tar acids and oils	Available as a liquid or a paste to disinfect surfaces, such as greenhouse staging, pots, seed trays, and tools. Not for use as a fungicide.
Winter wash	Plant oils	Limited control of aphid eggs and other overwintering pests on deciduous fruit trees and bushes.

Pesticide Resistance

Some pests and diseases develop resistance or build up tolerance, so that a once-effective pesticide no longer works. Continual use of the same pesticide can exert a powerful selection pressure by killing the more susceptible individuals, but leaving those with some tolerance to pass this characteristic on to the next generation.

In gardens, resistant pests are mainly found in greenhouses, where the enclosed environment and a pest's rapid reproduction rate can make pesticides ineffective against glasshouse whitefly, red spider mite, and aphids. Some diseases, such as grey mould, rose black spot, powdery mildews, and apple scab, can become tolerant of fungicides.

Reducing Resistance

Gardeners can reduce the likelihood of resistance occurring by using pesticides sensibly. Where possible, non-pesticide alternatives, such as biological controls (*see pp.92–93*) or the removal of badly affected plants, should be considered.

When pesticides are used, the continual use of the same product should be avoided. Note that active ingredients are often chemically related and it is likely that a pest or disease resistant to one will also tolerate others in the same group.

Phytotoxicity

Plants are sometimes damaged by pesticides; this is known as phytotoxicity. Read and follow carefully the manufacturer's instructions to ensure the correct dilution is applied. Some plants known to be harmed by a chemical will be listed by the manufacturer, but this information is not known for all potential plant and pesticide interactions.

The risk of damage occurring can be minimized by not treating plants already under stress as a result of exposure to bright sunlight, extremes of temperature, or dryness at the roots. Seedlings are more susceptible than mature plants, and petals are more likely to scorch than leaves. When in doubt, treat just one plant and watch for adverse reactions over the next week. Damaged plants may suffer a significant check in growth but will usually recover.

CONTROLS FOR EDIBLES

Edible plant	Name of synthetic control					
	CYPERMETHRIN	DELTAMETHRIN	LAMBDA-CYHALOTHRIN	ACETAMIPRID	TRIFLOXYSTROBIN + TEBUCONAZOLE	SURFACTANT AND NUTRIENT MIXES
Strawberries, grapes		X (strawberry only)	X			X
Blackcurrants, redcurrants, gooseberries		X (blackcurrant only)	X			X
Blackberries, raspberries		X	X			X
Apples, pears		X	X	X		X
Cherries, plums, and other stone fruit		X	X	X		X
Chestnuts, hazelnuts (inc. cobnut and filbert), walnuts			X			X
Beetroot			X			X
Carrots			X			X
Parsnips, turnips, and swede			X			X
Celeriac			X			X
Radishes		X	X			X
Tomatoes (indoors and outdoors) and aubergines (indoors only)		X	X	X (indoors only)		X
Peppers (indoors only)		X	X	X		X
Cucumbers and courgettes (outdoors only)		X	X			X
Cucumbers, courgettes, and summer squash (indoors only)					X	X
Sweetcorn		X	X			X
Cabbages, broccoli (inc. calabrese), cauliflowers and Brussels sprouts	X	X	X			X
Lettuces, curly endive, and other salad plants (inc. leafy brassicas) (outdoors only)		X	X	X (lettuce only)		X
Herbs (outdoors only)		X	X			X
Dwarf French beans, runner beans, and navy beans		X	X			X
Broad beans		X	X			X
Peas		X	X			X
Celery, fennel			X			X
Leeks		X				X
Spinach		X				X
Globe artichokes		X				X
Potatoes	X	X	X			

Pesticides and their Actions

Many pesticides are contact in action. This means they must cover the plant surface thoroughly so that the pest is either hit directly or picks up the pesticide from the treated surface. Systemic insecticides and fungicides are absorbed into the plant and may be transferred to parts that have not been directly treated. This movement may be within the leaves or up to the shoot tips. Systemic insecticides and fungicides will not move from the foliage down to the roots. Systemic fungicides generally only move within a single leaf. They kill fungal growth within plant

tissues and have a curative action, rather than a preventive role. Systemic insecticides are useful in controlling sap-sucking pests (see pp.82–83), including those concealed in distorted leaves that are difficult to reach with contact sprays.

A few pests can be controlled by pesticides that have been mixed with something palatable to form a poisoned bait. These include rat poisons, slug pellets, and ant baits. All such pesticides should be used with care around pets and children. The insecticides available to home gardeners generally control a wide range of

The Ways Chemicals Work

Different problems require chemicals to be applied in different ways. Sometimes, a combination of methods of application is needed to achieve complete elimination.

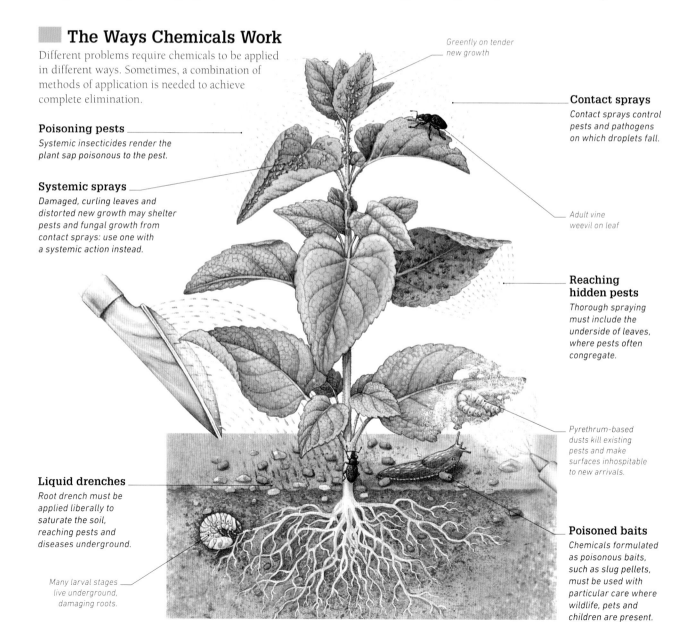

Greenfly on tender new growth

Contact sprays
Contact sprays control pests and pathogens on which droplets fall.

Poisoning pests
Systemic insecticides render the plant sap poisonous to the pest.

Systemic sprays
Damaged, curling leaves and distorted new growth may shelter pests and fungal growth from contact sprays: use one with a systemic action instead.

Adult vine weevil on leaf

Reaching hidden pests
Thorough spraying must include the underside of leaves, where pests often congregate.

Pyrethrum-based dusts kill existing pests and make surfaces inhospitable to new arrivals.

Liquid drenches
Root drench must be applied liberally to saturate the soil, reaching pests and diseases underground.

Many larval stages live underground, damaging roots.

Poisoned baits
Chemicals formulated as poisonous baits, such as slug pellets, must be used with particular care where wildlife, pets and children are present.

APPLICATION METHODS

Chemical type	Advantages	Disadvantages
Liquid Spray	Contact and systemic types available. Most have a wide range of applications. Economical.	Handling and accurate measuring of concentrate needed. Requires special equipment for application.
Liquid Drench	Contact and systemic types available. Economical.	Handling and accurate measuring of concentrate needed. Only available for pot plants and ants in lawns.
Dust	Used direct from pack, so little is wasted. It is easy to apply direct to the target.	Only contact types available. May leave unsightly deposit.
Ready-to-Use Sprays and Pump-Guns	No need to handle concentrate as contents already correctly diluted. Quick and simple to use.	Only suitable for treatment of relatively small areas of infection/infestation.
Pellets	In measured doses, so easy to apply correct quantity. Repellents in them to deter non-target animals.	May be eaten by pets, children, or wildlife. Can become less active after rain.
Aerosol	Quick, simple, and easy to use. No equipment needed.	Propellant may be potentially flammable and may prove phytotoxic to plants.

insects, including beneficial types if they are on the plants when they are sprayed. Such pesticides are said to have a broad spectrum of activity. There are others, available to professional growers, which are selective, such as pirimicarb, which control aphids but leave most other pests and beneficial insects unharmed.

Other pesticides may be relatively selective by having a short persistence. Organic insecticides, such as pyrethrum, fatty acids, and plant oils, are broad-spectrum treatments, but remain effective on the plant for less than 24 hours, so their impact on non-target organisms is limited.

Protecting Ponds
Fish are extremely susceptible to harm from pesticides, which should not be used on plants in or near ponds. Avoid chemical fertilizers also where they may leach into pond water.

Pesticides in the Food Chain
Birds may pick up insects and other pests treated with pesticides, either to eat themselves or to feed to their young. It is difficult to prevent this happening, but it is rare for them to eat sufficient amounts for harmful effects to be observed, and this is taken into account during the registration process.

Persistence of Pesticides
The rate at which insecticides and fungicides lose their effectiveness and break down into other less active compounds varies. This is largely due to the degree of stability of the chemical's molecular structure, but is also affected by factors such as temperature and exposure to sunlight.

Most pesticides currently available to gardeners remain active against pests or diseases for less than two or three weeks after application, and the period of protection is often much less. Organic insecticides, such as pyrethrum, plant oils, and fatty acids, remain effective on the plant for less than 24 hours. It is important, therefore, to spray plants thoroughly, and to apply the treatment at the correct time or times in order to get the best results. Pesticides used on fruits, vegetables, and culinary herbs often require an interval between treatment and harvesting. For the pesticides currently available to home gardeners, this harvest interval varies from nothing to two weeks.

Harm to Other Creatures
Many pesticides, in particular some insecticides, have the potential to be harmful to creatures other than garden pests. If used with care, and precisely according to the manufacturer's instructions, the risk will be low. Product labels will state clearly where risks exist and it is important to take all possible efforts to keep any creatures that could be harmed out of the area where a pesticide is being used. In most cases, once a spray has dried on the plant which has been treated, it no longer poses a threat to pets, children, or wildlife. Avoid applying pesticides when plants are in flower in order to protect pollinating insects. If treatment is essential on flowering plants, spray in the evening when bees and other daytime pollinators are less active, though this still risks affecting night-time pollinators.

Applying Pesticides

Garden pesticides are a useful tool for controlling pest and disease problems. If used with care and only when needed, they pose little threat to the environment. Changes in legislation and increasing concerns about safety, both to the environment and the user, mean that the range of pesticides available is constantly being reviewed. All registered garden products have been rigorously tested before being marketed.

Before spraying plants, always consider whether the situation really necessitates their use. Slight pest or disease problems, or those that occur towards the end of the season may not pose a sufficient threat to the plant.

Wherever possible, select a treatment that controls just the pest or disease. Formulations that contain both a fungicide and an insecticide are only appropriate when both the specified pest and disease are present at a level that warrants control.

Read the Label

Always consult the product label to determine precisely what the pesticide can be used for and whether it is formulated for use against the problem and on the plant in mind. Products carrying a label recommendation for use on edible crops will also specify a minimum interval that must elapse between application and consumption of the fruit or vegetable. Provided that this harvest interval is observed, it is safe to eat the crop. The timing, application rate, and frequency of application are also very important and again information about this is available

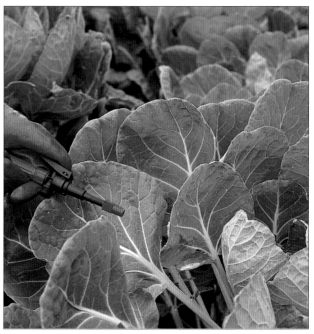

Healthy Vegetables
To make sure that vegetables and fruits are safe to eat after treatment (cabbage being treated for whitefly, above), keep to the harvest interval recommended on the product label.

on the label. Applying a pesticide at the incorrect rate, frequency, or time may prove ineffective, damaging, or simply wasteful, and so it is essential to follow the manufacturer's instructions carefully.

Rules for Safety

When applying pesticides there are a number of general rules that should be followed. Only use pesticides when necessary and always apply them according to the instructions. Avoid contact with your eyes and skin, and do not inhale sprays or dusts. Although there is no stated need for protective clothing, wearing rubber gloves when handling concentrates is advisable. Different products must not be mixed together unless recommended by the manufacturer.

Keep children and pets away and never eat, drink, or smoke while applying pesticides. Do not spray or dust in windy, hot, sunny, or completely calm weather, to minimize the threat of chemical drift affecting adjacent plants. Spray in the late afternoon, early evening or early morning. Consider carefully any potential risks to children if using slug pellets.

Equipment for Application

You may need to invest in a sprayer. This should be labelled clearly and kept separately, so that it is used for no other purpose. Use completely different equipment

▎ Hazard Information

The symbols in red lozenges shown below are standard in the UK and the EU and can be found on some garden pesticides; they indicate potential hazards, which can be avoided by correct use. In the USA, a system of signal words is used to indicate a pesticide's toxicity level to humans: CAUTION indicates slightly toxic to relatively non-toxic; WARNING indicates moderate toxicity; DANGER denotes highly toxic and is accompanied by POISON and the skull and crossbones symbol if oral, dermal, or inhalation risks are very high.

FLAMMABLE

HEALTH HAZARD

POISON

Mixing and Applying Pesticide Concentrates

1 Measuring Measure out the required amount of concentrate accurately. Avoid making up more than necessary and always read the instructions carefully. Wear rubber gloves and take care not to spill any of the concentrate.

2 Mixing Part-fill the sprayer with water and then carefully pour the concentrate in. Rinse out the measurer with water and add this to the sprayer. Top up to the required level with water and close all containers.

3 Applying If possible, spray both surfaces of the leaves. This is particularly important if contact action sprays are used. Spray at dusk to reduce the risk of scorching.

for the application of weedkillers, as even trace amounts of these can cause serious damage to garden plants. A small sprayer is suitable for most chemicals unless you have a very large garden. One that produces a fine spray is needed for controlling most pests and diseases. Insecticide or fungicide drenches or weedkillers can be applied using watering cans fitted with dribble bars.

It is important to clean all equipment regularly and, when rinsing it out after use, to spray all washings onto empty land or suitable plants. Never dispose of dilute or concentrated chemicals by pouring them down a lavatory or drain, into watercourses, or onto land where they may contaminate underground water.

Applying Pesticides in Enclosed Spaces

When sprays or dusts are used in enclosed spaces, such as greenhouses or conservatories, particular care is needed to avoid contact or inhalation. Temperatures are often higher and ventilation much lower than outside in the garden. Greenhouses or conservatories that open directly into the house pose further difficulties and wherever possible, doors or windows into the house should be closed tightly.

The risk of scorching plants is also greater, due both to the higher temperatures and to the magnifying effect that glass may have on sunlight. To minimize problems, apply pesticides in early morning or evening, in order to avoid the hotter, brighter parts of the day.

Storing Pesticides

It is very important to store pesticides safely, since even small amounts can sometimes be harmful to people, animals, and garden plants. Below are a few useful tips to help you store such chemicals correctly. Rules for safe storage also apply to the equipment with which you apply pesticides. It is worth marking items such as watering cans clearly so that they are not confused with equipment for general garden use.

Storage Tips

• Store pesticides in a safe place, preferably a locked cupboard inaccessible to children or animals.
• Keep pesticides in their original containers, with their tops tightly closed.

• Make sure all containers are clearly labelled and kept separately.
• Clear out cupboards annually, so that you do not accumulate old and withdrawn products. To dispose of them, wrap them well and contact your local council. Check for and get rid of leaking products similarly.

Safety Hazards

• Do not transfer a pesticide out of its original container into another one with an incorrect label or no label at all.
• Only buy sufficient pesticides for one to two years' use and do not store them for a long period of time. Pesticides become less effective once they have been opened and may go out of date.
• Do not keep surplus quantities of diluted sprays. Use them on suitable plants. Rinse out equipment as soon as you have finished treating plants.
• If you need to get rid of old pesticides, do not add them to household rubbish or pour them down a drain or toilet. Dispose of them safely without removing them from their original containers.

Preventive Measures

Well grown, vigorous plants are not immune to pests and diseases, but they are less likely to be badly damaged. When buying plants, bulbs, and seeds, consider paying more for good quality stock from reputable suppliers, and select plants likely to do well under the local soil and climatic conditions. If you have space to spare, it is worth keeping new purchases in a "quarantine area" for several weeks before planting. This will allow some pests and pathogens to become apparent, which may be too small or immature to see or have had their development suppressed by treatment with pesticides. Any pests or diseases that do appear may then be treated before they pose a threat to existing plants in your garden.

Resistant Cultivar
'Monclano' calabrese has been bred for its resistance to clubroot, a devastating soil-borne disease.

Resistant Cultivars

Many plant cultivars have been developed to withstand common disease strains or to be less susceptible to pest attack. These include some fruits and vegetables that virtually eliminate the risk of several serious problems. Resistant cultivars recommended for garden use are suggested in the *A–Z of Pests, Diseases, and Disorders* (see pp.105–199), but do check seed and plant catalogues regularly as the availability of resistant cultivars changes.

Care of Young Plants

Take care to ensure that planting or sowing is done properly. Sowing too early or late (when soil may be cold and wet), or planting pot-grown plants too deeply or with pot-bound roots, will result in plants that struggle and fail to make good growth. When growing your own plants, ensure that you raise strong, healthy specimens by following procedures correctly. It is important, for example, to thin, prick out, and pot on seedlings (*see facing page*) promptly. Crowded seedlings competing for limited resources will be weak and undersized, and may never make vigorous plants. Also, the younger a plant is, the more swiftly it recovers from being transplanted.

Maintaining Healthy Plants

Plant problems can arise at any time of the year but are more prevalent during spring to autumn, their peak growing period. It is a good idea to make regular inspections to detect problems before they become serious. Look for plants that are underperforming; this may show as slow growth, often with symptoms of discoloration, distortion, or damage that should be investigated as soon as possible.

Plants showing signs of attack by pests or diseases should not be ignored, since it is likely that other plants may also become affected. Treatment with a suitable pesticide can be very effective when the problem is detected and dealt with early on. In a group of plants, removal of the affected one may help stop the problem spreading. As plants grow and develop, thin out any

Choosing Healthy Plants and Bulbs

Vigorous growth

Selecting Healthy Plants
Choose plants that are free of any sign of pests or disease. Look for compact growth, with good leaf colour, and a well-developed root system. Avoid those with yellowing leaves, weak growth, or premature flowering.

Buying Bulbs
Since there is a good deal of variety in the quality and size of bulbs available to gardeners, they should be closely examined before buying. Bulbs should be as fresh and firm as possible (*see bottom right*) and without signs of mould or discoloured patches. The bulb shown top right has both diseased tissue and damaged outer scales, so it would not be a good choice. Undersized bulbs may not flower in their first year, so always choose the largest bulbs available of the chosen cultivar.

POOR EXAMPLE

Diseased tissue

GOOD EXAMPLE

Good Planting

1 Sowing Sowing into a seed tray is a good way of raising large numbers of seedlings, but they will soon need more space to allow them to develop. Overcrowding will restrict their growth.

2 Pricking out Select the stronger seedlings and carefully remove them from the tray. Handle them by gently holding the seed leaves rather than the stem, which is easily bruised.

3 Transplanting Space the seedlings out and plant out or put into pots once established. Avoid damping off by using potting compost, rather than garden soil, and tap water, rather than water from a butt.

that have become overcrowded. This improves their vigour and allows air to circulate around stems and leaves, thus reducing damp air conditions that favour germination and infection by fungal spores.

Feeding Plants

Most plants will benefit from supplementary feeding, to increase yields of fruits and vegetables and to increase the quality and quantity of flowers. Organic fertilizers include garden compost, farmyard manure, bone meal, dried blood, and seaweed extracts. These all provide a variety of nutrients, although not necessarily in the right balance required for optimum plant growth.

Man-made fertilizers can be "single nutrient" compounds, such as super phosphate (supplying phosphorus) or ammonium sulphate (nitrogen), but more widely used are general fertilizers that provide all the major nutrients. These are available as granules for adding to the soil, or in soluble form for watering in or spraying on the leaves (foliar feeding). Controlled-release fertilizer pellets will give container-grown plants the nutrients they need throughout the growing season.

Feeding is generally done during the spring and early summer, when plants are making rapid growth. Fertilizers applied in autumn or winter are largely wasted; dormant plants do not need feeding, and nutrients added then may be washed away by rain before spring.

Watering

Watering can be crucial in a garden, particularly at critical stages in a plant's development, such as the seedling stage and when flowers and fruits are being produced. In dry periods, it is important to water before plants show signs of drought stress, and also to water at regular intervals.

Watering is best done in the evening so that less water is lost by evaporation, and water should be directed into the soil, around the roots, not on the plant itself. Do not water plants in sun as the plant can be scorched. Covering the soil surface with a mulch will help retain moisture. Rain butts can be used to collect water from the roof. This source of water is suitable for most plants but is best avoided for seedlings and cuttings, as it can harbour fungal diseases that cause damping off and stem rots.

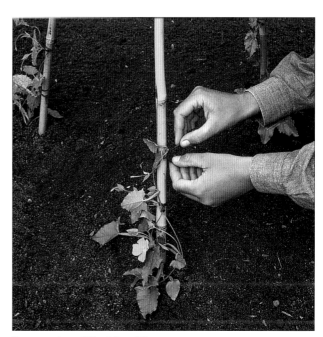

Supporting Climbing Plants
Plants that cannot support their stems need tying in to keep them off the ground, where growth will be vulnerable to insects and slugs. Even self-clinging climbers need help when young.

Drainage

Waterlogged soil can be just as damaging as dry earth to plants. A heavy soil, such as clay, has small spaces between the soil particles, making it moisture-retentive and difficult for water to drain through. Improve heavy soil texture by digging in plenty of coarse organic matter, such as compost, leaf mould, or manure, to improve aeration.

Thin soils over chalk are very alkaline. Water drains through easily, as do soil nutrients, so that acidic types of organic matter such as manures are needed to give bulk and counteract the alkalinity. Sandy soil, too, is very light, needing plenty of mulching and feeding, but it is easy to work, and warms up quickly in spring.

Weeding

Good weed control is important. Weeds compete with cultivated plants and if they gain dominance, make plants weak and more vulnerable to problems. Most weeds can be removed by hand or by hoeing, but in lawns and garden paths using a suitable weedkiller may be more convenient. Some weeds can be attacked by the same pests, viruses, and fungal diseases that occur on cultivated plants, and are a source of reinfestation after the cultivated plants have been sprayed. They also help some pests and diseases survive in the absence of cultivated host plants and thus diminish the value of crop rotation.

Removing Dead Plant Material
Unless features such as winter seed heads are required, dead or dying flowers and foliage, together with their stalks or stems, should be removed, as they can become sites for rots and fungal growth, which may spread to living plant material.

Removing Suckers to Keep Plants Strong
Rose suckers grow from the roots of the plant, since roses are grafted onto a rootstock. Gently clear away soil round the sucker to its union with the root, and cut or pull it off.

Removing Other Unwanted Growth

To keep plants strong, suckering growth from the roots needs removing, particularly on grafted plants such as roses and most fruit trees. These are usually grafted onto rootstocks, which will produce sucker growth of inferior quality to, but usually of greater vigour than, the named cultivar. If possible, the suckers should be traced back to the roots from which they arise and cut off at source. Cutting them off at soil level simply encourages more suckers. Avoid digging the soil around roses, cherries, and other plants prone to suckering as damage to the roots can stimulate the growth of suckers.

Dead, Diseased, and Damaged Growth

Pruning is generally done in autumn or winter, or after flowering, depending on the type of plant. It is, however, best to remove any dead, diseased, or damaged growth from a plant as soon as it is noticed, rather than waiting for the usual pruning time. Branches that have been broken by storms or other causes should be tidied up with a clean cut to speed callusing. Branches infected with fungal diseases, such as coral spot, canker, or silver leaf, should be cut out well below the signs of infection, to prevent the further spread of these diseases.

Routine Pruning

Not all plants require pruning but many woody plants do benefit from some form of thinning or reduction in size. This may be done on an annual basis – as with apple trees, in order to promote the development of spurs that will produce flower buds and hence fruits. Other plants may need infrequent pruning, with the intention of doing no more than removing the occasional branch or shoot

BAD CUT

Ragged tissue invites infection

BAD CUT

Budless stub will die back

BAD CUT

Too-sharply angled cut endangers selected bud

CORRECT CUT

Good and Bad Pruning Cuts
Make sharp pruning cuts close to a strong bud, with the cut sloping away from the bud. Avoid making ragged cuts or leaving stubs above the bud. Use a sharp knife or secateurs.

that is growing too far in the wrong direction. Always prune to prevent branches crossing each other, since in chafing together they may create wounds that are open to infection.

Pruning also provides an opportunity to thin shrubs and trees with a dense system of branches to provide a more open centre. This improves growth and the development of flowers and fruits by allowing sunshine

Thinning gooseberry bushes as here lets in sunshine and air

Pruning to Avoid Crowded Growth
Cut out some of the older stems in order to encourage new growth. Thin the stems if overcrowded to create a more open centre. This will help reduce the incidence of diseases.

in. There is also a better flow of air among the foliage and branches, which will help deter the germination of fungal spores. With spring-flowering deciduous shrubs, the removal of about one-third of the older stems each year will revitalize the plants by stimulating the development of young shoots.

Making Correct Cuts
Bad pruning can cause problems, rather than curing or preventing them, so it is important to make cuts correctly with clean, sharp tools. Invest in good quality secateurs to prune stems up to the thickness of a pencil; anything larger is better tackled with loppers or a pruning saw. Shoots should be cut cleanly back to a strong bud with a slanting cut (see left). Never leave a stump of stem above the bud as this will encourage dieback.

Do not cut a branch off absolutely flush with the trunk or branch from which it arises. There will be a slight thickening or collar at the base of the branch and the cut should be made just outside this point. If the collar is also removed, the growth of scar tissue (callus) over the pruning wound is impaired and can leave a large area of inner wood exposed.

It is sometimes necessary to remove large branches from trees. Before doing so, it is generally wise to seek the advice of a professional tree surgeon; major work such as this can result in an unstable or unbalanced tree.

Large branches are heavy, and it is often advisable for safety reasons to remove them in smaller pieces instead of making a single cut. This also reduces the danger of the branch dropping as it is being cut and tearing a strip of bark from the trunk. This risk can be further reduced by initially making a cut on the underside of the branch before completing the task by sawing down from the top.

Crossing Branches
Crossing branches rub against each other, wearing away the bark and creating entry points for disease.

Wound Paints
The use of proprietary wound paints on large pruning cuts is less fashionable today than in the past. There is considerable evidence to show that they can restrict the natural healing process that occurs around correctly-made pruning cuts. However, where trees and shrubs are known to be susceptible to diseases that infect fresh wounds, such as silver leaf or coral spot, it may be helpful to use a wound paint.

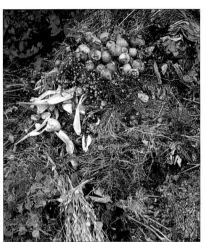

Compost Heap
The composting process will kill some pests and pathogens, but certain types of infested or diseased material must never be composted. The rotten apples shown here should be buried deep within the heap to help prevent the spread of brown rot spores.

Raking up Old Leaves
Dead leaves harbour slugs and encourage grey mould (*Botrytis*) and other diseases. Keep the garden tidy by collecting plant debris and putting it on the compost heap.

Disposal of Debris and Prunings
At the end of the growing season, tidy up plants and rake up fallen leaves to remove hiding places for pests and reduce the potential for disease carry-over. Composting garden debris is an excellent way to return nutrients and organic matter to the soil, but take care when composting unhealthy plant material as the temperatures reached in home compost bins are rarely high enough to kill many pathogens. Covering the surface of the compost heap with a deep layer of weeds, well-rotted manure, grass-clippings, or compost will help to seal in the material beneath and increase its temperature. Some of the less persistent foliar diseases, such as powdery mildews and rusts, are unlikely to survive for long in such conditions if material has been collected early enough, before the pathogens have had time to produce resilient spores or structures. However, plants with root pests or diseases and prunings from plants infected with canker or coral spot should not be composted. These should be binned or burnt, or taken to a council-run "green waste" recycling centre where the composting facilities reach far higher temperatures. Councils vary in what they are prepared to accept, so check for any restrictions before you set off.

Replanting and Crop Rotation
Growing the same or closely-related plants in the same piece of ground in successive years can lead to a build-up of problems. Changing the growing site does not guarantee protection: many insects have winged adults able to locate their host plant's new position, and many diseases are spread by wind-blown spores. There are, however, some less mobile problems, such as eelworms and root rots, which can be avoided by adopting a crop rotation system (*see box, right*).

There are other good reasons for changing the growing sites of plants. Roses and apples suffer from replant disorders. If replacement plants are put in the same place

from which an apple or rose has recently been removed, it is likely that the new plant will fail to thrive. Micro-organisms associated with the roots of the old plant seem to have an antagonistic effect on the new plant. Crop rotation in the vegetable garden also helps make better use of the available nutrients, as plants differ in their nutritional requirements.

Simple Crop Rotation

GROUP 1	GROUP 2	GROUP 3
Cabbage	Potato	Beans
Brussels sprouts	Tomato	Peas
Cauliflower	Carrot	Onions/Shallots
Kale	Parsnip	Leeks
Broccoli	Celery	Lettuce
Turnip	Parsley	Sweetcorn
Swede	Beetroot	Marrow
Radish	Spinach beet	Cucumber

FIRST YEAR	SECOND YEAR	THIRD YEAR
GROUP 1	GROUP 2	GROUP 3
GROUP 2	GROUP 3	GROUP 1
GROUP 3	GROUP 1	GROUP 2

A BED OF GROUP 3 CROPS, INCLUDING PEAS AND BEANS

A–Z of Pests, Diseases, and Disorders

This dictionary section contains entries for the most common pests, diseases, and disorders likely to be found in the garden. Some occur in varying forms on many plants, and have a general entry with cross-references. Many others are largely host-specific, and will be found under plant names. Suggested controls and preventive measures give a choice between pesticide and non-pesticide methods wherever appropriate.

A

Acer Gall Mites

SYMPTOMS In early summer, small, raised, reddish pimples appear on the upper leaf surfaces of maples and sycamores (*see p.17*), or a felt-like growth of yellowish-white or purplish-red hairs develops on leaf undersides.
CAUSE Microscopic mites that suck sap and secrete chemicals into the foliage that induce the abnormal growths. *Aceria cephalonea* causes pimple galls and *Aceria pseudoplatani* is responsible for felt galls.
CONTROL None of the pesticides available to gardeners is effective. Control is not needed, because apart from creating the galls, these mites have no harmful effect on their hosts. On small plants, light infestations may be controlled by removing affected leaves.

Acorn Gall Wasp

SYMPTOMS In late summer, oak acorns develop as sticky, ridged, yellowish-green structures (*see p.18*) instead of the familiar nut and cup. Later, the acorn (or knopper) galls become brown and woody and drop from the tree. In some years nearly all acorns on an oak tree are affected.
CAUSE A gall wasp, *Andricus quercuscalicis*, lays single eggs in developing acorns in early summer. The egg hatches into a small, white, legless grub, which secretes chemicals that cause the developing acorn to be converted into a gall. The grub pupates inside the gall, and the adult wasp emerges the following year.
CONTROL No control is available; the tree itself is unharmed by the gall wasp.

Acute Oak Decline

SYMPTOMS This infection is most commonly seen on mature sessile oak trees (*Quercus petraea*) and mature pedunculate oak trees (*Q. robur*) and their hybrids, but on occasion other *Quercus* species and younger oaks have been infected. Long cracks of 5–10cm develop in the bark; these secrete a dark liquid that may flow down the trunk (*see p.55*), or dry out and be washed off by heavy rainfall. These oozing cracks can appear at various heights on the trunk, and the tree can then die, sometimes in less than five years.
CAUSE While there is still much research to be done, the disease is believed to be caused by bacteria. It is thought that the bacteria may be spread by rain splash from the oozing areas or by insects such as the wood-boring beetle *Agrilus biguttatus*. For now, the disease remains restricted mainly to southern England and the Midlands.
CONTROL If only one tree is infected out of several growing in a garden, it may be advisable to fell it, to reduce the risk to other susceptible oaks in the area. An infected tree without others nearby could be left and simply monitored, as long as there is no reason to think it might be unsafe. All parts of an infected tree should be properly disposed of; they should not be chipped and used as mulch or soil conditioner, even if composted first.

Adelgids

Only conifers, particularly pines (*Pinus*), spruce (*Picea*), larch (*Larix*), Douglas fir (*Pseudotsuga menziesii*), and silver fir (*Abies*), are attacked by adelgids. These black aphid-like insects feed by sucking sap from the foliage, stems, and bark. Some species, particularly on spruce, cause galls at the shoot tips in which they develop, but on other

conifers they feed openly, although invariably covered by a fluffy, white waxy substance secreted from their bodies. Heavy infestations may cause some yellowing of the foliage, but generally the conifers will continue to make satisfactory growth. Control measures are limited, but are detailed in individual entries.
• *see also* Douglas Fir Adelgid *p.132*; Larch Adelgid *p.150*; Pine Adelgid *p.168*; Pineapple Gall Adelgid *p.168*.

Agapanthus Gall Midge

SYMPTOMS The flower buds on African lilies (*Agapanthus*) are distorted, become discoloured, and fail to open. Creamy yellow maggots up to 3mm long may be found inside the buds in summer.
CAUSE A tiny fly, *Enigmadiplosis agapanthi*, which lays eggs in the developing flower buds. When fully fed, the larvae go into the soil, where they overwinter as pupae. The adult flies are 2mm long with pale orange bodies and large black eyes.
CONTROL Remove and destroy infested flower buds before the larvae escape into the soil to help reduce infestations. Insecticide controls are likely to be ineffective since it is difficult to prevent the females from laying eggs, and the larvae are well protected inside the buds.

Alder Leaf Beetles

SYMPTOMS Holes and brown, dried-up patches appear on the leaves of alders (*Alnus*), and occasionally other deciduous trees, during spring to midsummer due to beetles and their larvae feeding on the foliage (*see p.45*). Heavy infestations can result in extensive defoliation.
CAUSE The adult beetles and larvae of the alder leaf

beetle, *Agelastica alni*. The black, soft-bodied larvae are up to 11mm long, while the dark blue adults are 6–7mm long. The adult beetles go into the soil to overwinter in late summer and emerge in late spring. At one time thought to be extinct in Britain, alder leaf beetles reappeared in 2004 and are now common in parts of England.
CONTROL Treatment is only feasible on small trees, which can be sprayed with a contact or systemic insecticide (*see p.94*) when damage is detected. Those that are too tall to be sprayed may have an unsightly appearance in midsummer, but they will survive.

Alder Sucker

SYMPTOMS Between mid-spring and early summer a fluffy, white waxy substance covering small, greenish-black insects appears on the foliage and shoot tips of alder trees (*see p.48*).
CAUSE The woolly wax is secreted from the bodies of the nymphs of *Psylla alni*. The adult stage is a winged aphid-like insect that does not produce this substance. Both nymphs and adults feed by sucking sap but cause little harm to the trees.
CONTROL Usually unnecessary. However, if the infestation is causing concern small trees could be sprayed with an insecticide (*see p.94*).

Algae

SYMPTOMS Powdery deposits develop on the plant's surface; these are commonly green or grey-green and can be rubbed off, leaving no discoloration on the plant tissue beneath. The plant's growth and development are not affected.
CAUSE Harmless microscopic, non-flowering plants that develop in films of moisture

on plant surfaces. They are particularly common on plants growing in damp situations. Leaves and trunks or stems are most commonly affected, especially those on the north-facing side of the plant.

CONTROL No control is needed as the plant is not damaged. If unsightly, algal deposits can usually be washed off with a sponge and soapy water. Improve air circulation, and reduce dampness to prevent regrowth.

Allium Leaf Miner

SYMPTOMS White legless maggots up to 10mm long tunnel through the leaves and stems of leeks, onions, shallots, and other edible alliums during spring and autumn. The dark brown barrel-shaped pupae are 2–3mm long and can be found by peeling back the outer leaves of leeks in summer and winter. The adult flies feed on sap from the leaves and cause a row of white dots along the leaf.

CAUSE The maggots of a leaf-mining fly, *Phytomyza gymnostoma*, which has two generations a year. This pest was first detected in Britain in 2003 and has since become widespread in England.

CONTROL There is no pesticide available for garden use on onions, leeks, and other edible host plants. The female flies can be prevented from laying eggs by practising crop rotation and covering new crops with insect-proof netting in spring and autumn. Crops sown or planted after April will avoid the first generation of flies.

• *see also* LEEK MOTH *p.152*.

Alternaria Leaf Spot

SYMPTOMS Grey or brown spots appear on affected foliage, which frequently show concentric rings of tissue dieback. In extreme conditions the spots may join together, causing extensive areas of leaf to be killed. In the majority of cases, however, the damage remains as distinct near-circular spots. Areas of pale or variegated foliage are likely to be the most readily affected, when the leaf may wither and die back. Plants commonly affected include potatoes, carrots, and brassicas, and also cinerarias.

CAUSE Various species of the fungus *Alternaria*. The spores are air-borne or carried by rain or water splash.

CONTROL Remove and dispose of affected leaves as soon as they appear. This should help to limit spread.

• *see also* BRASSICA DARK LEAF SPOT *p.119*; POTATO EARLY BLIGHT *p.172*.

American Gooseberry Mildew

SYMPTOMS Whitish-grey powdery fungal growth appears on the upper surface of gooseberry foliage (*see p.40*), stems, and fruits (*see p.15*). Severely affected leaves, especially the young growth towards the shoot tips, may become distorted and die off. Shoot tips may also die back. The mildew on the skin of affected berries turns pale brown or buff in colour as it ages, and can be scraped off. Infected fruit is still edible, but turns brown when cooked. Blackcurrant plants may also be affected.

CAUSE The fungus *Podosphaera mors-uvae* (syn. *Sphaerotheca mors-uvae*), which is encouraged by stagnant air around the branches and by excessive use of fertilizers with a high nitrogen content. The fungus overwinters on the branches and within the buds of plants.

CONTROL Prune out any branches that have been affected as soon as they are seen; also prune to thin growth and improve air circulation within the crown of the bush. Avoid excessive use of nitrogenous fertilizers; a general fertilizer is usually the preferable choice. Consider growing cultivars that show resistance to this infection, for example 'Invicta', 'Greenfinch' 'Hinnomaki Red', or 'Remarka' (gooseberries), or blackcurrants 'Ben Sarek', 'Ben Conan', and 'Ben More'.

Anemone Rust

See PLUM RUST *p.169*.

Anemone Smut

SYMPTOMS Anemone stems and leaves show dark, raised blisters, which burst to release dark brown to black spores. The plant may show some dieback, but is rarely killed. Anemone smut may also affect globe flowers (*Trollius*).

CAUSE The fungus *Urocystis anemones*, which can overwinter in infected plant debris and on other plant material on the soil surface. It is easily spread by rain splash.

CONTROL Remove affected parts promptly, before the blisters erupt. Destroy severely affected plants and choose a fresh site for growing new anemones or globe flowers.

Anthracnose

Various fungi cause patches or spots of discoloration to appear on infected plant tissue. Leaf, stem or, in beans, pod tissue may be killed, and in severe cases the whole plant may die back. Precise symptoms vary with the host plant and the particular fungus involved.

• *see also* BEAN ANTHRACNOSE *p.113*; PLANE ANTHRACNOSE *p.168*; WILLOW ANTHRACNOSE *p.197*.

Antirrhinum Rust

SYMPTOMS Concentric rings or circles of pustules full of dark brown spores appear on the underside of snapdragon (*Antirrhinum*) leaves (*see p.27*). A chlorotic (yellow) "halo" develops around each patch of spores, and the upper leaf surface may also become yellowed. Affected leaves die, and if left unchecked, the infection may spread to stems and flower buds. Severely infected plants fail to flower and may be killed.

CAUSE The fungus *Puccinia antirrhini*, which can overwinter on snapdragons if provided with sufficient warmth to survive the winter.

CONTROL Remove and dispose of severely infected plants promptly. Remove all snapdragons at the end of the season and do not grow them on that site for at least two years. Try growing cultivars that show a degree of resistance to the infection, such as the Tahiti series, 'Monarch', and 'Coronette'.

Ants

Several species of ants occur in gardens, including the black ant, *Lasius niger*, the yellow meadow ant, *Lasius flavus*, and red ants of various *Myrmica* species. They are often seen on plants infested with aphids and other sap-feeding insects. Ants collect the sugary honeydew excreted by these insects and may protect aphids by driving off predators. Fine heaps of soil appear on lawns (*see p.67*) and in flowerbeds, where soil excavated from the underground nests that the worker ants build is deposited on the surface. This can interfere with mowing and partly bury low-growing plants.

CONTROL Ants can be reduced in numbers but they cannot be eliminated from gardens. They are generally a nuisance rather than a significant pest. Many queen ants that fly into gardens are killed and eaten by ants from existing nests before they have an opportunity to establish new colonies. Rigorous attempts to eliminate nests may increase the survival rate of queens flying in from nearby gardens, and lead to more nests in the near future. Control measures should be confined to nests that are causing real problems. In gardens, nests in lawns can be treated with approved insecticide drench (*see p.94*). A pathogenic nematode, *Steinernema feltiae*, is available for watering into the soil in places where nests are situated. Ants in buildings can be killed with insecticide baits and dusts sold for this purpose, or nest entrances and ant pathways can be sprayed with aerosol lacquers that leave deposits of a crawling insect insecticide.

Aphids

Many species of aphid occur in gardens and few plants are never attacked. The more important aphid pests are described in more detail in individual entries. Generally, affected plants show reduced growth with varying degrees of leaf distortion. The upper leaf surface becomes sticky with the sugary honeydew excreted by aphids. A black sooty mould often grows on the honeydew, especially under damp conditions. Aphids may be green, pink, yellow, black, greyish-white, or brown, and are usually about 2mm in length, although some conifer and willow aphids are up to 6mm long. All nymphs shed their outer skins several times as they grow. These whitish cast skins accumulate on the upper surface of leaves below those where the aphids are feeding. Some virus diseases, especially of strawberries and raspberries, are spread by the aphids as they feed and move from one plant to another.

CONTROL Most contact insecticides, including organic types, will control aphids that are feeding in exposed positions. Where the aphids are feeding under curled leaves, a systemic insecticide will give better control. *See page 94* for more information.

• *see also* BEECH WOOLLY APHID *p.114*; BLACK BEAN APHID *p.116*; BLACKFLY *p.116*; CEDAR APHID *p.124*; CHERRY BLACKFLY *p.125*; CONIFER APHIDS *p.128*; CURRANT BLISTER APHID *p.129*; CYPRESS APHID *p.130*; GREEN SPRUCE APHID *p.142*; HONEYSUCKLE APHID *p.145*; JUNIPER APHID *p.149*; LETTUCE ROOT APHID *p.153*; LUPIN APHID *p.154*; MEALY CABBAGE APHID *p.155*; MEALY PLUM APHID *p.155*; MELON COTTON APHID *p.155*; PEACH APHID *p.164*; PEAR BEDSTRAW APHID *p.164*; PLUM LEAF-CURLING APHID *p.169*; ROOT APHIDS *p.177*; ROSE APHIDS *p.178*; ROSY APPLE APHID *p.180*; WATER LILY APHID *p.195*; WILLOW BARK APHID *p.197*; WOOLLY APHID *p.199*.

Apple and Pear Canker

SYMPTOMS Areas of bark sink inwards (*see p.52*), usually starting near a wound or bud. The bark becomes discoloured and then shrinks and cracks, forming concentric rings of flaky bark. The branch may become swollen around the cankered area. As the canker enlarges, the shoot may be girdled and so foliage and growth above it starts to die back. Stems and trunks may be attacked and very occasionally semi-mature fruits are infected and show rotting. In summer raised, white fungal pustules develop on the canker, and in winter raised, red fruiting bodies develop. Apples and pears are frequently affected; poplars (*Populus*), beeches (*Fagus*), willows (*Salix*),

hawthorn (*Crataegus*), and *Sorbus* may also succumb.

CAUSE The fungus *Neonectria ditissima* (syn. *Neonectria galligena*, *Nectria galligena*), which is spread mainly during the spring as wind-borne spores enter through wounded areas such as leaf scars, pruning cuts, scab infection, frost crack, or attack by woolly aphid.

CONTROL Prune out whole spurs or branches where possible. On larger branch or trunk infections, carefully pare away all infected and damaged bark and wood, cutting back to completely clean tissue, and treat the wound with a proprietary wound paint. Dispose of prunings and parings carefully. Improve general growing conditions, in particular soil drainage, since unsuitable conditions such as wet soil may aggravate the disease. On acidic soils, liming to increase soil pH may help to reduce risk of canker. Avoid growing apples known to be particularly susceptible, such as 'Cox's Orange Pippin', 'James Grieve', 'Worcester Pearmain', and 'Spartan'. The following cultivars seem to show a degree of resistance under most circumstances: 'Laxton's Superb', 'Newton Wonder', 'Bramley's Seedling', and 'Lane's Prince Albert', 'Katy', and 'Lord Derby'.

Apple Bitter Pit

SYMPTOMS The skin of the apple fruit may develop sunken brown spots, each less than 1mm in diameter (*see p.13*). The flesh is spoiled by numerous pale brown spots or freckles. Affected apples may have an unpleasant, slightly bitter taste. Large fruits are especially susceptible, as are fruits produced by heavy-cropping trees. Symptoms may appear while the fruits are on the tree, but often only develop during storage.

CAUSE Calcium deficiency in the fruit. This results in the collapse of small groups of cells within the flesh, so that brown flecks develop. Calcium levels in the soil may be perfectly adequate, but in dry conditions, the tree cannot take up the amount it needs. Large fruits and heavy-cropping trees are more prone to this disorder because of their greater demand for calcium. Too high a concentration of magnesium or calcium in fruits may also be a cause of bitter pit.

CONTROL Keep apple trees well watered and mulched to maintain soil moisture levels. Maintain even growth by feeding with a balanced fertilizer; avoid excessive use of high-nitrogen feeds. Spray developing fruits with calcium nitrate solution from early summer until early autumn. The cultivars 'Bramley's Seedling', 'Discovery', and 'Crispin' may be damaged by this chemical, so care should be taken when treating them.

Apple Blossom Weevil

SYMPTOMS Apple flower buds start to develop but the petals fail to open and turn brown (*see p.65*). If these "capped" blossoms are pulled apart, a creamy white grub, up to 7mm long, or the pupal stage, may be seen. Damaged blossoms cannot produce fruits.

CAUSE A weevil, *Anthonomus pomorum*, which lays eggs in the unopened buds in mid- and late spring. Adult weevils are brown with whitish markings on their upper surface and 4–6mm long.

CONTROL Light infestations have little impact on the yield of fruits. Spraying with a contact insecticide (*see p.94*) when flowers are at the green cluster stage is only worthwhile if heavy attacks have occurred in previous years. Do not spray when the blossom is open.

Apple Capsid

SYMPTOMS Ripe apples have raised yellowish-brown corky bumps or scabby areas on the skin (*see p.12*). Leaves at shoot tips may be distorted, with many small holes.

CAUSE Apple capsid, *Lygocoris rugicollis*, which is a bright green sap-sucking insect up to 7mm long. It injects saliva into the plant as it feeds on the foliage and immature fruits during early summer. This kills some plant cells and, as the developing leaves grow and expand, these dead areas tear into small holes. Fruits respond to capsid damage by developing raised bumps or scabby areas.

CONTROL The fruit damage is only skin-deep and light attacks can be tolerated. If damage has been extensive in previous years, spray with an insecticide approved for use on apples (*see p.95*) at petal fall. Do not spray when the blossom is open.

Apple Frost Damage

SYMPTOMS Russeting of the skin, usually on the most exposed face of the fruit. The more protected faces are usually unharmed. Damaged areas may develop a corky, roughened appearance and the fruit may become deformed and uneven in growth. Flowers and young foliage subjected to frost wither and die off rapidly.

CAUSE Frost on young, unhardened plant tissues. This is exacerbated if the plant is growing in a position where, following overnight frosts, it is then subjected to rapid warming by early morning sun.

CONTROL Rarely a recurrent problem, unless the tree is in an unsuitable position. Seasonal use of horticultural fleece or similar material draped over trees will prevent injury.

Apple Fruit Split

SYMPTOMS Shallow cracks occur, usually only just penetrating into the flesh and most commonly running round the stalk end of the fruit (*see p.14*). These cracks may heal over and a corky layer forms, but if they occur when the fruit is more mature, the fruit usually succumbs to secondary infections.

CAUSE An erratic supply of water is the most common cause. The splitting occurs when the tree puts on a sudden spurt of growth, when moisture becomes available after a period of drought.

CONTROL Keep trees regularly watered and apply a mulch to moist soil to try to retain soil moisture.

Apple Leaf Miner

SYMPTOMS Silvery-white or brown lines appear on the foliage of apple and cherry trees and ornamental *Malus* and *Prunus* (*see p.32*), particularly in late summer. If the leaf miners have finished feeding and are pupating, hammock-like cocoons of white silk will be attached to the undersides of leaves or on the bark.

CAUSE Small green caterpillars of a tiny moth, *Lyonetia clerkella*, live within the leaves, and eat out meandering mines or tunnels, which show as white or brown lines.

CONTROL Although leaf miners are sometimes abundant, this pest is active too late in the growing season to have a damaging effect on the tree's health or ability to produce fruit, so its presence can be tolerated.

Apple Powdery Mildew

SYMPTOMS Leaves are covered in a powdery white fungal growth (*see p.40*). Young leaves breaking from buds which were infected in the previous year often show the symptoms. If attacked early on, the foliage rarely reaches full size, but is distorted and dies prematurely.

CAUSE The fungus *Podosphaera leucotricha*, which overwinters in the apple buds or on the young stem growth (as silvery white patches on the bark). This fungus may also occasionally attack pears, quinces, and medlars, but does not cause serious damage. Disease spread and development is most rapid in hot summers and if there are heavy dews at night.

CONTROL Keep apples well watered, as dry soil seems to encourage the development of powdery mildew. Apply a mulch to the moist soil in order to preserve soil moisture. Prune out any severely infected growth and mildewed stems. Prune established trees to improve air circulation within the crown. Grow cultivars that show good resistance, such as 'Blenheim Orange', 'Discovery', 'Greensleeves', or 'Worcester Pearmain'.

Apple Sawfly

SYMPTOMS Apple fruitlets fall off in early to midsummer with a hole in the side (*see p.14*). This hole is filled with excrement produced by a white caterpillar-like larva, or maggot, with a brown head. Mature fruits may be misshapen, with a broad ribbon-like scar on the skin, often running up from the eye end (opposite the stalk) to the middle of the fruit (*see p.13*).

CAUSE A sawfly, *Hoplocampa testudinea*, lays its eggs at flowering time. After hatching, the larva bores into the fruitlet and at first tunnels just beneath the skin before boring into the core. Fruitlets damaged in this way drop off, and when fully fed, the larvae leave them (creating the exit hole) and go into the soil to pupate. However, if the larva dies before it can bore into the core, the fruit will reach maturity. The early feeding beneath the skin causes fruits to be misshapen with the scar described above.

CONTROL Pick off and destroy any fruitlets showing signs of damage before the larvae have a chance to escape into the soil. Spray at petal fall with a contact insecticide approved for use on apples (*see p.95*) if sawfly has affected a significant proportion of the crop in previous years. Do not spray when the blossom is open.

Apple Scab

SYMPTOMS Blackish-brown, scabby patches develop on the fruits, with similar but more greenish-grey spots on the leaves (*see p.28*). In severe cases the fruit may be almost entirely covered with scabby patches, and be small and misshapen. Infected fruits may crack or split and so become infected with secondary organisms such as brown rot. Affected leaves fall early and may also show extensive infection on the petiole (leaf stalk). A disease very similar to apple scab affects poplars.

CAUSE The fungus *Venturia inaequalis*, which overwinters on scabby patches on the young stems and also on fallen infected leaves. Scab can occur on ornamental and edible *Malus* species and cultivars and may also occur on *Cotoneaster* and *Sorbus*. It is usually most prevalent in damp seasons and on trees with crowded branches.

CONTROL Rake up and dispose of any affected leaves as soon as they fall. Prune out cracked or scabby shoots to limit the overwintering places of the fungus. Grow apple cultivars that show resistance to this infection, for example 'Discovery', 'Lane's Prince Albert', 'Gavin', 'Beauty of Bath', 'Suntan', or 'Sunset'.

Apple Scald

SYMPTOMS Reddish or brown areas develop on the skin of the fruit (see **p.12**), on surfaces that are subjected to the full impact of the sun. In extreme cases the flesh beneath may be damaged, but the discoloration is usually limited to the skin.
CAUSE The scorching or scalding is caused by strong sunlight on the tender skin of young apples.
CONTROL In temperate zones, scald is generally only seen in unusually hot, bright summers, so controls are not really justified.

Apple Sooty Blotch

SYMPTOMS Sooty, greenish-black spots develop on the fruit skin, usually on apples but occasionally on plums. They do not penetrate into the flesh, however, and can be scraped off. The fruit's development is not affected and it is not distorted.
CAUSE The fungus *Gloeodes pomigena*, which overwinters on young stems and which is most common in trees growing in shade, and in cool, damp conditions.
CONTROL Although common in some seasons, it is rarely widespread and usually only occasional. Increasing air circulation around the fruits will decrease the likelihood of sooty blotch developing, so fruit thinning and regular pruning are advisable.

Apple Sucker

SYMPTOMS Flattened, pale green insects, up to 2mm long, occur on apple blossom trusses in the spring (see **p.61**). Heavy infestations kill the flowers and prevent fruit setting.
CAUSE Sap-sucking, aphid-like insects, *Psylla mali*, that hatch from overwintering eggs as apples come into leaf. Most of the damage is caused by the nymphs during the blossom period. The adult stage resembles a winged aphid and is present after petal fall.
CONTROL Spray with an insecticide approved for use on apples (see **p.95**) before flowering, at the green bud stage. Do not spray when the blossom is open.

Apple Water Core

SYMPTOMS This disorder is also known as "glassiness" because affected fruits develop a glassy, almost wet appearance to the flesh. This is usually most apparent close to the core, but may occur throughout. Symptoms may decrease when apples are stored, but more often affected fruits become brown and start to disintegrate.
CAUSE Can be encouraged by unsuitable cultural conditions, including excessive pruning and thinning, high temperature, high light intensity and extensive damage or removal of bark on the trunk or branches.
CONTROL Avoid excessive thinning or pruning. Other factors are usually impossible to control.

Aquilegia Downy Mildew

SYMPTOMS Angular yellowish patches develop on the foliage of columbines (*Aquilegia* spp.) and increase in size and number until the entire leaf begins to curl and then dies. A delicate white, slightly purple fungal growth develops in patches on the lower leaf surface (see **p.28**), each corresponding to an area of yellowing on the upper surface. Purplish blotches may also develop on the plant stem, while infected flowers take on a water-soaked appearance and become distorted. Seed pods may still form if the plant is infected relatively late in the season and after flowering, but these invariably develop brown blotches and are unlikely to set seed.
CAUSE The fungus-like *Peronospora* species, which was first reported in Britain in 2013. It thrives in cool, damp weather in spring and autumn, and is spread on air currents, rain, or splash from watering. It may also be seed-borne on home-saved seed.
CONTROL There is no chemical available to control this infection. Avoid overhead watering and reduce the risk of infection by growing plants spaced well apart, for improved air circulation. Do not save seeds from infected or suspect plants, or any grown in the vicinity of an outbreak.

Aquilegia Gall Midge

SYMPTOMS Flower buds of columbines (*Aquilegia* spp.) are distorted and fail to open fully, with the petals turning brown (see **p.64**). Yellowish-orange maggots up to 3mm long may be found inside affected buds.
CAUSE A tiny fly, *Macrolabis aquilegiae*, which lays eggs in the developing flower buds in late spring. After feeding, the larvae go into the soil to pupate and overwinter.
CONTROL Remove and destroy infested flower buds before the larvae escape into the soil to help reduce infestations. Insecticide controls are likely to be ineffective since it is difficult to prevent the females from laying eggs, and the larvae are well protected inside the buds.

Aquilegia Leaf Miners

SYMPTOMS During the summer, the foliage of Aquilegia is disfigured with large, whitish-purple blotches or white sinuous lines (see **p.32**).
CAUSE The larvae of two species of leaf-mining flies. The large blotch mines are caused by larvae of *Phytomyza aquilegiae*, while the linear mines are made by larvae of *P. minuscula*.
CONTROL Aquilegia will still flower and grow, so control is not essential, particularly with the less damaging linear type of mine. If required, pick off infested leaves to prevent the larvae completing their feeding.

Aquilegia Sawfly

SYMPTOMS Pale green caterpillar-like larvae, up to 18mm long, feed on the leaf margins of aquilegias. They can quickly defoliate plants, leaving just the leaf stalks and flower stems. Several generations occur between mid-spring and early autumn.
CAUSE A sawfly, *Pristiphora rufipes*, which lays small batches of eggs on the undersides of leaves. The adults have black bodies 6mm long, two pairs of transparent wings and reddish-brown legs.
CONTROL Check for signs of leaf damage on aquilegias, especially in spring and early summer, when defoliation will have a more severe effect on the plant's health. Either remove the larvae by hand or spray with a contact insecticide (see **p.94**).

Arabis Mosaic Virus

SYMPTOMS Yellow mosaic-like flecking and discoloration on leaves, not only of arabis but of many plants. Infected plants are often stunted and distorted, and commonly fail to reach maturity.
CAUSE Arabis mosaic virus, which is one of the most common and widespread viruses infecting garden plants, both edible and ornamental.
CONTROL There is no control measure available. Affected plants should be removed and disposed of promptly.

Arbutus Leaf Spots

SYMPTOMS Numerous small spots develop on the leaves of strawberry trees (*Arbutus*). *Ruptoseptoria* causes discreet, irregularly-shaped leaf spots, which may enlarge and develop a dry, grey-white central area bordered by a dark purple band (*see p.26*). The vigour of the tree is not affected. *Elsinoe* causes purple spots or ringspots on the foliage, often followed by premature leaf fall and twig dieback.

CAUSE The fungi *Ruptoseptoria unidonis* (syn. *Septoria unidonis*) and *Elsinoe mattiroloanum*.

CONTROL Collect up and dispose of fallen leaves.

Aruncus Sawfly

SYMPTOMS Pale green caterpillar-like larvae, up to 20mm long, feed on the edges or central areas of the leaves of goatsbeard (*Aruncus dioicus*). Several generations occur between late spring and early autumn and plants can be stripped of foliage.

CAUSE A sawfly, *Nematus spiraeae*, sometimes known as the spiraea sawfly because *Aruncus* was at one time classified as a *Spiraea* species. This sawfly does not attack any of the plants currently known as *Spiraea*. The adults have black and yellow bodies 6mm long, two pairs of transparent wings and legs that are mainly yellow.

CONTROL Look for signs of leaf damage, especially in spring and early summer, when defoliation will have a more severe effect on the plant's health. Either remove the larvae by hand or spray with a contact insecticide (*see p.94*).

Ash Catkin Gall Mite

SYMPTOMS Instead of producing the usual ash seeds or keys, the flowers become converted into lumpy, brown, woody galls. These persist on the tree and can be a distinctive feature on infested ash trees in the autumn and winter after leaf fall.

CAUSE Microscopic gall mites, *Aceria fraxinivora*, which suck sap from flowers in the spring, and secrete chemicals that convert the flowers into galls. Individual trees growing close together can vary considerably in the extent to which they are affected, indicating that some trees are more susceptible than others.

CONTROL Apart from reducing the number of seeds produced by ash trees, this mite has no adverse effect on their growth. There is no effective pesticide control available to home gardeners.

Ash Dieback

SYMPTOMS The leaves of affected ash trees develop black blotches (*see p.24*), often on the main vein and the leaf base, and then wilt. The bark on stems and branches develops small, lens-shaped necrotic spots, which enlarge to the extent that they can form perennial cankers (*see p.52*). If the bark of one of these lesions is removed, the wood beneath can have a brownish or greyish discoloration, which is more extensive than the bark necrosis. Infection can develop rapidly; affected trees will then show extensive shoot and branch dieback. Infected trees may produce many shoots, known as "epicormic", from previously dormant buds beneath the damaged areas.

CAUSE A fungus with two distinct stages: *Chalara fraxinea* (the asexual stage) and *Hymenoscyphus fraxineus* (the sexual stage). Several species of ash are susceptible, in particular *Fraxinus excelsior* and *F. angustifolia*. The infectious spores are produced from apothecia (cup-shaped fruiting bodies) in fallen leaves and leaf stalks found in the leaf litter beneath ash trees. They are mainly active from early summer to mid-autumn and can travel up to 30km (19 miles) on the wind. These spores then land on living leaves, allowing the fungus to progress into the petioles and stems. The fungus can also be spread when infected ash plants, timber, or logs are moved from an infected area. In an attempt to reduce the spread of infection in Britain, a ban on the movement of ash trees, seeds, logs, and timber was applied in 2012.

CONTROL Report a suspected infection promptly to the appropriate plant health authority. Logs and fallen leaves (including leaf mould) from infected trees or those in the immediate vicinity should be disposed of as advised by the plant health authority. There is no chemical control available for use by gardeners.

Ash Sucker

SYMPTOMS In early summer the margins of ash leaves become purplish-red, thickened, and curled over. On the undersides of the curled leaves will be some flattened aphid-like insects, which secrete a white waxy material from their bodies.

CAUSE Ash sucker, *Psyllopsis fraxini*, is an aphid-like, sap-sucking insect. The damage is caused by the wingless nymphs. The adults, which resemble winged aphids, are present during the summer but have no noticeable effect on the tree.

CONTROL Tall trees cannot be treated and spraying is not usually required, even on saplings. Although the distorted leaves may cause concern and look unsightly, ash sucker does not have a serious impact on the tree's health or vigour.

Asparagus Beetle

SYMPTOMS Asparagus plants lose their foliage as adult beetles and their larvae eat the leaves and outer bark from the stems (*see p.57*). The damage to the bark causes stems to dry up and turn yellowish-brown above the point where bark has been removed from all round the stem. The adult beetles are 6–8mm long, black with six yellow blotches on their wing cases, and a reddish thorax. They emerge from the soil in late spring and begin laying eggs on the stems and foliage. The larval stage is creamy black in colour, up to 10mm long, with three pairs of legs towards the head end. There are two generations between late spring and early autumn.

CAUSE Adults and larvae of a leaf beetle, *Crioceris asparagi*.

CONTROL Burn old stems at the end of the year to destroy some of the overwintering beetles. Watch for signs of infestation from late spring onwards. Hand removal of adults and larvae is feasible on small asparagus beds; otherwise spray with an organic insecticide (*see p.94*). Asparagus flowers attract pollinators, so avoid spraying when plants are in flower if insecticide control is needed.

Aster Wilt

SYMPTOMS Rapid wilting of the plant, followed by death. The stem base is blackened and shrinks inwards. A faint, furry, pale pink or white fungal growth may develop on the affected areas.

CAUSE The fungus *Fusarium oxysporum* f. sp. *callistephi*, or fusarium wilt. It is most common in heavy or wet soils but can attack asters in any soil, and even in seed trays. The fungus is usually introduced on infected seedlings or plants, but may also be present in the soil. It may also be seed-borne.

CONTROL Promptly remove and dispose of infected plants and the soil in the vicinity of their roots. Use a fresh site for future plantings of asters and only plant perfectly healthy-looking plants. Discard any with dead roots or damaged stem bases. Avoid planting asters on heavy soils, or improve soil texture before planting. Grow China asters that show resistance, such as 'Duchess', 'Milady Mixed', 'Gala Mixed', 'Starlight', or 'Matsumoto Violet Stripe'.

Astrantia Leaf Miner

SYMPTOMS The foliage of *Astrantia* becomes discoloured by whitish-brown patches caused by larvae feeding inside the leaves. The mines are initially linear but later broaden into a blotch as the larvae mature (*see p.32*). The foliage is often also marked with small, whitish dots where the adult flies have fed by rasping the leaf surface.
CAUSE A small fly, *Phytomyza astrantiae*, which lays eggs on the foliage. After hatching, the larvae tunnel into the leaves, where they feed internally. When fully fed, they pupate inside the leaves. There are two generations during the summer, with damage occurring April– May and July–August.
CONTROL *Astrantia* will still flower and grow despite leaf miner damage, so control is not essential. If required, spray plants with a systemic insecticide (*see p.94*) when leaf-mining activity is detected. Avoid spraying while the plants are in flower.

Azalea Gall

SYMPTOMS Pale green, waxy growths develop in place of foliage, or occasionally in place of petals. Sometimes they have a pinkish-red colour. The galls later turn white as a powdery spore layer develops on their surface, and at the end of the summer they wither and turn brown. The infected plant continues to grow and flower normally. Symptoms may occasionally also develop on rhododendrons.
CAUSE The fungus *Exobasidium vaccinii*, which is common on outdoor and indoor Indian hybrid azaleas (*R. indicum*). Spores are carried on air currents and by insects, and months may elapse between infection and the appearance of symptoms.
CONTROL Remove galls before the spore layer develops.

Azalea Whitefly

SYMPTOMS White-winged insects, 2mm long, readily fly up from the undersides of the leaves of evergreen azaleas during the summer. The nymphs are flat, oval, whitish-green, scale-like creatures that live on the undersides of leaves. They are present for most of the year, as this pest overwinters in the nymphal stage. Both adults and nymphs excrete a sugary substance called honeydew that makes the upper leaf surfaces sticky and allows the growth of sooty moulds.
CAUSE A whitefly, *Pealius azaleae*, which feeds by sucking sap from the leaves of evergreen azaleas.
CONTROL The adult stage, which is reached in midsummer, is more susceptible to insecticides than the eggs or nymphs. When adult whiteflies are seen, spray the underside of the foliage thoroughly with a contact or systemic insecticide (*see p.94*). Avoid spraying while the plants are in flower.

B

Bacteria

SYMPTOMS The symptoms of bacterial infection vary with both the plant part and the type of plant infected, as well as the precise bacterium involved. Leaves usually become spotted, often with a yellow "halo" around the edge of the leaf spot. Fleshier plant parts such as stems, rhizomes, tubers, roots, and bulbs may discolour and disintegrate rapidly, often becoming slimy in texture and giving rise to unpleasant odours. Specific bacterial infections, such as bacterial canker and fireblight, have individual entries in this section.
CAUSE Various species of bacteria may be involved in general infections, probably the most common being species of *Erwinia*. Many bacterial infections occur initially on injured tissues but then spread into other areas of the host plant. Bacteria may be spread by rain splash, by insects and other animals, and also on pruning tools; some, such as crown gall, may spread through soil water movement.
CONTROL The precise method of control varies, but prompt removal of all infected areas and adjacent plant tissue is essential.
• *see also* BACTERIAL CANKER *p.112*; BACTERIAL SOFT 112 *p.104*; BEAN HALO BLIGHT *p.113*; FIREBLIGHT *p.135*; LEAF SPOT (BACTERIAL) *p.151*.

Bacterial Canker

SYMPTOMS Clearly defined areas of bark flatten and sink inwards (*see p.51*). Amber-coloured, resin-like ooze may appear, closely associated with injured bark. Only *Prunus* are commonly affected; symptoms are most obvious on cherry trees, far less so on plums. Buds at the tips of affected branches may fail to open, or leaves may appear but later wither and die back, due to the ringing effect produced by the expanding canker. Leaves on affected limbs are often yellowed and small in size. Foliage may develop shothole symptoms (*see p.182*).
CAUSE The bacterium *Pseudomonas mors-prunorum* (on plum and sweet cherry) and *Pseudomonas syringae* (on peach, apricot, cherry, and plum). Most infections occur during wet, windy weather in autumn, or in damp conditions in spring, when young stems and leaves are readily infected. Spring infections usually start on the leaves and then spread to the wood. Most wood infections occur through injured bark, such as that caused by pruning, frost crack, or leaf fall. Infection during summer is rare.
CONTROL Prune out infected areas during the summer months. and bin or burn infected prunings. Consider growing cultivars that show a degree of resistance, such as the cherries 'Merton Glory', 'Merton Premier', 'Merla', and 'Merpet', or plum cultivars 'Marjorie's Seedling' and 'Warwickshire Drooper'.

Bacterial Leaf Spot

See LEAF SPOT (BACTERIAL) *p.151*.

Bacterial Soft Rot

SYMPTOMS Roots, tubers, rhizomes, and fruits are generally the most susceptible areas of host plants. Discoloured areas develop and, as the tissue beneath deteriorates, the injured area may sink inwards. Infection usually spreads rapidly and can prove fatal or seriously damaging if its progress is not halted.
CAUSE Various bacteria may be involved, most commonly *Erwinia atroseptica* and its various pathovars or strains. The bacteria usually enter via wounds and are easily spread on pruning tools, by handling or by insects and other animals.
CONTROL Control is rarely possible as the infection spreads

so rapidly. It may, however, be possible to limit the damage caused by very prompt removal of infected areas together with the adjacent tissue. Disinfect pruning tools by using a garden disinfectant.

• *see also* BRASSICA BACTERIAL SOFT ROT *p.119.*

Banded Palm Thrips

SYMPTOMS Indoor plants, especially those that have relatively tough leaves, such as *Ficus*, *Schefflera*, *Monstera*, and palms, develop a silvery-brown discoloration of the upper leaf surface (*see p.36*). Close examination will reveal narrow-bodied insects up to 2mm long, either creamy yellow in colour or banded black and white.

CAUSE Sap-sucking insects, *Parthenothrips dracaenae*, which are creamy yellow as immature nymphs. The adults are darker in colour, with dark bands on pale wings. The wings are folded over the thrips' bodies when not in use, which gives the adults their striped or banded appearance.

CONTROL Spray the foliage thoroughly with a contact or systemic insecticide (*see p.94*). Two or three applications at fortnightly intervals may be needed to break the pest's life cycle. Damaged leaves will not regain their normal colour but new growth will stay clean once thrips have been controlled.

Bark Beetles

SYMPTOMS Many woody plants are affected by bark beetles, especially trees or shrubs already in poor health, that have dead limbs, or have recently died. The larvae feed between the bark and the wood, eating out a series of channels that can be seen if the bark is peeled away (*see p.49*). The tunnelling pattern is often distinctive for a particular

species of bark beetle. It consists of a maternal tunnel made by the adult female, in which eggs were laid, and larval tunnels that radiate away from the original tunnel. Fully-fed grubs pupate at the ends of the tunnels and adult beetles chew their way out through the bark.

CAUSE Many different species of bark beetle occur on conifers and deciduous trees and shrubs. Most are harmless, as they are secondary pests on plants that are already dying from other causes. The most notorious bark beetle is *Scolytus scolytus*, the elm bark beetle. This breeds in dying elm trees, including those infected with Dutch elm disease (*see p.133*). When adult beetles emerge in early summer they carry spores of the fungal disease on their bodies and can spread it to other elms.

CONTROL Prune out and burn affected branches, which will be weak or dying, and check to see if another problem, such as honey fungus (*see p.145*), is present. Insecticides are not effective against bark beetles, since for much of the year they are protected under the bark.

Basal Rot

See FOOT AND ROOT ROTS *p.137.*

Bay Sucker

SYMPTOMS During the summer, leaves at the shoot tips of sweet bay (*Laurus nobilis*) develop thickened, pale yellow margins that are curled over (*see p.40*). Often just one half of the leaf is affected, and later the damaged parts dry up and turn brown. Immature insects feed underneath the curled leaf margins but emerge when ready to moult to the adult stage. These fully grown nymphs are grey, flattened insects that secrete a fluffy, white waxy material from their bodies. The adults are winged, aphid-like insects, 2mm

in length, which may be seen on the shoot tips during summer.

CAUSE A sap-sucking insect, *Trioza alacris*, which has two or three generations between late spring and mid-autumn.

CONTROL Deal with light infestations by picking off leaves that have been affected. The pesticides currently available are unlikely to give effective control.

Bean Anthracnose

SYMPTOMS Sunken, brown longitudinal marks develop on the stems of affected dwarf or runner beans. Leaf veins may develop a red coloration and may subsequently turn brown, wither, and then die off. A reddish-brown spotting develops on infected pods and, when the weather becomes wet, a pink-coloured slimy fungal growth may develop.

CAUSE The fungus *Colletotrichum lindemuthianum*, which is usually seed-borne. The fungal spores are spread from young infected seedlings onto adjacent plants either by rain or by irrigation splash.

CONTROL Remove infected plants as soon as the fungus is detected, and do not save their seed. Grow cultivars showing some resistance, such as 'Aramis' or 'Rido Kenyan'.

Bean Halo Blight

SYMPTOMS Small, angular spots appear on the leaves of dwarf French or runner beans, at first appearing water-soaked, and then darkening, each one surrounded by a bright yellow "halo". Leaves then develop yellowing between the veins (interveinal chlorosis); the entire affected leaf may yellow and die. Growth is affected and yields are reduced. If pods or stems become infected, they develop grey, apparently water-soaked patches.

CAUSE The bacterium *Pseudomonas phaseolica*,

which attacks dwarf French and runner beans. Water splash (from rain or overhead watering) spreads the bacteria. Initial infection usually starts from infected seed.

CONTROL Pick off affected leaves promptly, and avoid overhead watering. Remove and burn all affected plants at the end of the season, discarding their seed. Consider growing the dwarf French bean cultivar 'Red Rum', which both shows some resistance to this infection.

Bean Rust

SYMPTOMS Numerous dark brown pustules develop on the underside of the leaves of French and runner beans. Occasionally stems and pods are infected. Rarely, the white cluster cup form of the fungus appears beneath the leaves. Severely infected leaves wither and in extreme cases cropping is reduced.

CAUSE The fungus *Uromyces appendiculatus*, which is most common in warm, damp summers.

CONTROL Remove affected leaves promptly.

Bean Seed Beetles

SYMPTOMS Rounded holes (where adult beetles have emerged) are seen in dry beans and peas kept for seed (*see p.17*). Several plants are vulnerable, but in gardens it is mainly broad beans that are affected. While the beetle grub is present within the seed, a pale circular patch on the seed coat indicates where it has eaten a cavity in the seed's cotyledon.

CAUSE Several species of seed beetle, up to 3mm long, can attack pea and bean seeds. On broad beans it is usually *Bruchus rufimanus*. This lays eggs in the bean pods as the seeds are

forming. The small grubs go unnoticed in bean seeds that are cooked for eating, but their presence becomes obvious in beans kept for next year's seed when the adult beetles emerge.
CONTROL There is no effective treatment available to home gardeners. Feeding damage by the larvae is to the seed's food reserve (known as the cotyledons) rather than the embryo, so damaged seeds can still germinate.

Bean Seed Fly

SYMPTOMS Germinating seeds of runner and French beans are eaten by white, legless fly larvae. Seedlings are sometimes killed before the shoots emerge from the soil, but more usually ragged leaves and stems damaged by the soil-dwelling larvae appear above soil level (see **p.16**). Such plants will usually survive although they may make slow growth initially, especially if the growing points have been killed, causing sideshoots to develop.
CAUSE Maggots of a fly, *Delia platura*, which is similar in appearance to a house fly.
CONTROL Slow-germinating seeds are most at risk, so avoid sowing under adverse conditions, such as when the soil is cold or very wet. The female flies are attracted to soil that contains fresh organic matter, so apply this in autumn rather than spring. To avoid bean seed fly problems, beans can be germinated in pots or seed trays, and transplanted after the first true leaves have expanded. There are no suitable insecticides for bean seed fly.

Beech Bark Disease

SYMPTOMS Usually seen in association with beech bark scale (see right). Small areas of bark (a few centimetres in diameter) die and may appear as slight depressions (see **p.54**)

They may then exude a black, tar-like substance formed from the tree's sap. As the infection spreads, larger areas of bark are killed and the leaves become yellowed. Leaves produced subsequent to the infection are often unusually small and yellow. Tiny red fungal bodies the size of pinheads develop on affected bark. Trees decline in vigour and may be killed, often snapping off part way up the trunk.
CAUSE The fungus *Nectria coccinea*, which infects beech trees via wounds in the bark created by beech bark scale. Young beeches are only rarely infected.
CONTROL Fell severely infected trees, to minimise spread and because the weakened state of their wood may render them dangerous.

Beech Bark Scale

SYMPTOMS A white powdery substance appears in crevices in the bark of beech trees, especially on the trunk and larger branches (see **p.51**). Small orange insects up to 1mm long live underneath this waxy exudation.
CAUSE A sap-feeding scale insect, *Cryptococcus fagisuga*, also known as the felted beech coccus. It has little direct effect on the tree, but it is believed that heavy infestations make trees more susceptible to a form of canker known as beech bark disease (see left).
CONTROL There are no effective insecticides for this scale insect.

Beech Gall Midges

SYMPTOMS In late summer, yellowish-brown, cylindrical structures, or galls, which may be hairy or smooth, develop on the upper surfaces of beech leaves. These galls, each containing a single fly larva,

readily break off and drop to the ground in the autumn.
CAUSE Small flies – gall midges – lay eggs in leaves in midsummer. The hairy galls are caused by *Hartigiola annulipes* and in some years are abundant. The larger smooth-surfaced galls are caused by *Mikiola fagi*, which is uncommon in Britain but can be frequent elsewhere in Europe.
CONTROL Apart from creating galls, these midges have no effect on the health of beech trees, so their presence can be tolerated.

Beech Leaf-mining Weevil

SYMPTOMS In spring, beech foliage develops many small, round holes eaten out by the adult weevils. These are black beetles, 2mm long, and jump like flea beetles when disturbed. The larval stage feeds as a leaf miner, making a brown linear mine from the centre of the leaf, where the egg was laid, to the leaf tip. A large, dried-up brown area develops at the leaf tip where most feeding occurs (see **p.39**).
CAUSE A weevil, *Orchestes fagi*, which in some years is so abundant that nearly every leaf is affected by larvae. This can be unsightly but the tree is able to withstand attack.
CONTROL Nothing can be done on mature trees because of the problem of spraying large specimens. Clipping beech hedges will remove many of the damaged leaves and new growth produced after late spring will not be attacked. Pesticide treatment is not worthwhile.

Beech Red Spider Mite

SYMPTOMS In late summer to early autumn, the foliage on beech (*Fagus sylvatica*) develops a fine mottling on the upper surface, especially along the

leaf veins (see **p.36**). Tiny, greenish-yellow mites, less than 1mm long, can be seen on the undersides of leaves, together with their white cast skins. Beech grown as hedges in sunny positions are more prone to damage than trees.
CAUSE A sap-sucking mite, *Eotetranychus fagi*. Infestations build up during the summer, but damage generally does not become noticeable until late summer.
CONTROL Spider mites are difficult pests to control. Fortunately this mite does not cause any lasting damage and has little impact until the end of the growing season. The application of a pesticide is not worthwhile.

Beech Woolly Aphid

SYMPTOMS Pale yellow insects covered with a fluffy, white waxy material (see **p.48**) form dense colonies on the shoot tips and undersides of beech leaves during late spring to midsummer, after which their numbers decline. The foliage becomes sticky with the aphids' sugary excrement, called honeydew, and this encourages the growth of sooty moulds. Winged aphids are present in midsummer and fly up when beech hedges are clipped. Because they are white, they are sometimes mistaken for whiteflies.
CAUSE A sap-sucking aphid, *Phyllaphis fagi*, a specific pest of beech trees.
CONTROL Tall trees cannot be sprayed effectively and so the aphid must be tolerated. On hedges look for signs of aphid activity in late spring and spray with a systemic or contact insecticide (see **p.94**) before heavy infestations develop.

Beefsteak Fungus

See FISTULINA HEPATICA *p.136*.

Bees

See BURROWING BEES *p.121;* LEAF-CUTTING BEES *p.151;* NECTAR ROBBING *p.159.*

Beet Leaf Miner

SYMPTOMS Leaves of beetroot, Swiss chard, and spinach beet develop large white or brown areas (*see p.31*) where internal tissues have been eaten by white maggots.
CAUSE Larvae of the leaf-mining fly, *Pegomya hyoscyami*, which has two generations during summer.
CONTROL With light infestations, remove affected leaves or crush the larvae inside the leaf mines. There are no effective insecticides approved for leaf miner control on this fly's host plants.

Beetles

In terms of named species, beetles are the most numerous insects in the world. They are characterized by their modified fore wings, which are hard opaque structures (the elytra) that form a covering for the abdomen. Some beetles, such as ladybirds (*see p.150*) and ground beetles (*see p.143*), are beneficial as predators of garden pests. Some beetles are damaging plant pests.
Root damage – *See* CHAFER GRUBS *p.125;* TURNIP GALL WEEVIL *p.186;* VINE WEEVIL *p.192;* WIREWORMS *p.198.*
Stem Damage – *see* BARK BEETLES *p.113;* SHOTHOLE BORERS *p.182.*
Foliage Damage – *See* ASPARAGUS BEETLE *p.111;* BEECH LEAF-MINING WEEVIL *p.114;* COLORADO BEETLE *p.128;* FIGWORT WEEVILS *p.135;* FLEA BEETLES *p.136;* LEAF WEEVILS *p.152;* LILY BEETLE *p.153;* MINT BEETLE *p.156;* PEA AND BEAN WEEVIL *p.163;* ROSEMARY BEETLE, *p.180;* VIBURNUM BEETLE *p.192;*
VINE WEEVIL *p.193;* WATER LILY BEETLE *p.195;* WILLOW LEAF BEETLES *p.197.*
Flower damage – *see* APPLE BLOSSOM WEEVIL *p.108;* POLLEN BEETLES *p.170.*
Seed and fruit damage – *see* BEAN SEED BEETLES *p.113;* NUT WEEVIL *p.159;* RASPBERRY BEETLE *p.175;* STRAWBERRY SEED BEETLE *p.187.*

Begonia Powdery Mildew

SYMPTOMS Greyish-white, powdery fungal growth appears in distinct spots or patches on the upper leaf surface (*see p.38*). Infected leaves either become dry and brown or, more commonly with the fleshy-leaved types, yellowed and soggy. Stems and flower-heads may also be affected. The disease is more common on indoor begonias; plants grown in the open are rarely affected.
CAUSE The fungus *Microsphaera begoniae*, which is encouraged by dry conditions around the plant's roots, and by moist air around the foliage.
CONTROL Pick off affected leaves immediately. Keep plants adequately moist, and improve air circulation to help reduce humidity. Unless the disease is noticed and tackled early, control is very difficult.

Berberis Sawfly

SYMPTOMS Sawfly larvae up to 18mm long eat the leaves on barberry shrubs (*Berberis* spp.) and *Mahonia* spp. The caterpillar-like larvae have black heads and white bodies marked with small black dots and larger yellow blotches. The adult sawflies have metallic bluish black bodies 7–10mm long and the wings are black. There are two generations of larvae between late spring and early autumn when plants can be stripped of leaves. The fully fed larvae go into the soil to pupate.
CAUSE A sawfly, *Arge berberidis*, which has become more widespread in Europe in recent years. It has been present in Britain since 2000. The most susceptible barberries are *Berberis thunbergii* and *B. vulgaris*.
CONTROL The larvae can complete their feeding and cause serious damage in as little as two-to-three weeks so check regularly for the presence of larvae. If seen, remove by hand or spray with a contact insecticide (*see p.94*). Avoid spraying plants in flower.

Big Bud Mites

SYMPTOMS Buds become abnormally enlarged and rounded, and fail to develop into shoots and leaves (*see p.59*). Each big bud will contain many hundreds of microscopic, elongate white mites. Dried-up big buds can be found at any time of year, but they become swollen over the winter months.
CAUSE Several species of gall mite cause this problem. The mites most frequently found on garden plants are *Cecidophyopsis ribis* on blackcurrant, *C. psilaspis* on yew, and *Phytoptus avellanae* on hazel.
CONTROL Big bud mites on yew and hazel will kill some buds, but generally not enough to affect the plants' growth, so they can be tolerated. Blackcurrant big bud mite is more serious, since apart from causing a loss of vigour, it can spread a virus-like disease called currant reversion (*see p.130*). The pesticides available to home gardeners will not control big bud mites. Removing big buds from lightly infested bushes in late winter will limit infestations. Scrap badly affected bushes after fruiting and replace in the autumn. 'Ben Hope' is a blackcurrant cultivar that is resistant to big bud mite.

Birch Witches' Brooms

SYMPTOMS Extensively branched twigs develop, clustered closely together to form "witches' brooms" scattered throughout the crown of the tree. The leaves on the brooms turn yellow and bear a layer of fungal spores on their lower surface. During the winter, the growths may, at first, be mistaken for birds' nests and can spoil the outline of the tree or hedge, but during the summer they appear as dense, leafy areas and so are less obtrusive. The growths are particularly susceptible to frost and so may be killed during the winter.
CAUSE The fungus *Taphrina betulina*, which is spread by water splash and on air currents. *Betula pendula* and *B. pubescens* are most frequently attacked. Similar witches' brooms may develop on birch as a result of a mycoplasma infection.
CONTROL Prune out the growths, if desired. Spread is very slow, but removal helps to limit the problem. No chemical control is available.

Bird Damage on Fruits

SYMPTOMS Apples, pears, and plums show bird pecks on the ripening fruit (*see p.12*). Wasps often extend the damage. Smaller fruits such as strawberries, cherries, and currants are eaten whole.
CAUSE Various birds, including blackbirds, thrushes, and starlings.
CONTROL Small trees and bushes can be netted or grown in a fruit cage. On large trees, bags made from muslin or old nylon tights drawn over the better fruit trusses will protect them from birds.

Bitter Pit

See APPLE BITTER PIT *p.108.*

Bitter Rot

SYMPTOMS Also known as gloeosporum rot, this is seen as circular brown, somewhat sunken lesions on the surface of apples and occasionally quinces and pears while in store or, occasionally, while still on the tree. On close inspection it may be possible to see many tiny fungal pustules arranged in concentric circles on the surface of the lesions. Occasionally a canker of old stems may also develop at the base of the spurs.
CAUSE The fungus *Glomerella cingulata*, which usually enters fruits through their lenticels (breathing pores in the skin).
CONTROL Remove infected apples. Check apples in store regularly and remove any showing symptoms.

Black Bean Aphids

SYMPTOMS Dense colonies of black insects, up to 2mm long, mass on the shoot tips and undersides of leaves (*see p.48*) of broad beans and other plants, including globe artichokes, nasturtiums, dahlias, and poppies, during the summer. Affected plants are weakened and bean pods fail to develop. The same pest may occur on viburnum and mock orange (*Philadelphus*) during the spring, causing leaf curling at the shoot tips.
CAUSE A sap-sucking aphid, *Aphis fabae*, and closely related species.
CONTROL Inspect susceptible plants at regular intervals and if aphids are seen, spray with with a systemic (not on beans) or contact insecticide (*see p.94*). Organic insecticides are effective if applied thoroughly before heavy infestations develop (*see p.89*). They can be used on edible plants but, because of their short persistence, may need more frequent use.

Black Leg of Cuttings

SYMPTOMS Before or shortly after roots start to form, the base of an affected cutting darkens and atrophies; the upper parts start to discolour and then die off.
CAUSE Various soil- and water-borne fungi, in particular those responsible for causing damping off (*see p.131*) in seedlings. They are usually introduced on dirty tools, pots, or trays, in unsterilized compost or by the use of non-mains water.
CONTROL Maintain good hygiene when taking cuttings. Use only clean or sterilized materials and tools, and mains water. Using a rooting powder will stimulate rapid rooting, which may in turn help to minimise black leg.

Blackberry Cane Spot

SYMPTOMS Silvery-grey, lens-shaped spots with purple margins appear on the canes. Pinprick-sized, black pycnidia, or fungal fruiting bodies, develop in the centre of each lesion. In severe cases canes may die back.
CAUSE The fungus *Elsinoe veneta*.
CONTROL Prune out severely infected canes.

Blackcurrant Gall Midge

SYMPTOMS Leaves of blackcurrants remain small and crumpled (*see p.40*), due to whitish maggots up to 2mm long feeding on leaves as they develop at shoot tips. Affected leaves dry up and die. Three generations occur during the summer, with symptoms first appearing in late spring at flowering time. Shoot tips may be killed, causing side-shoots to grow.

CAUSE The leaf distortion is induced by chemicals secreted by the larvae of a tiny fly, *Dasineura tetensi*. When the maggots pupate, the shoot may produce a couple of normal leaves before the next generation affects more new leaves.
CONTROL The first generation in late spring is potentially the most harmful. There are no effective insecticides available for controlling this pest. Some blackcurrant cultivars, such as 'Ben Connan' and 'Ben Sarek', are resistant.

Blackfly

Blackfly is a common name given to several species of black aphids.
• See also APHIDS *p.108*; BLACK BEAN APHIDS *p.116*; CHERRY BLACKFLY *p.125*.

"Bleeding" After Pruning

SYMPTOMS Sap oozes from pruning cuts or other wounds on the stems or trunks of trees, shrubs, and woody climbers. On occasion, copious quantities of sap may be produced rapidly, wetting the ground beneath. The sap may leak out either as the tree is pruned or shortly afterwards. Grape vines, walnuts, birch, *Acer* spp. (maples), hornbeam, magnolia, laburnum, mulberry, lime, and poplar trees are notably prone to bleeding.
CAUSE As spring approaches, the sap pressure rises dramatically in preparation for leaf break. The risk of bleeding is substantially higher if the stem or trunk is damaged, or if pruning is carried out between late winter and early spring. For some plants, sap pressure increases much earlier, making them particularly likely to show bleeding if they are pruned any time between midwinter and early spring.

CONTROL Though it may look dramatic, bleeding rarely has any significant effect on the plant in the long term. It is best avoided by carrying out minimal pruning and only at the correct time of year. Prune susceptible deciduous trees, shrubs, and woody climbers only when they are fully dormant and sap pressure is relatively low. The precise timing varies from plant to plant and, to a lesser extent, with the prevailing conditions. Take particular care with grape vines, walnut trees, and birch trees, as they are extremely prone to bleeding. There is no reliable way to stem the flow of sap on a bleeding plant. Wrapping, cauterising, or sealing a wound in an attempt to stop the bleeding is not recommended, and indeed may trap moisture and increase the risk of infection. Bleeding usually ceases once the leaves start to break.

Bleeding Canker

SYMPTOMS Yellow or brown substances ooze from patches of dead or dying bark on the trunks or large limbs of trees. The ooze may dry on the bark, leaving crusty deposits. Affected bark often cracks, allowing other decay organisms to enter. Stems completely ringed by the lesions will die back. Horse chestnuts, limes, silver birch, and apple trees are particularly prone to attack.
CAUSE Species of the fungus-like *Phytophthora* (*see p.167*).
CONTROL There is no cure for this disease. An inspection should be made by an experienced arboriculturist to determine the tree's health and safety.

Blind Shoots

SYMPTOMS Occasionally, flowering shoots of roses and other flowering shrubs develop normally, but fail to develop a terminal flower bud (*see p.56*).

The plant is perfectly healthy, and other shoots flower normally. **CAUSE** Unknown; possibly a cultural disorder or adverse weather conditions. **CONTROL** Prune back the blind shoot to a strong bud or pair of buds. Normal flowering shoots should grow from these buds.

Blindness of Bulbs

SYMPTOMS Bulbs produce healthy, normal foliage, but fail to flower. Either no buds are formed at all, or buds form, but are dry and virtually empty of petals (*see p.59*). **CAUSE** Occasionally due to large bulb fly infestation or basal rot (on *Narcissus*), but more often due to cultural problems. An inadequate supply of moisture and nutrients to the bulb causes it to decline in size and fail to produce buds, or a full complement of petals. Excessively crowded clumps (which are competing for moisture and nutrients), naturalized bulbs that have not been fed and bulbs growing in dry situations or suffering from drought are also likely to be affected. Narcissi are frequently affected, especially the multi-headed or double forms. **CONTROL** Divide and replant congested clumps, preferably into a fresh, well-fertilized site. Ensure an adequate supply of water during dry weather. Feed established clumps in the spring with a complete fertilizer and treat with a foliar feed while in leaf. Do not tie up or cut off bulb foliage for at least six weeks after flowering is complete.

Blossom End Rot

SYMPTOMS A sunken patch develops at the blossom end (base) of developing fruits. The base of the fruit becomes tough and leathery and darkens to brown or black as cells collapse. Not all fruits on a truss will necessarily be affected, and some trusses on a plant may remain perfectly healthy. This is most commonly seen on tomatoes, especially those growing in containers, and occasionally sweet peppers. **CAUSE** Dry conditions around the plant's roots, which prevent the plant from taking up sufficient calcium. The low calcium content of the fruits causes cells to collapse and discolour. This disorder is also encouraged by a very acidic growing medium. **CONTROL** Ensure an adequate, regular supply of moisture at all times, especially on fast-growing and heavy-cropping plants. If blossom end rot does develop, pick off affected fruits and improve the watering routine. Grow small-fruited varieties, such as cherry tomatoes, as these are rarely affected. Grow plants in open ground or in larger containers where the soil or compost is less prone to drying out.

Blossom Wilt

SYMPTOMS Blossom withers and dies shortly after appearing. Dead trusses of blossom remain on the tree. Infection may then spread to adjacent leaves, which wilt, turn brown, and die, remaining attached to the branch. Pinprick-sized, raised, buff-coloured fungal pustules develop on infected areas, and localized dieback may occur (*see p.56*). *Malus*, *Pyrus*, and *Prunus*, including cherry, plum, apple, apricot, and peach, may be attacked. Occasionally *Amelanchier* is affected. **CAUSE** The fungus *Monilinia laxa* (on all except apple) and *M. laxa* f. *laxa* (on apple) which may overwinter as cankers on infected stems, or pustules on infected flowers or foliage. The spores are wind-borne and spread is most rapid during damp weather.

CONTROL If feasible, prune out infected flower trusses, preferably before the infection spreads to the foliage or into the spur. Grow cultivars showing resistance, such as plums 'Marjorie's Seedling', 'Jefferson', and 'President', and apricots 'Moorpark' or 'Monique'.

Blue Mould on Bulbs

SYMPTOMS Bulbs in store are most likely to develop symptoms, but they may also occur while bulbs are in the ground. Reddish-brown lesions develop on the side of the affected bulb or corm, and pink fungal bodies may subsequently develop beneath the skin. A pale, bluish-green fungal growth appears on top of the lesion as it increases in size (*see p.20*). On bulbous irises this may be accompanied by rapid softening and rotting. **CAUSE** Various species of the *Penicillium* fungi. These fungi are all encouraged by over-warm, moist storage conditions and poor air circulation. The fungus is believed to enter the bulbs whilst they are in the ground, but the symptoms will only develop if storage conditions are unsuitable. **CONTROL** Avoid purchasing infected bulbs or those which are damaged, as these too are very susceptible. Store bulbs for as short a time as possible, and in cool but frost-free and well-ventilated conditions.

Bluebell Rust

SYMPTOMS Dark brown fungal pustules develop in elliptical, often concentric, formations on the leaves. Flower stems may also be attacked. Leaves yellow and wither prematurely and if infections are recurrent and severe, flowering may be

somewhat reduced. Scillas and grape hyacinths (*Muscari*) may also be affected. **CAUSE** The fungus *Uromyces muscari*. Infection is encouraged by damp growing conditions. **CONTROL** Rarely necessary, but remove infected foliage to limit the spread of the infection.

Blushing Bracket

See DAEDALEOPSIS CONFRAGOSA *p.130*.

Bolting

SYMPTOMS Affected vegetable plants produce flowers and seeds prematurely, before the crop has developed fully or been harvested (*see p.64*). **CAUSE** Various factors may be involved; these may include exposure to low temperatures at a certain growth stage (often induced by a cold, late onset of spring) which then induces flowering, or by excessively dry soil conditions. **CONTROL** Where feasible, avoid growing early varieties of susceptible crops, as these are usually more prone to the problem. Check seed catalogues for cultivars showing resistance to bolting.

Boron Deficiency

SYMPTOMS Symptoms vary depending on the crop affected, but root vegetables such as beetroot, turnip, carrot, radish, and swede are most susceptible. Cauliflower, cabbage, celery, pears, and strawberries occasionally develop symptoms. **Beetroot** – Roughened, cankered areas develop on the sides of the roots. In severe cases, the subsequent rotting spreads to cause a heart rot, which shows as brown ring development in the inner root tissues and the crown. Foliage is unusually small and may die.

Cabbage – Leaf distortion and hollow areas in the stems.
Carrot – Splitting to expose the greyish and sometimes split central core, as well as yellow and pink discoloration of the older foliage and splitting of the leaf stems.
Cauliflower – This shows curd browning, poor curd development, and roughening of the main stem, leaf stalks, and midribs.
Celery – Transverse cracks appear in the outer surface of the leaf stalk, after which the inner tissues turn reddish brown. Leaves may occasionally show distortion.
Pears – Distorted fruits with brown, hardened flecks in the flesh, often combined with slight shoot dieback.
Radishes – Splitting and dull skin, often with woody flesh.
Strawberries – Small leaves and general stunting, with puckering and yellowing at the leaf tips. The berries are small, pale in colour and with a characteristic "waist" close to the calyx.
Swede and turnip – These show "brown heart", which can be seen when the roots are cut across to reveal grey or brown discoloured areas, often in concentric rings, in the lower parts of the root.
CAUSE This disorder can often occur if the soil has been limed excessively or if it has been allowed to become very dry. Boron is readily leached out of the soil, so heavy rainfall may encourage this deficiency on light soils.
CONTROL For vegetables, apply about 35g per 20m² of borax before sowing or planting (mixing the borax with horticultural sand to make an even application possible). For pear trees, spray at petal fall with 70g borax in 22 litres of water (2½ oz in 5 gallons), together with a wetting agent.

Botrytis

See GREY MOULD *p.142.*

Box Blight

SYMPTOMS In autumn or winter brownish-black spots develop on the leaves of box plants, which gradually coalesce and may cover the entire leaf. On the lower surface patches of greyish-white fungal growth may be visible, and in damp weather white spore masses may develop. The fungus may develop in the stems, causing black streaking on the bark of young stems. Defoliation and then dieback occurs. During the growing season the infected box plant will re-grow from the live parts. The appearance of the plant may be badly spoiled. This infection has only been recognized in the United Kingdom since the mid-1990s and so far it has only been recorded on species of Buxus. It is known to attack the native *Buxus sempervirens*, in particular the cultivar 'Suffruticosa'.
CAUSE The fungus *Cylindrocladium buxicola* (syn. *Calonectria pseudonaviculata*), which is encouraged by high humidity and rain. The spores are carried by rain- or water-splash, although animals such as birds may also carry it. When box plants are kept tightly clipped, for example when used for topiary, the fungus is all the more likely to cause extensive problems because air circulation around the dense stem growth and foliage is restricted.
CONTROL Box blight spreads very rapidly under favourable conditions, so check plants regularly. As the fungus can form resting bodies in infected plant material, which then falls to the ground and remains viable, it seems likely that the fungus may be carried when plants, or soil on contaminated boots or gardening tools, are moved from one location to

another and it is essential to minimize this risk. Severely infected plants should be completely removed as soon as possible; the removal of only the infected stems is unlikely to prove adequate. The fungus can survive in fallen leaves for about six years, so any infected fallen leaves should be removed promptly and certainly before box plants are re-planted in the area. Remove dead leaves from within the centre of a plant as these may also be harbouring infection. All infected material should be bagged up and binned or burned, not composted. Avoid overhead watering and apply mulch beneath plants to reduce the risk of rain splashing spores up on to them. Disinfect secateurs and other pruning or clipping tools as you move from one box to another. Consider spraying with a suitable fungicide (see p.94), but this must be done in conjunction with the various cultural methods suggested. It is essential to ensure that any new box plants are completely healthy: buy from a reputable source and "quarantine" the plants by heeling them in at a damp spot for a minimum of three weeks; commercial nurseries may have used fungicides that supress the fungus but do not kill it, and once in the garden infected plants should show symptoms within about 10 days. Consider striking your own cuttings from plants known to be disease-free, or grow replacement plants that are not susceptible to box blight, such as *Berberis buxifolia* 'Pygmaea', *Berberis* x *stenophylla* 'Nana' or 'Irwinii', *Euonymus fortunei*, *Ilex crenata*, *Lonicera nitida*, or *Taxus baccata* 'Semperaurea'.

Box Red Spider Mite

SYMPTOMS A light whitish flecking develops on the

upper leaf surface of box (*Buxus* species).
CAUSE A tiny sap-sucking mite, *Eurytetranychus buxi*, that lives and feeds on the underside of the leaves. The mite overwinters on its host plants as eggs. These hatch in the spring and most of the feeding damage occurs during spring to early summer, although the leaf flecking remains visible throughout the year.
CONTROL The damage is generally not sufficient to affect the plants' growth, so this is a pest that can be tolerated. If control measures are required, the plants could be sprayed with organic insecticides in the spring (*see p.89*).

Box Sucker

SYMPTOMS The shoots of box are stunted, with closely-spaced cupped leaves that resemble cabbages (*see **p.40***). Droplets of liquid with a white, waxy coating can be tipped out of the shoot tips during late spring. This is excrement produced by the sucker's nymphal stages.
CAUSE The nymphs of a sap-sucking insect, *Psylla buxi*, which attack the new growth in spring. The nymphs are flattened, pale green insects up to 2mm long. Winged adults are present in midsummer, but they do not affect the plant's growth.
CONTROL If the box plant has grown to the desired height and is being clipped to keep it in shape, something that restricts shoot extension is not really a problem. Protect young plants that need to grow larger by spraying with a systemic or contact insecticide (*see p. 94*) in mid-spring.

Box Tree Moth

SYMPTOMS Box (*Buxus*) plants become discoloured due to caterpillars feeding on the foliage. The caterpillars spin silk webbing over the area where they are feeding (*see p.46*). Heavy infestations can cause extensive defoliation between mid-spring and mid-autumn.

CAUSE A moth, *Cydalima perspectalis*, which has two or three generations a year. The adult moth has white wings with a purplish-brown border and a wingspan of 40mm. Eggs are laid on the underside of box leaves. When fully grown, the larvae reach up to 40mm long. They have black heads and yellowish-green bodies, with a black and white band along the sides of the body. When fully fed, they pupate in silk cocoons on the plant. During the overwintering stage, newly hatched larvae conceal themselves between two leaves held together with silk threads.

CONTROL The caterpillars are protected to a certain extent by hiding themselves in the silk webbing. Forceful spraying with a contact insecticide (*see p.94*) when signs of caterpillar activity are seen is required.

Bracket Fungus

SYMPTOMS Fungal outgrowths develop on large limbs or the main trunk of affected trees (*see pp.53, 54*). They may appear throughout the tree, but are most common on the trunk, particularly towards the base. Some may be found growing from the buttress roots, or even attached to roots at some distance from the trunk. Some are annual, but most are perennial and quite woody in texture. Most have a tough leathery or woody upper surface, with spores being produced from pores on the undersurface. Most bracket fungi cause damage to trees – usually, extensive deterioration of the structural woody tissues. Infected trees often become dangerous because they are

weakened structurally, and may suffer wind snap or wind throw.

CAUSE Various fungi, some of the most common being *Fistulina hepatica*, species of *Ganoderma*, *Innonotus hispidus*, *Laetiporus sulphureus*, *Meripilus giganteus*, *Phaeolus schweinitzi*, *Phellinus pini*, *Piptoporus betulinus*, and *Sparrisis crispa*.

CONTROL Precise identification is essential, together with examination of the tree and consideration of its environment. As infected trees may be hazardous, employ an arboriculturist to examine the tree. Removal of the fungal brackets may limit spread, but will not cure the problem as the deterioration continues within.

• see also *DAEDALEOPSIS CONFRAGOSA p.130*; *FISTULINA HEPATICA p.136*; FOMES ROOT AND BUTT ROT *p.137*; *MERIPILUS GIGANTEUS* (GIANT POLYPORE) *p.155*; *PHELLINUS IGNARIUS p.166*; *PHELLINUS POMACEUS p.166*.

Brassica Bacterial Soft Rot

SYMPTOMS Patches of foliage or areas within the heart of the plant, often towards the stem base, discolour. Rapid disintegration of the tissues follows, usually accompanied by a foul smell.

CAUSE Various bacteria may be involved, but *Erwinia carotovora* is the most common. The bacteria usually gain entry through injured areas of the plant, when it has been attacked by pests such as slugs, snails, and caterpillars.

CONTROL Avoid initial injury, where possible. It is rarely possible to save an infected plant because the soft rot is generally only noticed when in a fairly advanced state, but prompt removal of affected parts when the first symptoms occur may prevent its development.

Brassica Dark Leaf Spot

SYMPTOMS Numerous round, brown-to-grey spots develop on leaves (*see p.26*), each with dark concentric rings. Infected areas of the leaf are killed and may drop away, leaving holes and a rather ragged-looking leaf.

CAUSE The fungus *Alternaria brassicola*, which is encouraged by wet weather and poor air circulation, and thrives on plants which are grown rather soft.

CONTROL Remove infected leaves promptly. Burn debris at the end of the season. Improve air circulation by spacing plants well and weeding between the rows. Avoid overhead watering and excessive use of high-nitrogen fertilizers.

Brassica Downy Mildew

SYMPTOMS Yellow blotches appear on upper leaf surfaces, each corresponding with a patch of greyish-white fungal growth on the lower leaf surface. Seedlings and young plants are more prone to attack than mature plants, but all stages can be susceptible. Severe infection may cause extensive leaf yellowing and the death of seedlings.

CAUSE The fungus *Hyaloperonospora brassicae* (syn. *Peronospora parasitica*), which is encouraged by mild, damp weather and humid conditions. It has a wide host range, attacking many cruciferous plants, including some ornamentals and weeds. The spores can be spread by wind to nearby plants, and infection of young plants may also occur through the roots from infection within the soil.

CONTROL Improve air circulation by wide plant spacing and regular weeding. Remove affected leaves as seen.

Brassica Light Leaf Spot

SYMPTOMS Discoloured circles of leaf tissue develop, becoming covered in fluffy, off-white fungal growth. In extreme cases, spots merge, causing extensive damage.

CAUSE The fungus *Pyrenopeziza brassicae*.

CONTROL Remove affected leaves promptly, and dispose of all debris at the end of the season. Rotate crops, or at least allow a two-year break between brassica crops.

Brassica Ring Spot

SYMPTOMS Purple-brown spots with yellow edges develop on leaves, the older, outer leaves being most susceptible. Well-defined concentric ringing of the tissue is usually visible, and the spot may be covered in numerous, black, raised, pinprick-sized, fungal fruiting bodies. This disease may occur on all brassicas but is most common and damaging on Brussels sprouts, where it may also attack the buttons.

CAUSE The fungus *Mycosphaerella brassicicola*, which is encouraged by damp weather conditions.

CONTROL Remove severely affected leaves and clear up all plant debris at the end of the season.

Brassica Wire Stem

SYMPTOMS Leaves on brassica seedlings, especially those of cauliflower and broccoli, become yellowed and narrowed, or may die back.

CAUSE A deficiency of molybdenum, which is most common in acid soils.

CONTROL Lime the soil to reduce acidity, and water affected plants or seed rows with sodium molybdate.

Broad Bean and Pea Rust

SYMPTOMS Both foliage and stems may develop discrete spots covered with dark brown, powdery spores on both leaf surfaces (see *p.27*). Each spot usually has a pale yellow "halo" around its perimeter. Infected leaves may die, and if the infection is severe it may cause defoliation. The infection is likely to be most prevalent and severe during damp or humid seasons, and in areas with poor air circulation.

CAUSE The fungus *Uromyces viciae-fabae*, which spends its entire life cycle on one host plant. The infection is most commonly found on broad beans, but may also develop on some closely related plants, such as vetches and sweet peas (*Lathyrus odoratus*). The fungus produces semi-dormant resting spores, allowing it to overwinter successfully on the remains of the plant. It can also overwinter on seeds.

CONTROL Pick off infected leaves promptly. Improve air circulation around the plants, and never water plants from overhead. Remove any other host plants that may be harbouring the infection, such as vetches and sweet peas. Dispose of all infected or suspect plant remains at the end of the season to prevent the fungus overwintering. Do not save seed from infected plants.

Broad Bean Chocolate Spot

SYMPTOMS Chocolate-brown spots develop on the upper surfaces of broad bean leaves (see *p.24*). Chocolate-brown streaks may also appear on the stems, pods, and flowers, and the seed coat may be affected in a similar way. Adjacent spots or streaks may coalesce to form large blotches. Plants that have been severely infected may die off, and even a very mild infection may reduce yields.

CAUSE The fungus *Botrytis fabae*, which is encouraged by damp or humid air. The fungus may overwinter on the remains of infected plants and may also be seed-borne.

CONTROL Grow beans on a properly drained site, avoiding the use of high nitrogen fertilizers (which will only increase the production of soft, infection-prone growth). Apply sulphate of potash so that growth will harden slightly. Space plants wider apart than normal – at least 23cm – to improve air circulation, and to ensure good weed control is maintained between the rows. This should help to minimize humidity and damp air around the plants. Common vetch can harbour the infection and allow it to carry over from one year to the next, so eliminate this weed where possible.

Broom Gall Mite

SYMPTOMS During early summer, buds on broom (*Cytisus*) fail to develop into normal shoots and leaves. They proliferate instead into green cauliflower-like growths that are covered in silvery hairs (see *p.55*). Later in the summer, these galled buds dry up and become greyish-brown. They remain on the stems over winter, and can persist on the plant for several years. Infestations increase in successive years until the shrub's flowering is significantly reduced.

CAUSE A microscopic gall mite, *Aceria genistae*, which lives and feeds within the affected buds.

CONTROL None of the pesticides available to home gardeners is effective against this mite. Remove badly infested plants and replant with clean stock in the autumn. Limit light infestations by picking off galled buds.

Brown Core of Primulas

SYMPTOMS Primula and polyanthus plants show signs of poor growth, and leaves may be yellowed. Plants wilt readily, especially during warm weather, but even when there is plenty of soil moisture available. Plants wither and die, usually within a single season. The root system deteriorates, leaving a reduced number of roots and little fine root growth. If roots are cut lengthwise, a distinct reddish-brown core of discoloration is seen running down them.

CAUSE The fungus *Phytophthora primulae*, which is soil-borne and builds up in the soil when susceptible plants are grown on the same site for a number of years in succession.

CONTROL Remove infected plants promptly. There is no soil treatment available, but growing fresh plants on a new site is usually successful. Take care not to spread soil from around infected plants to other parts of the garden.

Brown Rot

SYMPTOMS A soft, brown area develops on the fruit skin of tree fruits (see *p.13*) and also in the flesh beneath when it starts to deteriorate. The whole fruit is ultimately affected. Raised creamy white pustules appear on the rotted areas, in roughly concentric rings (see *p.14*). Infected fruits either fall off the tree, or may become "mummified" and dry and remain attached to the branches. Most tree fruits can be affected, especially apple, pear, plum, peach, and nectarine.

CAUSE The fungi *Monilinia fructigena* (on apples) and *M. laxa* (other host plants), which gain entry through injured skin. Bird pecks, wasp damage, codling moth exit holes, frost crack, cracking due to irregular growth, and fungal scab infections are all common causes of the initial injury. The spores are readily spread by insects or birds, by rain splash and by fruits coming into contact with a source of infection.

CONTROL Avoid, or minimize, possible causes of injury to fruits by taking appropriate control or preventive measures. Remove infected fruits promptly, as well as fallen ones, and dispose of them well away from the garden, or burn them. Prune out all mummified fruits, together with a short section of the spur to which they are attached.

Brown Scale

SYMPTOMS Brown, shell-like, convex objects, up to 5mm in length, are attached to the stems of many ornamental shrubs and fruit bushes (see *p.58*), especially those growing against walls or in other sheltered places. Honeydew excreted by the insects makes the foliage sticky and allows the growth of sooty moulds.

CAUSE A sap-sucking insect, *Parthenolecanium corni*, which usually has one generation a year. Its eggs are laid under the scale.

CONTROL The newly-hatched scale nymphs are the most vulnerable stage in the life cycle, feeding on the stems and undersides of leaves in early to midsummer. Spray ornamental plants in midsummer with a systemic or contact insecticide (see *p.94*). Fruit trees and bushes can be sprayed with approved insecticides, including organic types (see *p.89*). Plants in flower should not be sprayed.

Brown-tail Moth

SYMPTOMS Greyish-black, hairy caterpillars, up to 40mm in length, feed together in colonies under silk webbing on hawthorn, cherry, apple, blackthorn, blackberry, rose, and other trees and shrubs. The caterpillars have red and

white markings on their bodies and a pair of orange warts on the upper surface at the rear end. They overwinter as small larvae inside dense, white silk webs spun at shoot tips.
CAUSE Caterpillars of a moth, *Euproctis chrysorrhoea*, which during late spring and early summer can cause severe defoliation. In Britain, the moth is mainly found on the south and east coasts of England, but sometimes occurs further inland.
CONTROL Prune out the overwintering silk nests after leaf fall. Wear gloves to avoid contact with the hairs, as some people develop an intensely irritating rash when the caterpillar hairs come into contact with their skin. From mid-spring to early summer, spray infested plants with a contact insecticide (*see p. 94*), or carefully prune out infested shoots. Young caterpillars are more susceptible to sprays than the older larvae.

Bryobia Mites

SYMPTOMS Black-bodied mites, up to 1mm in length, with four pairs of pinkish legs, occur on the upper surface of leaves on ivy, primulas, polyanthus, gooseberry, and other plants during late winter to early summer. Leaves develop a fine, pale mottling on the upper surface.
CAUSE One of several species of sap-sucking mites in the genus *Bryobia*.
CONTROL Spray thoroughly with a suitable organic pesticide (*see p.89*) when bryobia mites are seen. Heavy infestations may need three applications at one-week intervals.

Bud Drop

SYMPTOMS Buds form, usually to full size, but drop while still apparently healthy. The plant is otherwise healthy.
CAUSE A period of dry weather while the flower buds are forming. On very susceptible plants, such as camellia and rhododendron, just a few consecutive days of drought at the end of summer when the buds are forming are enough to do damage. Plants growing in containers are particularly vulnerable because they are more prone to dry conditions around the roots. Feeding too late in the growing season (after early summer) may also induce bud drop the following spring.
CONTROL Water thoroughly and regularly and never let the soil or compost dry out completely. Apply a deep mulch around the root area.Feed no later than early to midsummer and avoid excessive feeding.

Bud, Dry

SYMPTOMS Buds form, but may be undersized and are usually dry and brown; the plant is otherwise healthy.
CAUSE As with bud drop (*see above*), a period of dry weather while the flower buds are forming. On very susceptible plants, such as rhododendron, just a few consecutive days of drought at the end of summer are enough. Plants growing in containers are particularly vulnerable because they are more prone to dry conditions around the roots.
CONTROL Water thoroughly and regularly and do not let the soil dry out completely. Apply a deep mulch around the root area.

Buff-tip Moth

SYMPTOMS Colonies of black-and-yellow chequered caterpillars, up to 50mm in length, feed together on leaves of deciduous trees and shrubs, including lime, oak, hornbeam, birch, hazel, rose, and cherry, from mid- to late summer. Unlike most moths with gregarious larvae, buff-tip caterpillars produce no silk webbing.
CAUSE Caterpillars of a moth, *Phalera bucephala*, which can defoliate branches on their host plants.
CONTROL Buff-tip moth is only likely to harm young trees or shrubs. If necessary, prune out infested shoots or spray with a contact insecticide (*see p.94*).

Bulb Scale Mite

SYMPTOMS Daffodil and amaryllis bulbs grown indoors for early flowering are much more likely to show symptoms of this pest than similar bulbs planted out in gardens. The whitish-brown mites are less than 1mm long and barely visible to the naked eye. Their feeding causes stunted and distorted growth. Flower stems and leaves have saw-tooth scarring along the edges and show a sickle-shaped curvature.
CAUSE Sap-sucking mites, *Steneotarsonemus laticeps*, which live in the neck area of dormant bulbs. These should not be confused with *Rhizoglyphus* species mites, which are larger, broader, pearly-white mites often found feeding on rotting bulbs. These are not pests, as they feed on damage initiated by some other cause.
CONTROL None of the pesticides available to home gardeners will control bulb scale mite. Infested bulbs should be thrown away.

Bullfinches

Bullfinches (*Pyrrhula pyrrhula*) feed on wildflower seeds for most of the year, but turn to flower buds when the seed supply runs out during winter. During late autumn to spring they will peck out the flower buds of plum (*see p.65*), pear, cherry, apple, gooseberry, almond, and forsythia. The outer bud scales are discarded and are easily seen when the ground below the plant is covered by snow or frost. Buds at the shoot tips often survive and so, in the spring, shoots which are mostly devoid of flowers will have clusters of blossom at the tips.
CONTROL Only netting will give complete protection; bird-repellent sprays are unreliable and may need frequent applications.

Burrowing Bees

SYMPTOMS Small, conical heaps of fine soil appear in lawns or on bare soil during late spring and summer (*see p.66*). They are distinguished from ant nests by the hole at the top of the heap where the bee enters its nest. Each nest consists of a vertical shaft in the soil; this shaft has several side-chambers, which the female bee provisions with nectar and pollen as food for her grubs.
CAUSE Many different species of solitary bee nest in this way in gardens. A common species is a ginger-brown bee, *Andrena fulva*, which is active in late spring. Although all burrowing bees are solitary, with each female constructing her own nest, the nests may be clustered together and the same piece of ground may be used by the bees in successive years.
CONTROL Control measures are not necessary or desirable, as burrowing bees cause no real damage to lawns, and they are of benefit as pollinating insects for fruit trees and bushes. The females have stings, but lack aggressive instincts and do not attack people.

C

Cabbage Moth

SYMPTOMS All types of brassica are damaged during the summer by yellowish-brown or green caterpillars, up to 42mm in length. They eat holes in the foliage (*see p.46*), especially the heart leaves of cabbages. Cabbage moth caterpillars also enter onion leaves, making holes and causing them to flop over.
CAUSE Caterpillars of a moth, *Mamestra brassicae*, which has one generation a year.
CONTROL Check brassicas regularly in summer and take control measures as soon as caterpillars are seen; they are difficult to deal with once they grow larger and go into the heart leaves. Remove caterpillars by hand or use a contact insecticide (*see p. 94*) against the young larvae. Insect-proof netting will prevent infestation if used in conjunction with crop rotation (*see p.104*).
• *see also* CABBAGE WHITE BUTTERFLIES (*right*).

Cabbage Root Fly

SYMPTOMS All brassicas and allied vegetables such as turnip, swede, and radish are attacked. Some related ornamentals, including stocks, alyssums, and wallflowers, are sometimes damaged. Affected plants grow slowly and tend to wilt on sunny days. Young brassica plants are particularly vulnerable, especially shortly after transplanting, when they can be killed. White, legless maggots, up to 9mm in length, eat the finer roots until just a rotting stump is left. The maggots also bore into the swollen roots of radish, swede, and turnip.
CAUSE Maggots of a fly, *Delia radicum*, which has three generations from mid-spring to early autumn, hatching from eggs laid in the soil.

CONTROL Established plants usually tolerate cabbage fly attacks: it is seedlings and transplants that need protection. There are no insecticides for use against cabbage root fly. Infestations can be reduced without using insecticides by placing squares or discs made of carpet underlay, cardboard, or roofing felt, about 10cm in diameter, on the soil around the stems of transplants. The female flies deposit their eggs on the discs, instead of in the soil; they then dry up before hatching. Growing brassicas, etc., under insect-proof netting, together with crop rotation (*see p.104*), will exclude the adult flies.

Cabbage White Butterflies

SYMPTOMS Holes are eaten in the leaves of brassicas and some ornamental plants such as nasturtium. Caterpillars of the large cabbage white butterfly are yellow with black markings. They have hairy bodies, up to 40mm long, and feed mainly on the outer leaves of brassicas. Small cabbage white butterfly larvae are up to 25mm long and feed mainly in the hearts of cabbages and other brassicas. They are pale green with a velvety appearance, unlike the smooth and largely hairless caterpillars of cabbage moth (*see above*).
CAUSE Large cabbage white, *Pieris brassicae*, and small cabbage white, *P. rapae*. Both have two generations a year, with caterpillars occurring from spring to early autumn.
CONTROL Check brassicas regularly from early summer to early autumn and take control measures as soon as caterpillars are seen. They are difficult to deal with once they have grown larger and have gone into the heart leaves. Remove caterpillars by hand or use a contact

insecticide (*see p.94*) against the young larvae. Insect-proof netting can prevent eggs being laid.

Cabbage Whitefly

SYMPTOMS Brassicas, such as Brussels sprout, kale, cabbage, cauliflower, and broccoli, have 2mm long, white-winged insects on the undersides of the leaves (*see p.48*). The adults are active throughout the year and readily fly up when disturbed. The immature nymphs are flat, oval, scale-like insects that remain immobile on the lower leaf surface. Heavy infestations make the foliage sticky with the whiteflies' sugary excrement, and sooty moulds will grow on this substance.
CAUSE The adults and nymphs of a sap-sucking insect, *Aleyrodes proletella*. This insect has several generations during the summer, and overwinters as an adult whitefly.
CONTROL Light infestations can be tolerated, but if large numbers of whitefly are present and the foliage is becoming soiled with sooty moulds, use a contact insecticide approved for use on brassicas (*see p.95*). Several treatments at weekly intervals may be necessary with organic insecticides.

Cactus Corky Scab

SYMPTOMS Buff to tan corky-brown markings develop on the skin and may become sunken (*see p.37*). Central areas often become grey or dark brown. Adjacent areas of the plant usually retain their normal green colour. Cacti and many succulents can be affected, especially *Opuntia* and *Epiphyllum*.
CAUSE Unsuitable growing conditions, in particular excessively high humidity or light levels.
CONTROL Improve growing conditions by moving the plant

to a slightly cooler position and improving ventilation to reduce humidity. Take care not to alter conditions too dramatically or rapidly or the plant may suffer even more.

Calcium Deficiency

SYMPTOMS These vary according to the plant and plant part affected. The most common symptoms are blossom end rot and bitter pit. Other symptoms may include: browning of the internal tissues of Brussels sprouts' buttons and the heads of cabbage; cavity spot of carrots; stunting and blackening of central leaves on celery; tip burn of lettuce, and the production of rolled leaves and spindly shoots on potato. On other plants the most common symptom is poor growth of young shoots and leaves, often combined with some curling and shoot or leaf deterioration.
CAUSE A deficiency of calcium, either in the soil or compost, or because of an inadequate uptake of calcium due to a low moisture content of the soil. This disorder is most common on acidic soils or growing media.
CONTROL Ensure an adequate supply of moisture and encourage soil moisture retention by mulching, where appropriate. Where possible on acid soils lime the soil to raise the pH level.
• *see also* APPLE BITTER PIT *p.108*; BLOSSOM END ROT *p.117*; CARROT CAVITY SPOT *p.123*.

Camellia Gall

SYMPTOMS Large, pale green swellings, often forked or hand-like, develop in place of the leaves (*see p.41*). Each gall may measure several centimetres in width and up to about 20cm in length. The surface soon becomes covered in a layer of spores, giving the gall a creamy white appearance. Small

shrivelled pieces of leaf may protrude from the tips of the galls. There are rarely more than one or two galls per plant. The overall vigour and flowering of the camellia is not affected.
CAUSE The fungus *Exobasidium camelliae*, which seems to be encouraged by damp weather. The spores are probably spread by insects and by rain splash.
CONTROL Remove galls, preferably before the spore layer develops.

Camellia Leaf Blight

SYMPTOMS Infected leaves develop discoloured patches and appear scorched (*see p.34*). Pycnidia appear on the damaged areas and may then exude black tendrils of spores in wet or humid conditions. Infected leaves may fall prematurely. Stem dieback occurs if the infection spreads from the leaf, down the petiole, and into the stem.
CAUSE The fungi *Pestalotiopsis guepini* and *Monochaetia karstenii* can both cause this disease. In either case, general lack of vigour or damage to the foliage increases the risk of the infection developing. Poor air circulation or generally damp conditions also favour the infection.
CONTROL Remove and dispose of any infected leaves or stems as soon as they appear. Disinfect secateurs and other pruning tools between each cut. As damp conditions encourage the spread of disease, always avoid overhead watering. Keep camellias adequately watered, mulched, and fed to encourage vigorous growth.

Camellia Petal Blight

SYMPTOMS Brown patches develop on the flowers and spread to the base. Infected flowers fall prematurely and a hard, black fungal resting body (Sclerotium) forms at the base

of the petals. These remain dormant beneath the tree and later produce structures that produce spores that land on petals of the next season's blooms. The sclerotia can remain viable for up to five years.
CAUSE The fungus *Ciborinia camelliae*, which is spread on air currents. Soil contaminated with the sclerotia can also spread the fungus.
CONTROL Remove fallen flowers promptly. Avoid moving contaminated soil; check treads of boots etc., too.

Camellia Yellow Mottle

SYMPTOMS Irregular blotching and mottling of camellia leaves, with bright yellow or near-white markings on the dark green surface (*see p.36*). Occasionally whole branches may bear completely white or strongly chlorotic leaves. The overall vigour of the plant seems not to be affected.
CAUSE An organism known as camellia yellow mottle, which is virus-like in its structure and behaviour. This organism is systemic within the plant, so propagation should not be carried out using material from any part of an affected plant.
CONTROL Prune out all affected stems.

Canker

SYMPTOMS The precise symptoms of the various cankers (see references below) vary with both the host plant and the causal organism. The bark is usually raised and often roughened and may split to reveal the tissues beneath. In some cases ooze may develop around the cankered areas. As cankers enlarge, they often ring the infected stem and so growth above the canker deteriorates and may die.

CAUSE Various fungal and bacterial organisms, most of which have a fairly restricted and specific host range.
CONTROL Remove affected areas as soon as possible and, except for bacterial canker and on edible plants, consider treating with an appropriate pesticide. Improve growing conditions.
• *see also* APPLE AND PEAR CANKER *p.108*; BACTERIAL CANKER *p.112*; MULBERRY CANKER *p.157*; NECTRIA CANKER *p.159*; SEIRIDIUM CANKER *p.181*; PHOMOPSIS CANKER *p.167*; POPLAR CANKER *p.170*.

Capsid Bugs

SYMPTOMS Leaves at the shoot tips of many plants, such as fuchsias, *Caryopteris*, roses, hydrangeas, currants, forsythias, dahlias, and chrysanthemums, are misshapen, with many small holes (*see p.45*). Flowers of dahlia and chrysanthemum open unevenly and those of fuchsia (*see p.65*) abort at an early stage.
CAUSE Pale green insects, such as the common green capsid, *Lygocoris pabulinus*, and tarnished plant bug, *Lygus rugulipennis*, both of which are about 6mm long when adult. They suck sap from the shoot tips, and secrete saliva that kills some of the plant cells. Later, as leaves develop from the damaged shoots, these dead areas tear into many small holes.
CONTROL Damage occurs any time between late spring and late summer. Check vulnerable plants during the summer and spray with a systemic or contact insecticide (*see p.94*) as soon as capsid bugs or signs of damage are seen. Avoid spraying plants in flower.
• *see also* APPLE CAPSID *p.109*.

Carnation Tortrix Moth

SYMPTOMS Pale green caterpillars up to 18mm long, with brown heads, feed between two leaves bound together with silk webbing, or under a curled leaf margin similarly held in place by silk. The caterpillars graze away the inner surfaces of their hiding places, causing the remaining damaged tissues to dry up and turn brown (*see p.34*). This pest occurs on a wide range of shrubs and herbaceous plants, both in gardens and in greenhouses.
CAUSE Caterpillars of a small, brown-winged moth, *Cacoecimorpha pronubana*. This moth has several generations a year, and its caterpillars can be found in any season on plants growing indoors or in sheltered places. Another species, the light brown apple moth, *Epiphyas postvittana*, attacks a wide range of plants in a similar manner.
CONTROL The hidden nature of the caterpillars makes them difficult to reach with insecticides. Some control can be achieved by looking for bound-up leaves and squeezing them to crush the caterpillars and pupae. If plants are too heavily infested for this to be feasible, spray with a contact insecticide (*see p.94*). Pheromone traps can be used to capture adult males of these moths in greenhouses (*see pp.86–87*).

Carrot Cavity Spot

SYMPTOMS Lens-shaped spots of discoloration develop on carrot roots. Cracks and associated craters develop beneath. Secondary organisms may then invade and cause rapid deterioration of the root.
CAUSE The precise cause is uncertain, but a deficiency of calcium is believed to be the main cause. This deficiency

causes groups of cells to collapse, so causing discoloration and crater formation.

CONTROL Ensure an adequate supply of moisture. Where feasible on acidic soils, lime the soil to raise the pH level to about 6.5.

Carrot Fly

SYMPTOMS The roots of carrots and related plants, such as parsley, parsnip, and celery, are tunnelled by slender, creamy-yellow maggots up to 9mm long. These cause rusty brown lines on the outside of the tap roots where tunnels close to the surface have collapsed (see **p.17**). Damaged roots are susceptible to secondary rots, which can affect their storage during the winter.

CAUSE Larvae of carrot fly, *Chamaepsila rosae*, which has two or three generations between late spring and late autumn. Carrot is the favourite host.

CONTROL Seedlings of carrots sown after late spring will miss the first generation of larvae; carrots harvested before late summer will miss the second generation. Growing susceptible plants on a fresh site under insect-proof netting prevents female flies getting close enough to lay eggs. There are currently no insecticides approved for carrot fly control. Some carrots, such as 'Resistafly', and 'Flyaway', are less susceptible to damage than most other cultivars.

Carrot Motley Dwarf Virus

SYMPTOMS Foliage of carrots and parsley develops distinct yellow and pink hues (see **p.37**), and is stunted and sometimes twisted. Leaf stalks may show pronounced twisting. Affected plants are stunted, and carrots form little if any edible root. Root tips and root hairs die back. Similar foliage discoloration may be seen when these plants are suffering from a deficiency of magnesium, but the other symptoms described are not seen.

CAUSE Carrot motley dwarf virus is usually found together with carrot red leaf virus, and it is these two viruses, in combination, which are responsible for the symptoms. The viruses are both spread by aphids.

CONTROL Dig up and dispose of infected carrots. Attempt thorough aphid control to minimize the risk of infection development and spread.

Caterpillars

Caterpillars are the larval stage of butterflies and moths. There is much variety in their colour, size and degree of hairiness, but they generally have an elongate tubular body shape with a distinct head capsule. They have three pairs of jointed legs at the head end and two to five pairs of clasping legs (prolegs) on their abdomen. Larvae of sawflies look similar but have more than five (usually seven) pairs of prolegs. Most pest species feed on leaves, sometimes gregariously under silk webbing. Others live in soil and feed on roots, or bore into stems, or feed inside leaves as leaf miners, or feed on fruits, berries, and seeds.

Root damage – *see* CUTWORMS *p.130.* SWIFT MOTHS *p.188.*

Stem borers – *see* CURRANT CLEARWING MOTH *p.130;* LEOPARD MOTH *p.153.*

Leaf miners – *See* APPLE LEAF MINER *p.109;* HORSE CHESTNUT LEAF MINER *p.146,* LABURNUM LEAF MINERS *p.149;* LEEK MOTH *p.152;* LILAC LEAF MINER *p.153;* PYRACANTHA LEAF MINER *p.173.*

Leaf damage – *see* CABBAGE WHITE BUTTERFLIES *p.122;* CABBAGE MOTH *p.122;* CARNATION TORTRIX MOTH *p.123;* CHINA MARK MOTH *p.126;* DELPHINIUM MOTH *p.131;* ELEPHANT HAWK MOTH *p.134;* MULLEIN MOTH *p.157;* VAPOURER MOTH *p.192;* WINTER MOTH *p.198;* YELLOW-TAIL MOTH *p.199.*

Gregarious leaf feeders, often with silk webbing – *see* BOX TREE MOTH *p.118;* BROWN-TAIL MOTH *p.120;* BUFF-TIP MOTH *p.121;* COTONEASTER WEBBER MOTHS *p.128;* GYPSY MOTH *p.143;* JUNIPER WEBBER MOTH *p.149;* LACKEY MOTH *p.150;* OAK PROCESSIONARY MOTH *p.160;* SMALL ERMINE MOTHS *p.184.*

Fruit, seed, and berry damage – *see* CODLING MOTH *p.127;* PEA MOTH *p.163;* PLUM MOTH *p.169;* TOMATO MOTH *p.190.*

Caterpillars can be removed by hand, especially on small plants. Many caterpillars feed after dark and are easier to find by torchlight on mild evenings. An alternative treatment is to use a contact insecticide (see **p.94**), when signs of caterpillar feeding are seen.

Cats

Despite their value as pets, cats can be a nuisance in gardens. They soil flower borders and vegetable plots and may damage plants as they attempt to bury their droppings. Their urine can scorch foliage, and the bark of trees and shrubs is damaged when cats use them as scratching posts.

CONTROL In localities where cats are numerous, it is difficult to stop them being a problem. Proprietary animal repellent sprays and powders often fail to give reliable long-term protection. Devices that emit ultrasonic sounds that are virtually inaudible to humans but sound loud to cats can encourage them to go elsewhere. However, as cats become used to the noise some may turn a deaf ear to it. Cats looking for toilet areas are attracted to loose dry soil or mulch-covered flower beds. Close planting helps to cover up the soil surface with vegetation, making it less accessible.

Cavity Decay

SYMPTOMS Once exposed to the elements and open to infection by fungal spores, damaged areas on trees may soon deteriorate and form a cavity. The decay may then spread further into the trunk or limb, which may result in sufficient structural damage to render the tree dangerous.

CAUSE Initial injury may result from storm damage, poor pruning, or other physical injury.

CONTROL Prevent cavity formation wherever possible. When cavities develop, the rotting wood should not be removed and the healthy wood exposed: this was once believed to be the best method of dealing with wood rotting, but it has since been shown that this action simply promotes the breaching of the barriers that trees form in response to injury, and more extensive rotting subsequently develops. Trees with cavities should be examined regularly by a professional arborist to determine their condition and safety.

Cedar Aphid

SYMPTOMS The foliage and stems of cedars become thickly coated with a black, sooty mould that grows on the sugary honeydew excreted by aphids (see **p.48**). Heavy infestations cause the foliage to turn yellow in colour and drop off in midsummer.

CAUSE A greyish-brown, sap-sucking aphid, *Cinara lapportei*, up to 2mm in length, which forms dense colonies at the leaf bases in late spring to early summer.

CONTROL Spray small trees with a systemic or contact

insecticide (*see p.94*) when aphids or sooty moulds are detected. Damage to large trees has to be tolerated but, fortunately, infestations resulting in defoliation are infrequent occurrences.

Celery Leaf Miner

SYMPTOMS The leaves of celery and related plants such as parsnip, parsley, and lovage develop brown dried up patches (*see p.33*). This is where the interior of the leaf has been eaten by one or more white maggots up to 7mm long. Heavy attacks give the plants a scorched appearance, with celery plants producing thin, bitter-tasting stems.
CAUSE Larvae of a leaf-mining fly, *Euleia heraclei*. It has two generations a year, with damage occurring from late spring to late summer.
CONTROL With light infestations, pick off the damaged portions of the leaves. There are no suitable insecticides for celery leaf miner.

Celery Leaf Spot

SYMPTOMS Pinprick-sized spots develop on the leaf surface (*see p.25*), then even smaller black fungal fruiting bodies develop on the spots. The infection may spread to the leaf stalks, and severely affected leaves die off.
CAUSE The fungus *Septoria apiicola*, which is spread by handling, by insects, and by water splash. In dull, damp weather, or on damp, misty nights, the spores are spread very readily. The infection may also be seed-borne and the fungus may overwinter on infected crop debris.
CONTROL Grow plants with care to ensure that they are vigorous and so better able to compensate for any losses. Always use disease-free seed which has been treated against leaf spot, from a reliable

company. Practise crop rotation (*see pp.100–104*) to minimize the risk of disease carry-over from year to year. Destroy all infected plant debris at the end of the season.

Centipedes

Centipedes are yellowish-brown animals with many body segments, each of which has one pair of legs. This distinguishes them from the superficially similar millipedes which have two pairs of legs per segment. Centipedes also have longer antennae on their heads, and are more active animals, which run away if disturbed. They are predators of small soil-dwelling animals and may help control soil pests.

Chafer Grubs

SYMPTOMS Plants may suddenly wilt. Those species of chafers that eat grass roots cause yellow patches in lawns (*see p.66*). Heavy infestations attract foxes, badgers, rooks, magpies, and crows, which scratch out the loosened turf to feed on the grubs during late summer to spring.
CAUSE Plump, white-bodied grubs with brown heads and three pairs of legs that live in the soil and feed on roots (*see p.19*). Their bodies are curved like the letter C. They can kill small plants and gnaw cavities in root vegetables. Several species of chafer occur in gardens. The largest species is the cock chafer or maybug, *Melolontha melolontha*, which has larvae up to 40mm long. It mainly occurs in cultivated soil such as flowerbeds and vegetable gardens. Lawns are attacked by garden chafer, *Phyllopertha horticola* and Welsh chafer, *Hoplia philanthus*. These two species are particularly associated with well-drained soils and are smaller, with grubs

up to 20mm long. When fully fed, the grubs pupate in the soil and emerge as adult beetles in late spring to early summer.
CONTROL There are no pesticides available to home gardeners for control of chafer bugs. Infestations in flower borders and vegetable gardens are generally light, and it is feasible to search for the grubs in the soil near plants that have suddenly wilted. Problems in lawns are sporadic, but can be devastating when they do occur. A pathogenic nematode, *Heterorhabditis bacteriophora*, can be watered into lawns in summer while the soil is moist and in the temperature range 14–20°C (57–68°F). The nematodes enter the bodies of the grubs and kill them by releasing bacteria that cause a fatal infection.

Cherry Blackfly

SYMPTOMS Leaves at the tips of shoots become curled in late spring, and dense colonies of black aphids are visible on the leaf undersides (*see p.39*). In midsummer some of the affected leaves often turn brown and dry up.
CAUSE A sap-sucking aphid, *Myzus cerasi*, which overwinters as eggs on fruiting cherries and some ornamental species such as *Prunus avium* and *P. padus*. Japanese flowering cherries are generally not attacked.
CONTROL Look for aphids on new growth from mid-spring onwards and spray ornamental and fruiting cherries with a systemic or contact insecticide (*see p.94*) before leaf curling becomes extensive. A plant oil winter tree wash applied in early winter to control overwintering eggs can reduce the need to spray in spring.

Cherry Leaf Scorch

SYMPTOMS In the winter infected trees stand out because

their branches remain covered with dead, brown, shrivelled leaves. The buds and stems, however, remain alive. Close examination of the lower surface of infected leaves with a hand lens reveals tiny reddish-brown blisters. These produce the fungal spores and so, once the spores have been shed, small indentations are all that is left on the leaf. In the summer yellow-margined brown blotches develop on the leaves, which then brown completely and die.
CAUSE The fungus *Apiognomonia erythrostoma*. The spores from the over-wintering dead leaves infect new foliage in the spring. Little is known about the precise host range, but it is most widely seen on *Prunus avium* and its cultivars.
CONTROL Despite the premature leaf browning, this disease seems to have few, if any, serious adverse effects. No fungicidal controls are available, but it may be worth removing infected leaves (perhaps with a garden vacuum). Infected leaves must not be composted

Cherry Leaf Spot

SYMPTOMS In the summer the leaves develop purplish spots, often followed by shothole symptoms or yellow mottling. The lower leaf surface may be marked with white flecks. Leaf fall is common, and both twigs and fruit crop levels and quality may be affected.
CAUSE The fungus *Blumeriella jaapii*, which is widespread in much of Europe but has only recently been reported in the UK. The fungus overwinters on fallen leaves then produces spores which infect new foliage in the spring. Rain and high humidity encourage spread of the infection.
CONTROL Rake up and dispose of fallen leaves; do not compost them. There are reports

to suggest that *Prunus* 'Kanzan' and *P*. 'Shirofugen' show a useful level of resistance.

Chicken of the Woods

See LAETIPORUS SULPHUREUS *p.150*.

China Aster Fusarium Wilt

See ASTER WILT *p.111*.

China Mark Moth

SYMPTOMS Water lily pads and other floating leaves of pond plants have oval holes up to 15mm in length (*see* **p.45**). The removed leaf pieces are spun together with silk by a small caterpillar to form a protective case around it. The caterpillars become fully grown in summer.
CAUSE Caterpillars of the China mark moth, *Elophila nymphaeata*. The adult moths have brown wings with white blotches and they rest on pond-side vegetation.
CONTROL Insecticides cannot be used in ponds as they will kill fish and other pond life. Search for the caterpillars' leaf cases, which are often on the undersides of floating leaves, and remove them by hand.

Chironomid Midges

SYMPTOMS The leaves of water lilies and other plants with floating foliage, such as water hawthorn, frog-bit and *Potamogetum* species, have a nibbled appearance around the edges. The foliage goes yellow and rots. Slender, transparent larvae, up to 10mm long, can be found feeding in damaged tissues.
CAUSE Maggots of certain chironomid flies (*Cricotopus* species). These seem to be a particular problem in new ponds or recently cleaned

established pools. Usually, once the pond has settled down and regained its complement of predatory insects, the midge population is kept at a non-damaging level.
CONTROL Insecticides cannot be used because of the danger to fish and other pond life. Remove damaged leaves and consider introducing fish if there are none in the pond.

Chlorosis

SYMPTOMS Yellowing of plant tissue, most commonly in the foliage. The loss or deterioration of the green pigment (chlorophyll) may also allow other pigments, which are usually masked by the green, to be seen. In addition to yellowing, red or orange discoloration may also develop.
CAUSE Many quite unrelated factors may be involved. Most common are deficiencies of iron and manganese (so-called lime-induced chlorosis), nitrogen and magnesium. Waterlogging, low temperatures, virus infection, or weedkiller contamination are also fairly commonly responsible. The precise patterning apof the chlorotic symptoms and their distribution within the plant may help to determine which factor is responsible.
CONTROL Take appropriate action to remedy any mineral deficiencies, sources of contamination, or cultural problems.

Chocolate Spot

See BROAD BEAN CHOCOLATE SPOT *p.120*.

Chrysanthemum Eelworm

SYMPTOMS Foliage on chrysanthemums, Japanese anemones, penstemons, and

many other herbaceous plants progressively turns brown and dries up from the base of the plants upwards. Spread of the microscopic eelworms within the leaf is restricted by the larger leaf veins, therefore recently infested leaves have brown wedges or islands separated from green healthy parts by the leaf veins. Eventually the entire leaf will turn brown or black.
CAUSES Microscopic nematodes or eelworms, *Aphelenchoides ritzemabosi*, which feed within the foliage. Although present on suitable host plants all year, infestations develop most rapidly during damp weather in late summer and autumn as the foliage starts to age.
CONTROL None of the pesticides available to home gardeners is effective. Destroy badly affected plants. Dormant chrysanthemum stools dipped in hot water will produce shoots for cuttings that are free of eelworm. Cut down the old stems and wash soil from the stools before immersing them for five minutes in water held at 46°C (115°F). The amount of heat is critical, as too much will damage the plant, while too little will allow eelworms to survive.

Chrysanthemum Leaf Miner

SYMPTOMS Sinuous white or brown lines develop in the leaves of chrysanthemums (*see* **p.37**), gerberas, cinerarias, marguerites, pyrethrum, and other related composite flowers. If an infested leaf is held up to the light, a grub or its pupa may be seen at the end of its tunnel. Small, pale, slightly raised round scars appear on the upper surface of leaves at shoot tips where the adult flies have been feeding.
CAUSE Maggots of a leaf-mining fly, *Chromatomyia*

syngenesiae, which has two generations during the summer on garden plants, but can breed continuously throughout the year in greenhouses.
CONTROL Deal with light infestations by picking off mined leaves, or crush the grubs at the ends of their tunnels. If necessary, spray with a systemic insecticide (*see p.94*). Avoid spraying plants in flower.

Chrysanthemum Petal Blight

SYMPTOMS Water-soaked lesions or smaller, brown spots develop on the petals of chrysanthemums (*see* **p.63**) and occasionally other members of the Compositae family, and on anemones. Infected flowers deteriorate rapidly.
CAUSE The fungus *Itersonilia perplexans*, which is encouraged by wet weather and by humid conditions. It is less common than ray blight (*see below*).
CONTROL Remove infected plant parts or whole plants immediately. Improve air circulation and attempt to lower humidity.

Chrysanthemum Ray Blight

SYMPTOMS Red-brown spots develop on the petals; as the spots increase in number and size, the petals are killed off and the whole flower head starts to decay. Rotting of shoot tips and brown spots on the foliage may also be seen. Occasionally infections occur at the stem base, causing the whole plant to be stunted and discoloured, and eventually to die. Similar symptoms may be caused by chrysanthemum petal blight (*see above*) or grey mould (*see p.142*).
CAUSE The fungus *Didymella ligulicola*, which can overwinter on infected plants and plant debris. Conditions of high

humidity, especially common in glasshouses, encourage disease development and spread. **CONTROL** Remove and dispose of infected plants immediately. Decrease humidity by opening vents and windows to improve air circulation.

Chrysanthemum White Rust

SYMPTOMS Pale buff or cream-coloured raised, warty pustules develop on the lower leaf surface (see **p.63**). The upper surface is marked with yellow pits, each pit corresponding with a pustule. As pustules age they become pale to tan-brown, and leaves may become distorted and die. **CAUSE** The fungus *Puccinia horiana*, which is encouraged by moist, humid atmospheres. It may persist from year to year on over-wintering plants and stools, and can infect indoor and outdoor chrysanthemums. **CONTROL** Remove infected leaves promptly. Destroy severely infected plants and do not propagate from them. Improve air circulation to decrease humidity.

Clematis Green Petal

SYMPTOMS The flowers of clematis open to reveal a distinct green coloration. This is especially clear on pale-flowered or white-petalled clematis. It is usually only the first flush of flowers that is affected. **CAUSE** The most common causes are adverse growing or cultural conditions at a critical stage in the flower's development. Green flowers frequently develop a few weeks after there has been an unusually cold spell of weather. If cold damage is responsible, subsequent flushes of flowers are perfectly normal. However, if the problem lasts

throughout the season, and reappears the following year, then the development of green pigments in the petals could be due to the presence of a phytoplasma infection. **CONTROL** Protect plants against excessively cold conditions with horticultural fleece or similar material. If the plant has a phytoplasma infection it should be destroyed, as this is infectious and there is no cure available.

Clematis Wilt

SYMPTOMS Leaves and shoots wilt and die back. Wilting usually starts at the shoot tips, leaf stalks blacken where they join the leaf blade and young foliage wilts. New, healthy shoots may appear from below soil level or from nodes beneath the wilted area. A small spot of discoloration may develop under the lowest pair of wilting leaves. Waterlogging, graft failure, and insect, or slug damage may cause wilting too, and indeed are more frequently responsible for general wilting symptoms, but in these cases affected clematis do not show the other symptoms. **CAUSE** The fungus *Calophoma clematidina* (syn. *Ascochyta clematidina, Phoma clematidina*). **CONTROL** Prune out affected stems, cutting right back into healthy tissue, even cutting below soil level if necessary. Mildly infected and otherwise vigorous plants should reshoot. Some clematis seem to show useful levels of resistance to fungal wilt, including *Clematis alpina, C.* 'Avant Garde', 'Bonanza', 'Confetti', *C. macropetala, C. montana, C. orientalis*, and *C. tangutica*.

Click Beetles

Click beetles are elongate brown beetles, 8–20mm long. If a click beetle is placed on its

back it will right itself by jumping into the air with an audible clicking noise, hence its common name. Click beetles cause no damage as adult insects, but several species are the adult stage of soil pests known as wireworms (see p.198).

Clubroot

SYMPTOMS Roots become swollen, and the whole root system may become distorted (see **p.21**). The foliage is pale, chlorotic, or pinkish in colour and wilts readily, especially during hot weather, but may even wilt when the soil is moist. Plants fail completely or produce a poor crop. Members of the Cruciferae may be attacked, in particular brassicas. Certain ornamentals, including wallflowers, stocks, and candytuft, and also some weeds, including shepherd's purse, charlock, and wild radish, are susceptible. **CAUSE** The fungus-like organism *Plasmodiophora brassicae*, which is soil-borne and can remain viable in the soil for in excess of 20 years, even in the absence of susceptible host plants. Infection occurs via the root hairs. When swollen roots disintegrate, they release many spores. The pathogen is easily introduced into a garden, on infected plants, on soil adhering to boots, tools, and wheelbarrows and in soil, manure, or garden compost. Acidic and water-logged soils are most prone to clubroot. **CONTROL** Improve soil drainage and lime the soil in order to deter the pathogen. Burn all infected plants immediately, certainly before the roots start to disintegrate. Avoid spreading the spores around the garden on tools or boots, or by moving soil that is potentially infective. Raise your own plants, rather than buying them in; if this is not possible,

make sure that you only purchase plants from a reputable source and check them carefully before planting. Practise good weed control. If you have no choice but to grow crops of brassicas on infested soil, raise plants in proprietary compost in individual pots that are at least 5cm in diameter. Plant out only once plenty of roots have formed. Choose resistant vegetable cultivars such as calabrese 'Trixie', Chinese cabbage 'Harmony', kale 'Tall Green Curled', swede 'Marian', cauliflower 'Clapton', cabbage 'Kilaxy' or 'Kilatron', and Brussels sprouts 'Chronos' or 'Crispus'.

Codling Moth

SYMPTOMS In mid- to late summer, fruits of apple and pear are tunnelled by small white caterpillars with brown heads (see **p.13**). By the time the fruit is ripe, the caterpillars have usually finished feeding and have sought overwintering sites under loose flakes of bark. The caterpillar feeds in the core and makes an exit tunnel leading either to the eye end or elsewhere on the fruit surface. **CAUSE** Caterpillars of a small moth, *Cydia pomonella*, which is generally more widespread and troublesome than the other apple fruit pest, apple sawfly (see p.101). The latter attacks apples at the fruitlet stage while codling moth damages ripening fruits in late summer. **CONTROL** Apply a suitable contact insecticide (see p.94) in early summer with a second treatment three weeks later. Pheromone traps (see pp.88–89) are available for detecting when adult codling moths are active and likely to be laying eggs. Accurate timing of insecticides is important, since it is necessary to kill the young caterpillars before they have bored into fruits. Pheromone traps may catch enough male moths to reduce the mating success of females on isolated apple trees and so may reduce the number of maggoty apples.

Collar Rot

see PHYTOPHTHORA *p.167*.

Colletotrichum

SYMPTOMS Leaf and occasionally stem spotting, developing into lesions that may become sunken. Stems may become ringed, causing the whole plant, or parts of it, to die back. **CAUSE** Various species of the fungus *Colletotrichum*, some of which are believed to be seed-borne, and which can all be spread by water splash. **CONTROL** Remove affected plant parts promptly, and entire plant if stem infections are present. Do not save seed from infected plants.

Colorado Beetle

SYMPTOMS The foliage of potato, tomato, aubergine, and pepper plants is eaten by pale yellowish-orange beetles, up to 10mm in length, with five black stripes on each wing case. Rotund, orange-red grubs with black heads and two rows of black spots along the sides of the bodies also attack leaves and can cause extensive defoliation. **CAUSE** A leaf beetle, *Leptinotarsa decemlineata*, which originates from North America but is now found throughout Europe, except in Britain and Ireland. **CONTROL** Strains of the beetle that have gained resistance to some insecticides can occur. If Colorado beetles are found in Britain, they must be reported to the Animal and Plant Health Agency, which will organize the necessary measures to eradicate them. Elsewhere, infested plants can be sprayed with an insecticide approved for this purpose when the adults or larvae are seen.

Conifer Aphids

SYMPTOMS Colonies of greyish-black or brown insects develop on the shoots, branches, and sometimes trunks of conifers. Spruce (*Picea*), pine (*Pinus*), juniper (*Juniperus*), cypress (*Cupressus*), and Leyland cypress (× *Cuprocyparis leylandii*) are liable to be attacked by the aphids which, at 3–6mm long, are significantly larger than most aphids on other plants. The foliage and stems become sticky with honeydew excreted by the aphids from late spring to midsummer. A black, sooty mould grows on the honeydew, and heavy infestations can cause foliage to turn yellow and dry up. **CAUSE** Several aphid species, mainly *Cinara* species. **CONTROL** Only small trees and hedges can be treated effectively. Spray with a systemic or contact spray (*see p.94*) in mid- to late spring before heavy infestations build up.
• *see also* CEDAR APHID *p.124*; CYPRESS APHID *p.130*; GREEN SPRUCE APHID *p.142*; JUNIPER APHIDS *p.149*.

Conifer Red Spider Mite

SYMPTOMS Foliage becomes finely mottled and yellowish-brown in midsummer, and may fall prematurely. **CAUSE** Careful examination with a hand lens will show tiny mites, *Oligonychus ununguis*, yellowish-green with dark markings, crawling among the foliage on fine silk webbing. Orange-red spherical eggs may also be seen on the leaves and stems. Spruce is the most susceptible conifer, especially *Picea albertiana*, but this mite also attacks juniper, thuya, cedar, and cypress.

CONTROL Watch for signs of infestation from late spring onwards. If necessary, spray thoroughly with an organic pesticide (*see p.89*) on two or three occasions at weekly intervals.

Conifer Woolly Aphids

See ADELGIDS *p.106*.

Conifers, Needle Rusts of

See NEEDLE RUSTS OF CONIFERS *p.159*.

Copper Deficiency

SYMPTOMS Blue-green or slightly yellow discoloration of the foliage, occasionally followed by dieback. Symptoms are not easy to recognize or differentiate from other problems and are fairly rare, being most likely on peaty soils. **CAUSE** A deficiency or poor availability of copper in soil. **CONTROL** Treatment with a complete fertilizer may help to rectify the problem, but other possible causes for the symptoms shown should be investigated too.

Coral Spot

SYMPTOMS Bright, coral to orange raised pustules appear on dead, woody stems and branches (*see p.56*). The pustules may only appear after the stem has been dead for several weeks. Many different plants may be affected, but currant bushes (*Ribes*), elaeagnus, magnolias, and maples (*Acer*) are very susceptible. Dieback occurs, and if the infection spreads right down into the crown, the whole plant may be killed. **CAUSE** The fungus *Nectria cinnabarina*, which may also live on dead logs, twigs, woody debris, and old pea sticks. Spores are produced all year, and water, either as rain or irrigation splash, is the main method of dispersal. The fungus enters a plant via a wound or colonizes a dead snag left by physical injury or poor pruning. **CONTROL** Prune out all dead or dying stems promptly, cutting well back into sound, healthy wood. Remove wood bearing the pustules. Dispose of garden debris that may be harbouring the disease. Do not compost or pass through a shredder for use as mulch.

Cotoneaster Webber Moths

SYMPTOMS In early summer, the stems of *Cotoneaster horizontalis* are covered with silk webbing spun by numerous dark brown caterpillars up to 12mm long (*see p.53*). The leaves dry up and turn brown where they have been grazed by the caterpillars. Less extensive webbing and feeding damage also occurs in late summer, caused by the newly-hatched caterpillars before they hibernate for the winter and resume their activities in spring. **CAUSE** Caterpillars of two moths, *Scythropia crataegella*, which also attacks hawthorn, and *Acrobasis suavella*. Both have one generation a year and their caterpillars pupate among the silk webbing. **CONTROL** Prune out light infestations, otherwise spray with a contact insecticide (*see p.94*) when webbing is seen. Avoid spraying plants in flower.

Crane Flies

Crane flies, or daddy-long-legs, often have soil-dwelling larvae – leatherjackets – some of which feed on roots and the stem bases of small plants. The adults do not feed.
• *see also* LEATHERJACKETS *p.152*.

Crown Gall

SYMPTOMS Rounded swellings or galls – that are off-white when young and harden and turn brown with age – develop on woody or herbaceous plants. Galls are very variable in size, ranging from less than one centimetre to more than 30cm in diameter. They usually develop in small groups. Crown gall may appear on the roots or stem base (the crown of the plant). Less frequently, symptoms of an aerial form may show at some distance up the stem or trunk (*see* **p.54**). Galls on woody plants are persistent, those on herbaceous plants, are softer and more short-lived. They soon disintegrate, often allowing entry of secondary organisms in the process. The plant's overall vigour is rarely affected. However, if galls ring the stem or trunk, or cause extensive disruption of water and nutrient flow, the whole plant may decline. If stem or root splitting occurs due to the erupting out of the galls, secondary organisms may gain entry and cause dieback..
CAUSE The bacterium *Agrobacterium tumifaciens*, which is soil-borne, and capable of living saprophytically (without a living host) in the soil. The bacteria gain entry via wounds and then stimulate rapid proliferation of cells, so forming the galls. Gall growth is greatest when the plant is growing rapidly, so infections that occur at the very end of the growing season may only produce symptoms the following year.
CONTROL Improve soil drainage. Check all plants carefully before planting. Keep root and stem injury to a minimum. Remove infected plants and grow root crops (except beetroot and swedes) on the site for at least a year to minimize the risk of subsequent infections, as these crops are not susceptible to crown gall.

Crown Rot

SYMPTOMS Deterioration and rotting of the tissues at the crown (the junction of the stems and roots) of the plant. Herbaceous plants are very prone, but other woodier hosts, such as wisterias, may also succumb. It may then cause deterioration of the stems and foliage and the whole plant can die.
CAUSE Various soil- and water-borne fungi and bacteria, often several in combination.
CONTROL If dealt with very promptly it may be possible to save an infected plant. Remove affected areas, cutting well back into healthy tissue.

Cuckoo Spit

SYMPTOMS Blobs of white frothy liquid appear on the stems of many plants in early summer (*see* **pp.56, 61**). A yellowish-green insect lives in the froth where it feeds by sucking sap.
CAUSE The nymphal stages of froghoppers, such as *Philaenus spumarius*. Once they have reached the adult stage in early to midsummer, they stop producing froth and become less noticeable.
CONTROL Spraying is often unnecessary, since cuckoo spit is only a temporary nuisance on plants, and the nymphs cause little harm unless they are feeding at the shoot tips, where they may cause some distorted growth. Picking off the nymphs by hand is usually sufficient to control infestations.

Cucumber Foot and Root Rot

see FOOT AND ROOT ROTS *p.137.*

Cucumber Gummosis

SYMPTOMS Discrete sunken grey-black spots, often with a dense, furry, grey fungal growth, develop on the fruit. A viscous liquid may ooze from these lesions, and severely infected fruits may split. Stem and leaf lesions may also develop but are less common.
CAUSE The fungus *Cladosporium cucumerinum*, which is encouraged by cool, damp conditions.
CONTROL Remove infected areas promptly. Improve ventilation and, if feasible, raise temperatures. Grow resistant cultivars such as 'Femspot'.

Cucumber Mosaic Virus

SYMPTOMS Plants are stunted and show distinct yellow mosaic patterning (*see* **p.35**), usually on stunted, deformed foliage (*see* **p.37**). Flowering is reduced or non-existent, and plants may even die completely. If fruits are produced on cucurbits (such as marrows, courgettes, or cucumbers), they are small, pitted, and strangely dark green with bright yellow blotches. Not only is their texture hard but the fruits will be inedible. On ornamental plants, flower breaking (changes in colour on clearly defined areas on the petals) may occur.
CAUSE The cucumber mosaic virus, which has an extremely wide host range, including many common garden plants, food crops, and weeds. Virus particles are readily spread by handling and by sap-feeding pests, in particular aphids.
CONTROL There are no cures available and infected plants should be destroyed. Practise aphid control and minimize the risk of infection spreading. Remove weeds. Avoid unnecessary handling of plants and always handle healthy plants before touching infected or suspect ones. After handling infected plants, ensure that you wash your hands thoroughly. Grow resistant cultivars where available, such as aubergine 'Bonica', courgette 'Defender', and marrow 'Tiger Cross'.

Currant and Gooseberry Leaf Spot

SYMPTOMS In early summer, very small dark brown or black spots develop on the leaves of currants, gooseberries, and, occasionally, ornamental currants (*Ribes*). The spots may enlarge and join together, and the infected leaf turn brown. Affected leaves fall prematurely and the bush may be extensively defoliated by midsummer. In extreme cases, fruits shrivel.
CAUSE The fungus *Drepanopeziza ribis*. It produces spores from the leaf spots that are spread by wind and rain. The fungus overwinters on the leaf spots on fallen leaves.
CONTROL Regularly remove and dispose of fallen leaves to prevent the fungus overwintering. Feed and maintain plants well to encourage new growth; spray with foliar feed in spring. Consider growing more resistant cultivars, such as blackcurrants 'Ben Connan', 'Big Ben', 'Ben Hope', or 'Black Reward', and redcurrant 'Rondom'.

Currant Blister Aphid

SYMPTOMS Leaves at the shoot tips become crumpled, with raised areas that have small pale yellow insects on the undersides. The affected leaves are yellowish-green or have a reddish coloration (*see* **p.41**), especially on redcurrants.
CAUSE Sap-sucking aphids, *Cryptomyzus ribis*, which are active on the foliage between bud burst and midsummer. After that they develop wings and migrate to the wildflower known as hedge woundwort (*Stachys sylvatica*). There is a return migration in autumn

when overwintering eggs are laid on currant bushes.

CONTROL Although unsightly, currants are not greatly harmed. To prevent infestations, spray with a contact insecticide (*see p.94*) between bud burst and flowering. A plant oil winter tree wash applied in early winter to control overwintering eggs can help reduce the need to spray.

Currant Clearwing Moth

SYMPTOMS Red- and black-currant stems are tunnelled by white caterpillars, up to 15mm long, with brown heads (*see p.49*). Their feeding activities sometimes cause shoots to die but more usually the presence of a caterpillar is detected when a weakened stem snaps. This is likely to happen during windy weather, especially when stems are heavy with berries. A blackened tunnel can be seen in the centre of attacked stems.
CAUSE Caterpillars of a clearwing moth, *Synanthedon tipuliformis*. The name clearwing is derived from the fact that the moth's wings are largely devoid of coloured scales and are therefore transparent, except at the wing tips. This feature, combined with a black and yellow-striped abdomen, gives this day-flying moth a wasp-like appearance.
CONTROL There is no effective pesticide treatment, as it is not possible to reach caterpillars inside stems. Fortunately, this pest does not usually damage all of the stems on a bush. During the winter, tunnelled stems can sometimes be detected by gently bending them, causing them to snap at the weak point. These stems should be cut back to sound wood and the prunings burnt to destroy the caterpillars.

Currant Reversion

SYMPTOMS Plants produce leaves with slight yellowing,

and unusually small main veins. The leaves are also smaller and have fewer lobes than healthy ones. Flowering and ultimately cropping of the plant is also reduced.
CAUSE The blackcurrant reversion virus, which is spread largely by big bud mite (*see p.115*) and may also be transmitted by grafting. Symptoms of mite infestation, such as swollen buds, may also be present.
CONTROL Remove infected plants, as the problem will not decline. Such plants will be a source of infection for other healthy bushes. Always buy blackcurrants from a reputable source and ensure that they are certified free from pests and diseases on purchase. The cultivar 'Ben Gairn' is resistant to reversion.

Cushion Scale

SYMPTOMS Leaves of holly, camellias, rhododendrons, *Trachelospermum*, and *Euonymus* develop sooty mould on the upper surfaces, especially during winter. Flat, oval, yellow-brown scales are attached to the undersides of the leaves, and in early summer the females deposit eggs under broad, white, waxy bands (*see p.27*).
CAUSE Sap-sucking scale insects, *Pulvinaria floccifera*.
CONTROL Spray the undersides of the leaves thoroughly with a systemic or contact insecticide (*see p.94*) in midsummer to control the more vulnerable newly-hatched scale nymphs.

Cutworms

SYMPTOMS Small plants in flowerbeds and vegetable gardens wilt and die. Examination of the plants will show that the roots just below soil level have been severed, or the outer bark has been gnawed

away. Root vegetables, such as potatoes and carrots, may have cavities eaten in them. Caterpillars, often creamy brown or sometimes greenish-brown in colour and up to 40mm long, may be found in the soil near recently damaged plants (*see p.19*). At night these caterpillars will feed above soil level on the foliage of low-growing plants.
CAUSE The soil-dwelling caterpillars of several moths, including large yellow underwing (*Noctua pronuba*), lesser yellow underwing (*N. comes*), turnip moth (*Agrotis segetum*), heart and dart moth (*A. exclamationis*), and garden dart (*Euxoa nigricans*).
CONTROL Cutworms tend to work their way along row crops such as lettuce, so it is worthwhile sifting the soil near a damaged plant to find the culprit. These caterpillars are more troublesome in weedy plots, so good cultivation does help. There are currently no insecticides for controlling cutworms. A soil drench of pathogenic nematodes is available.

Cypress Aphid

SYMPTOMS Conifers, such as *Cupressus* species, *Chamaecyparis lawsoniana*, × *Cuprocyparis leylandii*, and *Thuja*, develop sooty mould on the foliage and stems. This grows on the sticky honeydew excreted by the aphids, which are brownish-black and up to 4mm long. Attacks during early summer cause shoots to dry up and turn yellow-brown in midsummer. On hedges, the damage is generally worse on the lower parts, with some green growth left at the top.
CAUSE A sap-sucking aphid, *Cinara cupressivora*, which is active during summer.
CONTROL Look for aphids on the stems from late spring

onwards and if seen, spray immediately with a systemic or contact insecticide (*see p 94*). Once foliage has started to discolour, the process cannot be reversed, but in time plants may recover. Damaged trees and hedges should be fed with a general fertilizer in spring and kept watered during dry spells to help them regain vigour.

D

Daedaleopsis confragosa

SYMPTOMS Otherwise known as the blushing bracket, this is a common bracket fungus. It attacks deciduous trees, in particular alders, birches, and willows. The bracket-shaped, semi-circular fruiting bodies measure between 5 and 12cm in diameter (*see p.53*). The upper surface is smooth or slightly ridged, and reddish-brown in colour. The pores are circular but become elongate and red. The bracket has a corky, almost rubbery texture. It causes a white rot of the woody tissues and is generally associated with wounded or weakened trees.
CAUSE The fungus *Daedaleopsis confragosa*.
CONTROL Remove brackets to prevent spores spreading to other parts of the tree, or to other susceptible trees. Affected trees should be inspected by a professional arboriculturist to determine the extent of decay, and the overall health of the tree.

Dahlia Smut

SYMPTOMS Pale brown, elongate spots with yellow margins develop on leaves in midsummer. The lower, older leaves are affected first, but younger leaves may then be affected too. The spots darken and increase in size and may join together, often killing the leaf.

CAUSE The fungus *Entyloma calendulae* f. *dahliae*, which may live in the soil around infected plants and on plant debris. It can persist in the soil in the absence of living dahlias for about five years.

CONTROL Dispose of severely infected dahlias, and plant replacements on a fresh site. The fungus does not seem to be present in the tubers, so provided any affected leaves are removed and the tubers appear clear of infected debris, they can be stored and should produce healthy plants the following year.

Damping Off

SYMPTOMS Seedlings flop over, often showing discoloration and deterioration around the stem base (*see* **p.50**). Initially the basal lesions may appear water-soaked. The infection spreads rapidly, causing patches of seedlings to die off. An entire tray of seedlings may be killed within a few days. Fluffy white fungal growth may be seen on the surface of the dead and dying seedlings. If infection occurs before germination is complete, the seedlings only emerge patchily.

CAUSE Several soil- and water-borne fungi and fungus-like organisms, in particular species of *Pythium*, *Phytophthora*, and *Rhizoctonia*. Infection is encouraged by over-wet compost and prolonged raised temperatures. Seedlings that are not given adequate light, and which are from seed sown too densely, are particularly prone to infection. Damping off is most common when hygiene is poor – if pots, trays, and implements are not properly scrubbed and cleaned, if unsterilized compost is used, or if non-mains water is used.

CONTROL Observe strict hygiene and only use clean trays, pots, and implements and proprietary or sterilized compost. Do not use anything other than mains water, especially not water collected in a butt. Sow seedlings thinly and ensure they have adequate light and are not kept too warm for any longer than is needed to encourage germination.

Deer

Several species of deer cause problems in gardens. The most widespread is roe deer (*Capreolus capreolus*), but muntjac (*Muntiacus muntjak*), fallow deer (*Dama dama*), and red deer (*Cervus elephus*) also cause problems in some areas. A wide range of herbaceous and woody plants is eaten, with damage mainly occurring between dusk and dawn. Sometimes it is blooms rather than foliage that is eaten, as with tulips, but more usually whole shoots are lost. On woody plants, such as roses, the stems are partly bitten through cleanly but there remains a ragged edge to the stem (*see* **p.59**). This is because deer have no front teeth in their upper jaw and they cut off woody shoots by biting and tugging. The trunks of trees and side branches can be damaged by male deer rubbing their antlers against the bark, a habit known as fraying.

CONTROL It is difficult to exclude deer from gardens. They are agile animals good at jumping over obstacles, and a fence about 2m high is necessary to keep roe deer out. Proprietary animal repellents often give disappointing results; the effects tend to be short-lived and they may simply encourage deer to feed on something else in the garden. Deer are inquisitive feeders, attracted to new plantings, so give new plants some protection with netting. Deer can eat almost anything, but it is a good idea to look at other gardens in the locality to see which plants tend to be left alone.

Delphinium Black Blotch

SYMPTOMS Black blotches develop on leaves (*see* **p.29**), rapidly enlarging and spreading to the stems, petioles, and flower buds. Entire leaves may turn black and die off.

CAUSE The bacterium *Pseudomonas syringae* pv. *delphinii*, spread from the soil or by rain or watering splash.

CONTROL Remove and dispose of infected plants promptly and avoid over-head watering. Plant new delphiniums on a fresh site.

Delphinium Leaf Miner

SYMPTOMS The leaves of delphinium and monkshood develop large, brown, dried-up areas extending from the tips towards the leaf stalks (*see* **p.32**). White maggots up to 3mm in length may be found feeding within the mined areas.

CAUSE Larvae of a leaf-mining fly, *Phytomyza aconiti*, which has two generations during the summer months.

CONTROL For light infestations, the damaged parts of leaves can be picked off. If necessary spray with a systemic insecticide (*see p.94*).

Delphinium Moth

SYMPTOMS During the spring, leaves at the shoot tips of delphinium and monkshood are spun together with silk threads by small green caterpillars heavily marked by black spots. Later the caterpillars feed openly on the expanded foliage. They are greenish-white by the time they pupate in a pale yellow cocoon spun on the underside of leaves during late spring to early summer. The mature caterpillars are up to 40mm long, but often escape notice as they remain inactive on the underside of the foliage during the day.

CAUSE Caterpillars of a moth, *Polychrysia moneta*, which has one generation a year and overwinters as young larvae in the old plant stems.

CONTROL It is the young larvae feeding at the tips of new shoots in the spring that cause most harm. Look for bound-up leaves and remove the caterpillars by hand. For heavy infestations, spray with a contact insecticide (*see p.94*).

Delphinium Viruses

SYMPTOMS Yellowing of the foliage, usually as mosaic patterns, mottling, or vein-clearing (yellowing immediately adjacent to the leaf veins). Infected plants may be stunted and fail to flower properly, often dying within a year or so of the first sign of the symptoms.

CAUSE Various viruses may be involved, either singly or in combination. The most common is cucumber mosaic virus (*see p.129*). The viruses involved are not all spread in the same manner, but common means of spread include handling, and vectors that include sap-sucking pests, such as aphids, and soil-inhabiting nematodes.

CONTROL Remove infected plants promptly and avoid handling other plants after handling one showing symptoms. Do not propagate from infected plants.

Dianthus Anther Smut

SYMPTOMS Anthers of pinks and carnations (*Dianthus*) become swollen and darken as they fill with masses of black-purple spores. These are released in great quantity as the anthers rupture. Stems become stunted.

CAUSE The fungus *Microbotryum violaceum*, which appears to be present throughout the plant, although only producing symptoms on the anthers and flower stems.
CONTROL Dispose of infected plants promptly. Do not propagate from any part of an infected plant. Remove all plant debris and scrub down the greenhouse if infected plants were grown under glass. Do not grow *Dianthus* on the site for at least five years.

Dianthus Fusarium Wilt

SYMPTOMS Carnations (*Dianthus*) start to fade, and the leaves wilt and develop a greyish or straw-coloured discoloration. The leaves may also develop a purple-red discoloration. The whole plant dies, but the symptoms rarely occur over the whole plant simultaneously.
CAUSE The fungus *Fusarium oxysporum* f. sp. *dianthi*, which has thick-walled resting spores that enable it to remain dormant in the soil for many years, even in the absence of carnations in the area.
CONTROL Remove infected plants promptly and grow new plants on a fresh site. Do not propagate from infected plants.

Didymella Stem Rot

SYMPTOMS Black-brown, sunken blotches develop on the stems of tomatoes (*see p.49*), and occasionally aubergines, most commonly at soil or ground level, often combined with the development of adventitious roots (roots formed above soil level). The older leaves turn yellow. A blackish-brown rot, usually at the calyx end, develops on the fruit. Infected areas later develop numerous tiny, black, pinprick-sized fungal fruiting bodies on the surface.

CAUSE The fungus *Didymella lycopersici*, which can be seed-borne, but is usually spread from plant to plant by rain or watering splash, or by handling. The fungus can overwinter on debris from infected plants or in the soil.
CONTROL Remove and dispose of infected plants promptly; do not compost them. Clear up all crop debris at the end of the season and grow tomatoes on a fresh site. As aubergines may also occasionally be affected, choose a fresh site for these, too.

Dieback

SYMPTOMS The stems of plants start to die back. Dieback may be seen first either at the tip, the base, or part-way up the stem, it rarely affects all stems simultaneously. Dark blotches or sunken patches may develop at the point where the deterioration starts. Leaves wilt, yellow, and die off. The symptoms may spread down into the base of the plant, or its crown, so causing the whole plant to die.
CAUSE Various fungi, some of which can actively invade healthy stems, and others which only enter the plant through wounds or other points of damage. Dieback, especially from stem tips downwards, may also be caused by cultural problems, including poor establishment of young plants, drought, and waterlogging.
CONTROL Prune out affected stems to a point well below any signs of damage. Improve cultural conditions to ensure increased plant vigour and plenty of replacement growth.

Dogs

Although not pests, dogs can cause problems in gardens. Their urine can scorch foliage, with bitches causing circular brown patches in lawns, and dogs

damaging conifers and other plants that they cock their legs against.
CONTROL Spraying the foliage with water to dilute the urine can prevent damage, but this needs to be done soon after the dog has urinated. Proprietary animal repellent products are available from garden shops but these do not usually give reliable long-term protection.

Dogwood Anthracnose

SYMPTOMS Spots are seen on the leaves and bracts of *Cornus* in late spring or early summer; infected leaves discolour and fall prematurely. These symptoms usually appear first towards the base of the tree, and then progress upwards. Cankers may develop on shoots and branches. Minute, raised, dot-like fungal fruiting bodies may appear on infected areas, especially during damp weather. As the disease develops, extensive dieback may occur and the tree may even be killed. The precise range of dogwoods affected is still under investigation, but *Cornus florida* and *C. nuttallii* are known to be particularly susceptible. *Cornus sanguinea*, *C. alba*, and *C. kousa* are also affected, but on these, dieback is restricted to weak shoots. Native dogwoods seem to be unaffected.
CAUSE The fungus *Discula destructiva*, which is encouraged by cool, damp weather and which is most prevalent in areas of high rainfall. This fungus has only been reported in the UK since the late 1990s.
CONTROL Prune out infected stems immediately to try to save the affected plant. Rake up and bin or burn fallen infected leaves.

Douglas Fir Adelgid

SYMPTOMS Small, fluffy white balls of waxy threads appear

on the undersides of leaves on Douglas fir (*Pseudotsuga menziesii*) during the summer. Heavy infestations may cause some yellowing of the foliage, which can also be discoloured by the growth of sooty moulds on the pest's excrement.
CAUSE *Gilletteella cooleyi*, a blackish-brown sap-sucking insect, up to 2mm long, with an aphid-like appearance. It can be found on Douglas fir throughout the year, and may also occur on its alternative host plant, Sitka spruce (*Picea sitchensis*), where it causes an elongate swollen gall at the shoot tips. The spruce generation is not normally a problem in gardens.
CONTROL Large trees cannot be sprayed adequately and the presence of adelgids has to be tolerated. On small trees, some protection can be given against over-wintering adelgid nymphs by spraying with a systemic or contact insecticide (*see p.94*) during mild, dry weather in late February. At that time they are not protected by a waxy coating and are therefore more vulnerable to sprays.

Downy Mildew

SYMPTOMS Yellow or otherwise discoloured areas develop on upper leaf surface. Each corresponds to a slightly fuzzy, greyish-white or, occasionally, purplish fungal growth on the lower leaf surface. The infection may spread, causing large areas or even the entire leaf to discolour and die. A wide range of plants may be affected, but downy mildew is especially common on young plants and those growing in moist or humid environments. The symptoms may be confused with those of powdery mildew infection.
CAUSE Several different fungus-like organisms, in particular species of *Peronospora*, *Bremia*, and *Plasmopara*. These

include *Peronospora parasitica* (downy mildew of cruciferae), *Peronospora destructor* (on onion), *Peronospora farinosa* f. sp. *spinaceae* (on spinach), *Peronospora viciae* (on pea), *Peronospora sparsa* (on rose), *Bremia lactucae* (on lettuce), and *Plasmopara viticola* (on grape).

CONTROL Remove infected leaves promptly and improve air circulation around plants by increased spacing and good weed control. Increase ventilation in a greenhouse. Avoid overhead watering. Many downy mildews produce their spores at night and this, combined with their preference for damp conditions, means that evening watering should be avoided. Downy mildews also produce resting spores within the infected plants and these may be released into the soil when the plant rots down: removal of infected material is therefore essential.
• *see also* BRASSICA DOWNY MILDEW *p.119;* GERANIUM DOWNY MILDEW *p.139;* HEBE DOWNY MILDEW *p.143;* LETTUCE DOWNY MILDEW *p.153;* ONION DOWNY MILDEW *p.161;* PANSY DOWNY MILDEW *p.163.*

Drought

SYMPTOMS Symptoms vary depending on the plant affected and whether the drought has been recurrent or occasional. Common symptoms include poor growth and stunting, wilting of the foliage (*see p.43*) and, in extreme cases, other plant parts such as flower stems, stems, and flowers (*see p.62*). Prolonged drought may also cause poor flowering, bud drop, and the formation of small fruits. Drought may also cause blossom end rot (*see p.117*). Drought followed by sudden watering or rainfall may cause splitting or cracking of fruits and stems.

CAUSE Most commonly, inadequate rainfall or watering; occasionally, an inability of the plant to take up adequate moisture because of a damaged root system. Very free-draining soils, such as thin soils over chalk or soils with a high sand content, are particularly prone to drying out. Plants with restricted root runs, such as those growing in containers, are also particularly prone.

CONTROL Take any measures possible to ensure that the soil or compost never dries out completely. Where feasible, grow plants in open ground, or protect containers from the effects of direct sunlight. Improve soil moisture retention by incorporating bulky organic materials such as compost into light soils. Apply a mulch to improve soil moisture retention and reduce loss by evaporation.

Dry Rot of Bulbs and Corms

SYMPTOMS Foliage turns brown and dies, often with a clear dark discoloration just above soil level. Crocus, freesias, gladioli, and montbretias (*Crocosmia*) are particularly affected. Whole plants may die back. The corms develop numerous small sunken lesions, each of which bears many pinprick-sized black dots or fungal fruiting bodies.

CAUSE The fungus *Stromatinia gladioli*, which can overwinter on stored corms and which may persist in the soil as small, hard, black resting bodies (sclerotia). These can remain viable in the soil for many years even in the absence of host plants.

CONTROL Always check bulbs are healthy and free from the small black sclerotia before planting. Remove any infected plants immediately. Plant replacements, and any others that are susceptible, on a fresh site.

Dutch Elm Disease

SYMPTOMS Areas of the tree's crown show wilting and leaf yellowing. Affected leaves then turn brown and die. Dieback of twigs and branches usually follows and young shoots may curl over to form "shepherds' crooks". The whole tree may be killed by the end of the first summer. If the bark on affected stems and branches is peeled away, longitudinal, discontinuous brown staining is apparent. If stems or branches are cut across, they may show complete or discontinuous rings of brown staining in the current year's growth. Feeding damage caused by adult elm bark beetle (*see p.104*) may be seen in the crotches of young stems. The "galleries" of tunnels caused by the beetle's larvae show beneath the bark on larger stems and trunks. The infection may also be spread between root grafts, between trees in close planting or in hedgerows, or when young elms grow from the suckers of a parent plant. Symptoms are much more obvious and the disease more damaging during hot, dry weather, as these conditions exacerbate the problems caused by this 'vascular wilt' fungus blocking the tree's vascular (water transport) system.

CAUSE The fungus *Ophiostoma novo-ulmi*, which may cause this disease on elms and on *Zelkova*.

CONTROL Remove infected trees and dispose of by burning.

E

Earthworms

Earthworms (*see p.69*) are generally beneficial animals that improve soil condition. Their tunnels provide aeration and drainage pathways through the soil, and worms help incorporate organic matter by pulling dead leaves into the soil. They can,

however, be a nuisance when they excrete small piles of soil onto the surface of lawns (*see p.66*). These worm casts can spoil the appearance of a fine lawn, as well as making the surface muddy and slippery. When lawns are mowed, worm casts become smeared, and this creates patches of bare soil that lawn weeds can colonize. Not all earthworms produce worm casts. The culprits are usually the *Allolobophora* species of earthworms. They are mainly a problem during spring and autumn when warm, moist soil conditions encourage worm activity in the upper layers of the soil; drought or cold sends worms deeper and reduces surface casting.

CONTROL There are no pesticides approved for earthworm control available to home gardeners. Worm casts can be dispersed by brushing the lawn with a besom broom or using an up-turned wire rake. This is best done when worm casts are dry and crumbly. During wet periods, walking on the lawn should be kept to a minimum to avoid smearing the worm casts.

Earwigs

Earwigs, *Forficula auricularia*, are up to 14mm long, tawny brown in colour and easily recognised by the pair of pincers on the rear end of the body. They eat ragged holes in the young foliage and flowers of many plants (*see p.65*), especially clematis, dahlias, and chrysanthemums. During the day they hide in dark crevices or amongst the foliage and petals of plants. They emerge and feed after sunset, so a torchlight inspection is often needed. Earwigs are omnivorous and will also eat aphids in fruit trees, which can be beneficial.

CONTROL Vulnerable plants can be given some protection

by placing earwig traps made of flower pots inverted on bamboo canes and placed level with the plant tops (see p.90). Inspect the pots daily and destroy any earwigs found inside. If earwigs are numerous and still cause damage, spray the plants with a contact insecticide (see p.94) at dusk on a mild evening.

Eelworms

Eelworms or nematodes are mostly microscopic worm-like animals, some of which are important plant pests. Chrysanthemum eelworm (see p.126), narcissus eelworm (see p.158), onion eelworm (see p.161), phlox eelworm (see p.166), potato cyst eelworms (see p.171), and root knot eelworms (see p.177) all feed within their host plants, but there are others, such as *Pratylenchus*, *Longidorus*, *Trichodorus*, and *Xiphinema* species, that live in the soil and attack the root hairs of plants. The last three are able to transmit some virus diseases of ornamentals and soft fruits. Not all eelworms are pests however; the vast majority feed on bacteria, fungi, and other microorganisms, while some are predators. Slugs (see p.183), chafer grubs (see p.125), leatherjackets (see p.152), and vine weevil (see p.193) can be controlled by beneficial eelworms.
CONTROL There is no effective pesticide control for eelworms available to home gardeners. Parts of gardens from which virus-infected plants have been removed should not be replanted with the same types of plant, otherwise they may quickly be infected by the eelworms.

Elaeagnus Sucker

SYMPTOMS The shoot tips are infested with small insects that excrete a sticky honeydew onto the foliage. The adults are about 2–3mm long and yellowish brown with transparent wings strongly patterned with black markings. The immature nymphs are creamy orange with darker markings. Sucker nymphs have broad, flattened bodies with prominent wing pads on the older nymphs. Heavy infestations cause die back and leaf fall.
CAUSE A sap-sucking insect, *Cacopsylla fulguralis*, which originates from eastern Asia and became established in Britain in 2002. It has several generations between spring and early autumn but the worst damage occurs during the spring when heavy infestations build up on the new growth.
CONTROL Inspect the new growth on *Elaeagnus* plants and spray with a systemic or contact insecticide (see p.94). Many *Elaeagnus* spp. and hybrids are susceptible to this pest but *Elaeagnus angustifolia* and *E. multiflora* seem to be resistant.

Elephant Hawk Moth

SYMPTOMS Various garden plants, but particularly fuchsias, are eaten (see p.46) by green or, more usually, brownish-black caterpillars. When fully grown these are up to 80mm long and as thick as a man's finger. They have two pairs of eye spot markings on the front end of the abdomen. When the caterpillar is disturbed it retracts its head into its body, causing the front to swell into a false head and making the eye spots more prominent. In this state the caterpillar can be mistaken for a small snake.
CAUSE Caterpillars of a hawk moth, *Deilephila elpenor*, which are active during mid- to late summer. Feeding is generally done at night, and the caterpillars rest on the underside of leaves during the day. Despite their size, the caterpillars often go unnoticed until they have completed their feeding and are wandering about looking for somewhere to pupate.
CONTROL Damage to garden plants is infrequent. If the caterpillars are found they can be removed by hand and perhaps transferred to one of their wildflower hosts, such as rose bay willow herb or Himalayan balsam.

Elm Gall Mite

SYMPTOMS Elm leaves develop hard yellowish-green swellings on the upper surface (see p.29) during mid- to late spring that persist throughout summer.
CAUSE A microscopic gall mite, *Aceria campestricola*, which begins feeding on the foliage in the spring. Chemicals injected into the leaves by the mites induce the galls to develop. The mites spend the summer months feeding and breeding within these hollow structures before emerging in late summer and overwintering under bud scales.
CONTROL Garden pesticides will not control gall mites. Fortunately, apart from creating galls, this mite does little harm to elm trees so can be tolerated.

Enchytraeid Worms

These small animals are related to earthworms and are also sometimes known as "white" or "pot" worms. They have white bodies that are up to 20mm long and 1mm thick. They can be mistaken for eelworms or nematodes; but where these have completely smooth, unsegmented bodies, enchytraeid worms have segmented bodies, visible with the aid of a hand lens. Enchytraeids cause no harm as, like earthworms, they feed on decaying organic matter in the soil. They are sometimes abundant in soil that has been heavily dressed with farmyard manure or similar composts. They may also be found among the roots of a dead plant but would not be responsible for the plant's condition.

Escallonia Leaf Spot

SYMPTOMS Circular or irregularly shaped purplish-black blotches develop on the leaves, usually with a greyish centre. Tiny, black pycnidia may be visible on these central areas. The infected leaf then starts to yellow, spreading from the edges of the spots throughout the leaf (see p.24). Leaves fall prematurely, and although the plant will usually produce a replacement flush of leaves, these too are likely to become infected. If infection is severe, many branches may be completely devoid of leaves. New symptoms usually appear during the summer.
CAUSE It is thought that two different species of the fungus *Mycosphaerella* may be responsible. The fungi involved seem to spread and cause more damage during wet weather.
CONTROL Consider pruning back hard the infected plant in an attempt to stimulate healthy replacement growth. Keep the plant well watered and fed to encourage this new growth. It is thought that the fungus may overwinter on fallen leaves, and in this way carry over into the next growing season, so collect and dispose of leaves promptly.

Eucalyptus Gall Wasp

SYMPTOMS Eucalyptus leaves develop many slightly raised swellings about 1mm in diameter that are pinkish brown in spring. Similar markings can be caused by oedema (see p.160). The latter causes less regularly-shaped swellings that are solid. The insect-induced galls are hollow and contain a tiny white grub. Heavy infestations can result in excessive leaf fall in early summer.

CAUSE A tiny gall wasp, *Ophelimus* species, that originates from Australia but has become established in several European countries. It has been present in south east England since 2005 and will become more widespread. The tiny black adult wasps emerge from the galls in early summer and lay eggs in the new foliage. The galls gradually develop and mature the following spring.

CONTROL Eucalyptus trees are often too large to be sprayed effectively. On small trees, it is possible that spraying with a systemic insecticide (*see p.94*) in early summer may control the young larvae.

Eucalyptus Sucker

SYMPTOMS Grey or yellowish-orange insects, up to 2mm in length, cluster at the shoot tips of eucalyptus. The adults resemble winged aphids and are the overwintering stage. They lay yellow eggs at the shoot tips in early spring, and all stages in the life cycle can be found together during the summer. The nymphal stages are wingless, flattened insects that secrete wispy white wax filaments from their bodies. Heavy attacks cause stunted growth, and sooty moulds grow on the suckers' excrement.

CAUSE A sap-sucking insect, *Ctenarytaina eucalypti*, which originates from Australia but is now widespread in European gardens.

CONTROL Infestations on mature trees have to be tolerated because of the difficulty of spraying thoroughly. Small trees can be treated with a systemic or contact insecticide (*see p.94*).

Euonymus Scale

SYMPTOMS This is a specific pest of evergreen euonymus,

particularly *Euonymus japonicus*. The foliage becomes heavily coated with elongate, whitish scales about 2mm long (*see p.35*), which are the males. The females mainly cluster on the stems and are brownish-black, pear-shaped objects, up to 3mm in length, which are well camouflaged against the bark. Heavy infestations cause yellowish discoloration of the foliage, plants lack vigour and die back.

CAUSE A sap-sucking scale insect, *Unaspis euonymi*. In Britain it is mainly confined to southern England, but is spreading further north.

CONTROL The newly-hatched scale nymphs are the most vulnerable stage and are present in early summer and early autumn. Spray at those times with a systemic or contact insecticide (*see p.94*) if the scale is present.

F

Fairy Rings

See Turf Fairy Rings p.191.

Fasciation

SYMPTOMS Shoots or flower stems become enlarged and flattened, developing a ribbon-like, often slightly curled appearance. Leaves, buds, and flowers develop normally or in a reduced size on the distorted stem (*see p.62*). Sometimes many stems are affected simultaneously, but more commonly the majority of the plant remains unaffected. Where weather is involved, stems on the more exposed areas of the plant are more likely to be affected.

CAUSE Early injury to the growing point, for example that caused by frost or insect attack, mechanical injury such as hoeing, slug damage, or *Rhodococcus fascians* infection

(*see* Leafy Gall). It has been suggested that viral infection may also be involved on occasion. Chemical contamination may also cause fasciation.

CONTROL Although rather peculiar in appearance, fasciation is harmless and does not affect the overall growth and development of the plant. Prune out affected stems, if desired.

Felt Gall Mites

SYMPTOMS Various trees and shrubs, particularly alder, beech, birch, crab apple, lime, maple, and sycamore, develop a dense coating of hairs on the underside of the leaves. These hairs are usually creamy white, but on purple-leaved trees such as copper beech, some maples, and crab apples, the hairy patches will be reddish-pink (*see p.33*). Later in the summer the hairs dry up and become brownish. On some plants, such as walnut and grape vine, there is an upward bulging of leaves where felt galls have developed on the underside.

CAUSE Microscopic mites of various species which are usually specific to a particular plant species. The mites feed by sucking sap from the foliage and while doing so they secrete chemicals that induce the growth of hairs that form felt galls.

CONTROL Light infestations on small trees can be checked by picking off affected leaves. No pesticides are available for controlling gall mites, so heavier infestations have to be tolerated. Fortunately, apart from creating the galls, these mites have little impact on the host plant's growth.

• See also Acer gall Mites p.106; vine erinose mite p.193; walnut blister mite p.194.

Felted Beech Coccus

See Beech Bark Scale p.114.

Figwort Weevils

SYMPTOMS Small beetles, which are mainly black and mottled with brownish-white markings, infest the shoot tips of figworts (*Scrophularia*), phygelius, *Buddleja*, and verbascum. The beetles have a snout-like projection on their heads and there is usually a black circular mark where the wing cases meet. Both the adults, which are about 4mm long, and their brownish-yellow slimy larvae feed on foliage and flowers.

CAUSE Several species of weevil, *Cionus* and *Cleopus* species, which have two generations between late spring and late summer. When fully fed, the larvae pupate inside brownish cocoons on the shoot tips and these can be mistaken for seed pods.

CONTROL Inspect plants for signs of infestation and if necessary spray with a contact insecticide (*see p.94*). Avoid spraying plants in flower.

Fireblight

SYMPTOMS Flowers wilt and wither, then die back, followed by adjacent leaves and stems shortly afterwards (*see p.51*). The plant may show extensive dieback and be killed within a few seasons. Infections usually occur scattered throughout the crown of the tree or shrub, healthy stems often growing side by side with infected ones. Infection commonly occurs at flowering time. The bark sinks inwards on young stems and, if pared back, a foxy-red discoloration is visible on the wood. Bacterial ooze may be produced from affected areas. The symptoms may be confused with those of blossom wilt (*see p.117*), but the infection only occurs in pome-fruit (apple-like) bearing members of the plant family *Rosaceae*, which include apples, pears,

Cotoneaster, Crataegus, Photinia, Pyracantha, and *Sorbus.*

CAUSE The bacterium *Erwinia amylovora,* which is encouraged by warm, wet weather. The bacteria usually enter the plant via the open flowers, but infection may also take place through stem injuries. Bacteria may be carried by rain splash and spread on pruning tools. In certain areas of the British Isles (Northern Ireland, Isle of Man, and the Channel Islands) fireblight is still a notifiable disease, so cases in gardens should be reported to the Animal and Plant Health Agency.

CONTROL Prune out affected stems to a point at least 60cm into healthy wood. Dispose of all prunings by burning, and sterilize all tools after use on an infected stem. If the infection is widespread, or the plant small, it may be better to remove it entirely. Consider growing one of the 'Saphyr' pyracanthas as these show a good degree of resistance.

Fistulina hepatica (Beefsteak Fungus)

SYMPTOMS Soft bracket fungi, usually growing singly and measuring 5–30cm in diameter, develop on branches, stems, and tree stumps between late summer and late autumn. They are tongue-shaped or become semi-circular as they enlarge. Cream-coloured when young, they become reddish-brown and very juicy as they mature, the flesh often being marbled and resembling a slice of raw steak, giving rise to their common name. If cut or otherwise injured, blood-red juice is produced. The fruiting bodies produce spores, mainly from pores on the lower surface of the fruiting body. The wood on affected trees develops a streaky brown discoloration, called "brown oak" and later

cubical cracking occurs. The fruiting bodies disintegrate with the onset of cold weather and frosts.

CAUSE The fungus *Fistulina hepatica,* which may enter trees via wounds and may cause the wood to become so brittle that it is weakened.

CONTROL Remove fruiting bodies in order to limit spore spread. The fungus itself will continue to remain active within the tree. An inspection of the tree to determine its health and safety is advisable.

Flatworms

There are many flatworms native to Britain; most live in water but some occur in soil. They are mostly small animals and cause no damage to plants. Two species from Australasia have become established in Britain and are causing concern because they prey on earthworms and can reduce them to very low numbers. The loss of earthworms leads to impaired drainage and aeration of the soil, with consequent effects on plant growth. The New Zealand flatworm (*Arthurdendyus triangulatus*) is up to 17cm long when fully extended, and is purplish-brown on its upper surface, although the colour fades towards the sides of the body. It is widely distributed in Ulster and Scotland and is spreading southwards, particularly down the western side of Britain. The Australian flatworm (*Australoplana sanguinea*) mainly occurs in southwest England and is up to 10cm long. It is orange-brown in colour. Both species are mainly nocturnal and can be found during the day curled up under stones, logs, or other objects that are in close contact with the soil. These animals are flat to the point of being ribbon-like and have entirely smooth, unsegmented bodies. They can

be confused with soil-dwelling leeches, which are flattish but have fine lines running across their bodies to indicate their segmented body structure. Leeches also have a sucker underneath the rear end of their bodies; flatworms do not.

CONTROL There is currently no control for flatworms and it is likely that they will become more widespread. They do not move far by their own efforts but can easily be transported among the roots of container-grown plants. When receiving or purchasing plants, inspect the rootballs carefully. Immersing the pots in water for a couple of hours will encourage flatworms to emerge and be seen. Infested plants should be rejected or repotted in clean compost after all the old soil has been washed from the roots to remove flatworms and their shiny black egg capsules.

Flea Beetles

SYMPTOMS Small beetles, mostly 2mm long but sometimes 3–4mm, readily jump off foliage when disturbed. They are shiny black, sometimes with a yellow stripe running down each wing case, or can be metallic blue or yellowish-brown in colour. They eat small, rounded holes in the upper leaf surface (*see p.45*) which often do not go all the way through the leaf. The remaining damaged tissues dry up and turn brownish-white. Heavy attacks can kill seedlings and check the growth of older plants.

CAUSE There are many different species of flea beetle. On brassicas, including radish, swede, and turnip, it is mainly *Phyllotreta* species and these also attack wallflowers, alyssum, stocks, aubrieta, and nasturtiums. Potatoes and ornamental *Solanum* species are sometimes damaged by *Psylliodes affinis.* The damage is caused by

the adult beetles, with the larvae living in the soil and feeding on roots. On fuchsias, both adults and larvae of a metallic blue species, *Altica lythri,* feed on leaves at the shoot tips. Flea beetles are active between mid-spring and late summer.

CONTROL Protect seedlings by sowing when weather and soil conditions will allow rapid germination and growth through the vulnerable seedling stage. Keep young plants watered during dry spells. If necessary, control with a contact insecticide (*see p.94*). If used on food plants make sure the host is listed on the label.

Flies

Flies are characterized by having only one pair of wings; the hind wings are reduced to small club-shaped structures called halteres. There are a number of fly pests of garden plants and in all cases damage is caused by the larval stage rather than the adults.

Root and seedling pests – *see* BEAN SEED FLY *p.114;* CABBAGE ROOT FLY *p.122;* CARROT FLY *p.124;* FUNGUS GNATS *p.138;* LEATHERJACKETS *P.152;* ONION FLY *p.161.*

Bulb pests – *see* NARCISSUS BULB FLIES *p.158.*

Foliage pests – *see* ALLIUM LEAF MINER *p.107;* ASTRANTIA LEAF MINER *p.112;* BEECH GALL MIDGE *p.114;* BEET LEAF MINER *p.115;* BLACKCURRANT GALL MIDGE *p.116;* CELERY LEAF MINER *p.125;* CHIRONOMID MIDGES *p.126;* CHRYSANTHEMUM LEAF MINER *p.126;* DELPHINIUM LEAF MINER *p.131;* GLEDITSIA GALL MIDGE *p.141;* HOLLY LEAF MINER *p.144;* LABURNUM LEAF MINERS *p.149;* SEMPERVIVUM LEAF MINER *p.182;* VIOLET GALL MIDGE *p.193.*

Flower pests – *see* AGAPANTHUS GALL MIDGE *p.106;* AQUILEGIA GALL MIDGE *p.110;* HEMEROCALLIS GALL MIDGE *p.143.*

Fruit pests – *see* PEAR MIDGE *p.165;* SPOTTED-WING FRUIT FLY *p.185.*

Fluted Scale

SYMPTOMS Stems and foliage become encrusted with orange-brown sap-sucking insects that secrete white waxy filaments from their bodies. The mature females have a shell-like appearance and are about 4mm long. The females are perched on the edge of a mound of white waxy material in which the eggs are deposited. These egg masses have a distinctive grooved or fluted appearance. All stages in the life cycle can be found together throughout the year.
CAUSE A sap-sucking insect, *Icerya purchasi*, which is also known as the cottony cushion scale. This used to be an uncommon pest found only in heated glasshouses but it is now becoming established out of doors on patios and in sheltered inner city gardens, especially in London. A wide range of plants is attacked, especially *Acacia* spp., *Choisya*, and citrus plants. Heavy infestations soil plants with honeydew and sooty mould, with a severe loss of vigour in the plants.
CONTROL On small plants remove the scales and egg masses before the eggs hatch. Spray infested ornamental plants with a systemic insecticide (*see p.94*). On citrus plants with edible fruits, use an organic spray (*see p.89*). Several applications of organic sprays may be required at one week intervals.

Fomes Pomaceous

See PHELLINUS POMACEUS *p.166.*

Fomes Root and Butt Rot

SYMPTOMS A dark reddish-brown, woody-textured bracket fungus (*see p.53*), usually measuring 5–10cm across but occasionally up to about 30cm in diameter, develops on the trunk or butt of trees, usually at ground level. Infection is usually seen on conifers but occasionally deciduous species such as birch, beech, alder, and oak are attacked. The lower surface of the bracket is creamy white and it is from the minute pores on this undersurface that the spores are released. The fungus causes extensive deterioration of the woody tissues within the tree, causing rotting and deterioration of the wood. Trees are rendered unsafe, and may occasionally be killed. The fruiting bodies are fairly frost-resistant and usually perennial.
CAUSE The fungus *Heterobasidion annosum* (syn. *Fomes annosus*), which may infect stumps through cut surfaces, and then infects nearby trees by root contact.
CONTROL Trees are rendered unsafe, and because of the ease with which the infection spreads, should be felled promptly. Remove stumps, together with the soil in the immediate vicinity, to minimize the risk of spread.

Foot and Root Rots

SYMPTOMS Deterioration of the tissues around the stem base causes the upper parts of the plant to wilt, discolour and die back. Soft-stemmed plants, in particular seedlings and cuttings, are commonly affected, but most annual, biennial, and herbaceous perennial plants may show symptoms. Tomatoes, cucumbers, and melons, especially those growing in greenhouses, are frequently affected. The tissues at the stem base may darken, atrophy, become discoloured (usually darkening) and, occasionally soften. Root rotting may occur simultaneously (*see p.21*).
CAUSE Various fungi, often similar to or the same as those responsible for damping-off disease of seedlings (*see p.131*). Other fungi, including various species of *Fusarium* and *Verticillium*, may cause both wilt and foot rot symptoms.
CONTROL Once foot rot symptoms have developed, there is no cure available for infected plants. It may, however, be possible to prevent spread of the infection by very prompt removal of any infected plants, together with the compost or soil around their roots. Unhygienic cultivation techniques, including the use of unsterilized compost, dirty trays, pots, and tools and the use of non-mains water, are often responsible for introducing or spreading this type of infection.

Forsythia Gall

SYMPTOMS Numerous galls develop along the length of stems (*see p.56*). The whole or just certain areas of the plant may show the symptoms. Individual galls are usually up to 10–15mm in diameter, rough-surfaced and commonly near-spherical; occasionally several may be joined or fused together. They are the same colour as the forsythia stems. Overall growth and development of the plant is rarely affected but the plant's shape and outline can be spoiled.
CAUSE Unknown.
CONTROL If the forsythia becomes disfigured, prune out affected stems to a point several inches beneath the lowest gall.

Foxes

Foxes are becoming increasingly at home in urban gardens and often make their dens under garden sheds or decking. They cause damage by digging up or trampling plants. Dog foxes use their faeces and urine to mark out their territories and this can create an unpleasant nuisance. Fox urine can scorch the foliage of plants, especially conifers, and has a distinctive pungent smell.
CONTROL It is difficult to prevent foxes from coming into a garden. They are agile animals that are as good as cats when it comes to scaling walls and fences, but have the additional ability to dig underneath obstacles. Proprietary animal repellents have little lasting effect. In gardens where foxes are a problem, the use of bonemeal or dried blood should be avoided. Foxes dig up plants where these fertilizers have been used because the smell makes them think there is food beneath.

Froghoppers

see CUCKOO SPIT *p.129.*

Frost Damage, Apple

See APPLE FROST DAMAGE *p.109.*

Frost Damage (Buds and Flowers)

SYMPTOMS Buds, whether tightly closed or partially opened, discolour, usually turning brown and sometimes becoming soft and squashy. Petals of open blooms discolour, usually turning brown, and either retain their shape or wither (*see p.62*). The buds or flowers on the more exposed areas of the plant are usually the worst affected. Frost may only damage buds or flowers at certain developmental stages, killing some, damaging others and leaving some untouched. On fruiting plants, such as plums, there may be subsequent partial or total loss of crop.

CAUSE Frost, which is most likely to cause damage on slightly tender plants or those which have put on precocious growth. Late frosts, occurring once flower buds have opened, can also be very damaging. Plants which are subjected to the warming of early morning sun following overnight frosts are especially prone to injury.

CONTROL Choose planting positions carefully, especially for early flowering plants, or those known to be readily frost-damaged. Provide protection using horticultural fleece jackets or sheets.

Frost Damage (Foliage and Stems)

SYMPTOMS Foliage, usually towards the shoot tips and particularly on the more exposed parts of the plant, turns brown or black or appears scorched (*see p.43*) and may wilt, wither, and die back. Young stem growth adjacent to the frosted leaves may also die back. Isolated patches on otherwise sound stems may develop, but this is less common.

CAUSE Frost, which is most likely to cause damage on slightly tender plants or those which have put on precocious growth. Late frosts, occurring once flower buds have opened, can also be very damaging. Plants subjected to the warming of early morning sun following overnight frosts are especially prone to injury.

CONTROL Choose planting positions carefully, especially for plants known to be readily frost-damaged. Provide protection using horticultural fleece jackets or sheets. An autumn dressing of sulphate of potash helps to harden stem growth and so may help to minimize frost damage.

Fruit Tree Red Spider Mite

SYMPTOMS The foliage of apples and plums loses its glossy green colour and becomes increasingly dull and yellowish-green (*see p.34*). There is a fine pale mottling of the upper surface and large numbers of tiny mites and spherical eggs (*see p.57*) occur on the lower leaf surfaces. Heavy infestations can develop in hot, dry summers and result in early leaf fall.

CAUSE Sap-sucking mites, *Panonychus ulmi*, which are blackish-red in colour and less than 1mm long. This mite is generally not a problem on unsprayed trees where predatory mites and insects keep it at a low level. Spraying often kills beneficial mites and insects, allowing spider mites to breed unchecked. Overwintering eggs are laid in crevices on the underside of branches and can be so numerous that the bark has a reddish hue.

CONTROL Avoid unnecessary spraying against other pests as this may lead to spider mite problems. If spider mite needs controlling, spray thoroughly with an organic pesticide (*see p.89*). Three applications at weekly intervals may be needed.

Fuchsia Gall Mite

SYMPTOMS Growth at the shoot tips becomes grossly distorted and often has a reddish yellow colour. The flowers are also deformed or fail to develop. New growth comes to a halt once a heavy infestation has developed.

CAUSE A microscopic gall mite, *Aculops fuchsiae*, that originates from South America but has spread to the US and Europe. It was first found in England in 2007. The mites can be picked up and spread by bees when they visit the flowers.

CONTROL Cold winters may kill the mites on hardy fuchsias

but the mites will survive in greenhouses and in mild districts. There are no effective pesticides available to home gardeners so affected plants should be destroyed to prevent the pest spreading.

Fuchsia Rust

SYMPTOMS Numerous patches of pale orange or deep yellow fungal pustules develop on the lower leaf surface, older leaves towards the base of the plant usually showing symptoms first. The upper leaf surface shows corresponding areas of yellow or tan-brown discoloration (*see p.32*), followed by necrosis. Infected leaves fall prematurely, leaving leggy plants which may be killed.

CAUSE The fungus *Pucciniastrum epilobii*, which is encouraged to spread and infect by moist or overcrowded growing conditions and high humidity. Ornamental and weed species of *Epilobium* (willowherbs) may also be infected and so may harbour the fungus. In addition, the fungus may develop on the foliage of its alternate hosts the firs (*Abies*) but infection is much more common on fuchsia than on either of the other genera.

CONTROL Remove infected leaves promptly. Check willowherbs growing in the vicinity and remove them or control infections as appropriate. Consider spraying with a suitable systemic fungicide (*see p.94*), but as fuchsias are often damaged by chemical sprays, first test the effect by spraying on a couple of leaves and waiting for a week or so to see if there are any adverse effects.

Fungus Gnats

SYMPTOMS Greyish-brown flies with slender bodies, 3–4mm long, run over the soil surface of pot plants or fly

slowly and alight on leaves. The larvae are whitish maggots, up to 5mm long, with black heads (*see p.19*). They live in the soil and feed mainly on decaying organic matter, such as dead roots, but may damage seedlings or the base of soft cuttings.

CAUSE Several species of fungus gnats occur in greenhouses and on house plants. Most are harmless but some *Bradysia* species damage fine roots on seedlings and tunnel in the stems of soft cuttings, such as pelargoniums.

CONTROL The adult flies can be trapped by hanging sticky yellow strips close to soil level. A predatory mite, *Hypoaspis miles*, which feeds on the maggots can be purchased by mail order from biological control firms. Another biological control is a pathogenic nematode, *Steinernema feltiae*, which is watered into the potting compost. On established plants, fungus gnats are a nuisance rather than a pest, and so their presence can be tolerated. Maintain good hygiene in the greenhouse and remove dead leaves and flowers from the soil surface.

Fusarium Bulb Rot

See GLADIOLUS CORM ROT *p.140*; NARCISSUS BASAL ROT *p.158*.

Fusarium Wilt

SYMPTOMS Soft-stemmed plants wilt, either in part or in their entirety. Woodier-stemmed plants may retain their overall shape, but foliage on affected stems wilts and withers. Internal staining of the vascular tissue (conducting elements in the stems and roots) may occur and is usually brown or black. This may be apparent just beneath the bark on the stems or within the central area of the roots. Death almost always follows, either rapidly or over several seasons, depending on the plant affected. Cool, damp conditions may encourage very pale pink or white, slightly fluffy fungal growth from the infected tissues.

CAUSE Various types of the fungus *Fusarium*, most commonly *F. oxysporum*. The fungi involved are usually fairly host-specific, attacking only one genus or a few very closely related genera of plants. The fungi are responsible for the whole or partial blocking (usually by gum-like formations) of the vascular tissues, so the symptoms are very similar to those caused by drought. Dry soil conditions may encourage rapid wilt symptoms, but unlike plants suffering from drought, those affected by fungal wilts do not recover permanently when watered. The fungus may persist in plant debris and may also be capable of remaining viable in the soil for several years, even in the absence of any suitable host plants. In *Fusarium oxysporum*, the specialized spores responsible are known as chlamydospores.

CONTROL There is no chemical control available. Remove infected plants promptly to minimize spread. Where feasible, remove soil or compost from the immediate vicinity of the roots. Do not grow the same or closely related plants on the site again, or at least for a minimum of five years.

G

Gall Midges

Gall midges are tiny flies, usually no more than 2–3mm long, and most of them have larvae that feed on plants. The maggots are orange-white and many secrete chemicals that affect the development of plant parts, causing them to produce galls that enclose the larvae.

These midges lay their eggs on a particular part of their host plant, such as developing leaves, shoot tips, flower buds or fruits, and the galls that subsequently develop have a distinctive and characteristic shape that enables the pest to be recognized. Not all members of the gall midge family are pests; one species, *Aphidoletes aphidomyza*, has larvae that feed on aphids and is used to give biological control in greenhouses.
• *See also* AGAPANTHUS GALL MIDGE *p.106*; AQUILEGIA GALL MIDGE *p.110*; BEECH GALL MIDGES *p.114*; BLACKCURRANT GALL MIDGE *p.116*; GLEDITSIA GALL MIDGE *p.141*; HAWTHORN GALL MIDGE *p.142*; HEMEROCALLIS GALL MIDGE *p.143*; PEAR MIDGE *p.165*; VIOLET GALL MIDGE *p.193*.

Gall Mites

These are microscopic animals that can be found on many plants, particularly trees and shrubs. The mites have tubular bodies with just two pairs of legs. Although too small to be seen without magnification, the effect that gall mites have on their host plants is often very characteristic. They feed mainly on foliage or in buds and they secrete chemicals that induce the plant to make abnormal growths. On foliage, gall mites cause raised structures, thickened and curled leaf margins, blisters, excessive growth of hairs, or discoloured patches. Bud-feeding gall mites cause swollen buds which later dry up or proliferate into cauliflower-like structures, or shoots develop as witches' brooms. With a few exceptions, most gall mites do little harm to their host plants and the presence of galls can be tolerated.
• *See also* ACER GALL MITES *p.106*; BIG BUD MITES *p.115*; BROOM GALL MITE *p.120*; ELM GALL MITE *p.134*; FELT GALL MITES *p.135*; FUCHSIA GALL MITE *p.138*; LIME NAIL GALL MITE *p.154*; MOUNTAIN ASH BLISTER MITE *p.157*; PEAR LEAF BLISTER MITE *p.164*; PLUM LEAF GALL MITE *p.169*; RASPBERRY LEAF AND BUD MITE *p.175*; RED BERRY MITE *p.176*; VINE ERINOSE MITE *p.193*; WALNUT BLISTER MITE *p.194*.

Gall Wasps

Gall wasps are black or brown insects up to 5mm long, which are mainly associated with oaks and roses. The gall structures are produced by the plant in response to chemicals secreted by the larval stages, which are white, legless grubs. Although galls are sometimes numerous, they have no harmful effect and so control measures are unnecessary.
• *See also* ACORN GALL WASP *p.106*; EUCALYPTUS GALL WASP *p.134*; OAK GALL WASPS *p.160*; ROBIN'S PIN CUSHION *p.177*; ORIENTAL CHESTNUT GALL WASP *p.188*.

Ganoderma

SYMPTOMS Large, brown or tan-brown, tough fungal brackets develop, most frequently towards the base of the trunk. The lower surface of the bracket is usually a cream or pale brown colour and perforated by numerous closely packed, tiny pores, through which the spores are released. Initially, brackets may be rounded in outline, but they become increasingly bracket-shaped and flattened as they enlarge. They are perennial and woody in texture. Large quantities of dark brown spores are produced and may cover the vegetation beneath. Affected trees may initially show little sign of deterioration, but the woody tissues within the trunk may be seriously weakened, rendering the tree potentially dangerous. The tree may become thin in the crown, with a sparse covering of leaves, and show signs of dieback.

CAUSE Various species of the fungus *Ganoderma*.

CONTROL Remove the bracket to prevent further dissemination of spores. As the damage will continue within the tree, it must be inspected by a reputable tree surgeon or arborist to determine its overall safety. In most instances the tree has to be felled.

Geranium Downy Mildew

SYMPTOMS Pale green or light brown, often angular areas of discoloration develop on the leaves (see *p.37*) of *Geranium* species and hybrids. Severely affected leaves wither and die. Off-white, slightly fuzzy fungal patches develop on the lower leaf surface, corresponding with the discoloured areas. Symptoms may be confused with the much whiter fungal growth which occurs on upper leaf surfaces and stems and is due to powdery mildew (see *p.175*).

CAUSE A species of the fungus *Peronospora*, which is encouraged to spread by moist, humid conditions and poor air circulation.

CONTROL Remove infected leaves promptly. Try to improve air circulation around the plants. Avoid watering or wetting the foliage of plants in the evening.

Geranium Sawfly

SYMPTOMS This pest attacks hardy *Geranium* species and hybrids, not pelargoniums. Rounded holes appear in the foliage and greyish-green caterpillar-like larvae, up to 12mm long, may be found on the underside of leaves, although the larvae often drop to the ground when disturbed.

CAUSE Larvae of a sawfly, *Ametastegia carpini*, which has

two generations between late spring and late summer.
CONTROL By late summer, the foliage may be extensively holed but this has little effect on the plant's growth, so control may be unnecessary. If required, spray the undersides of the leaves with a contact insecticide (see p.94).

Geum Sawflies

SYMPTOMS Three species of sawfly occur on geums. One has leaf-mining larvae up to 4mm long that feed within leaves, causing the damaged areas to dry up and form large brown blotch mines. The other two have pale green larvae up to 13mm long with short, forked hairs on their upper surface. They feed on the young leaves in early summer and can cause severe defoliation.
CAUSE The leaf-mining sawfly is *Metallus lanceolatus* and it has two generations between late spring and late summer. The others are *Claremontia waldheimii* and *Monophadnoides rubi*, both of which have a single generation, with larvae active in late spring to early summer.
CONTROL Pick off light infestations, or spray infested plants with a contact insecticide (see p.94). Geum leaf-mining sawfly larvae are difficult to control with the pesticides available to home gardeners.

Ghost Spot

See GREY MOULD p.142.

Giant Polypore

See MERIPILUS GIGANTEUS (GIANT POLYPORE) p.155.

Gladiolus Core Rot

SYMPTOMS A moist rot and blackening affects the central core of the corm. The base of the corm looks normal but is soft if squeezed. The rot spreads outwards until the whole corm becomes soft and brownish-black. Fluffy, white fungal growth develops beneath the outer scales. Black sclerotia (resting bodies) of the fungus develop on the surface in among the white fungal growth. Symptoms are usually noticed in storage. If a slightly diseased corm is planted, it may grow but produces only a few yellowed and stunted leaves, often with brown spots and red margins.
CAUSE The fungus *Sclerotinia draytonii*, which spreads rapidly during damp weather. The spores are spread by rain and water splash. The fungus may survive in the soil for many years, as the sclerotia are very resilient.
CONTROL Remove and dispose of plants showing foliage symptoms. Check all corms thoroughly before storing, and again at the end of storage before replanting. Allow plants to dry off naturally but quickly after lifting. After lifting, treat corms with sulphur dust. Store corms in a cool, dry place. Plant corms on a fresh site each year.

Gladiolus Corm Rot

SYMPTOMS Concentrically ridged brown markings develop on the corm, usually at its base and whilst in storage (see p.21). Crocuses and bulbous irises may also be affected. The corm dries out, becoming mummified but not developing fungal growth. If plants grow from infected corms, the foliage shows yellow flecking and later striping towards the tips of the outer leaves (see p.21). Discoloration spreads downwards and the leaves turn brown and die. Roots of affected corms blacken and may die back.
CAUSE The fungus *Fusarium oxysporum* f. sp. *gladioli*, which persists in the soil and usually enters the plant through the basal plate or roots of the corm.
CONTROL Dispose of plants showing foliage symptoms. Check corms before storing, and again before replanting. Dry plants off after lifting. Store corms in a cool, dry place. Plant on a fresh site each year.

Gladiolus Dry Rot

SYMPTOMS Corms develop numerous slightly sunken lesions which may join together. Rotting occurs and causes dark brownish-black areas of discoloration, often accompanied by black fungal resting bodies (sclerotia). Infected corms may grow, but the foliage dies off and a dark brown rot occurs at the base of the foliage, later becoming covered in tiny sclerotia.
CAUSE The fungus *Stromatinia gladioli*, which can persist as sclerotia both on the corms and in the soil.
CONTROL Remove and dispose of plants showing foliage symptoms. Check all corms thoroughly before storing, and again at the end of storage before replanting. Dry plants off naturally after lifting and store only those corms that appear healthy. Store in a cool, dry place. Plant on a fresh site each year.

Gladiolus Hard Rot

SYMPTOMS Small, purple-brown spots form on leaves and develop grey centres bearing numerous pycnidia. Dark, reddish-brown to black areas appear towards the base of the corms. Affected areas of the corm often sink in and enlarge. Gladioli, crocus, acidantheras, and freesias may be attacked.
CAUSE The fungus *Septoria gladioli*, which is encouraged by wet weather and is most severe on gladioli growing in poor soils.
CONTROL Dispose of affected corms; plant only those which seem healthy.

Gladiolus Scab and Neck Rot

SYMPTOMS Small, red-brown spots develop on leaves and these darken as they enlarge. Leaf tips become shrivelled. Pale yellow to golden, sunken lesions develop on the corms and produce a resin-like, bacterial ooze. In addition to gladioli, freesias and crocuses are commonly affected by this disease.
CAUSE The bacterium *Burkholderia gladioli* (syn. *Pseudomonas*), which is encouraged by wet weather.
CONTROL Do not plant affected corms. Dispose of any showing symptoms and once the disease has developed, do not grow susceptible plants on that site for at least five years.

Gladiolus Thrips

SYMPTOMS The foliage of gladioli develops a fine, pale mottling with tiny black excrement spots during mid to late summer. Flowers also have pale flecking (see p.63) and in severe attacks, flower buds fail to open.
CAUSE Sap-sucking insects, *Thrips simplex*, which tend to be more troublesome in hot, dry summers. Elongate, narrow-bodied insects, up to 2mm in length, live in the flower buds and at the bases of the leaves. Adult thrips are brownish-black while the immature nymphs are pale yellow.
CONTROL Watch for signs of infestation and spray with a systemic or contact insecticide (see p.94). Avoid spraying open flowers, which can damage them, particularly if treated in sunny weather, and is a danger to pollinators. In autumn burn infested foliage and seed heads. Some thrips overwinter on corms; store in a cool, frost-free place, otherwise thrips continue feeding over winter and weaken the corms.

Glasshouse Leafhopper

SYMPTOMS A coarse, pale mottling occurs on the upper leaf surface (*see p.34*) of many indoor plants. This insect also attacks some garden plants including primulas, polyanthus, foxgloves (*Digitalis*), and nicotiana. The adults are pale yellow with greyish markings, 3mm long and broadest at the head end, tapering to a point behind. Adults jump off plants when disturbed and so it is easier to find the less active, creamy-white, wingless immature nymphs. Both adults and nymphs live on the underside of leaves.
CAUSE Sap-sucking insects, *Hauptidia maroccana*, which breed continuously throughout the year in greenhouses and in sheltered places out of doors.
CONTROL Spray the underside of the leaves with a systemic or contact insecticide (*see p.94*) when leaf spotting is seen. Damaged leaves remain discoloured, but new growth stays green once the pest is controlled.

Glasshouse Red Spider Mite

SYMPTOMS Leaves develop a fine pale mottling (*see p.36*) and the foliage becomes dull green at first, but later increasingly yellowish-white. These symptoms can be confused with a mineral deficiency but examination of the lower leaf surface, preferably with a x10 hand lens, will reveal tiny spider mites and their spherical eggs. The mites are less than 1mm long and have four pairs of legs; they are yellowish-green with two dark markings towards the head end of the body. They only become orange-red in the autumn and winter when the adult females are hibernating in sheltered places. When plants are heavily infested, mites crawl

over a fine silk webbing spun between the leaves and stems (*see p.62*). Leaves dry up (*see p.38*) and fall prematurely, so that only young leaves at the shoot tips remain.
CAUSE Sap-sucking mites, *Tetranychus urticae*, which attack a wide range of indoor and greenhouse plants. They also cause problems out of doors in warm dry summers, especially to beans, raspberries, strawberries, currants, roses, and many other ornamental plants.
CONTROL Red spider mites breed rapidly under warm conditions, and some strains have gained resistance to some pesticides. Biological control with the predatory mite, *Phytoseiulus persimilis*, gives good results if introduced before a heavy infestation has developed. It also requires warm daytime temperatures and an absence of pesticides if it is to become established. *Phytoseiulus* can be used on outdoor plants during the summer. Alternatively, plants can be sprayed with an organic spray (*see p.89*). Several applications will be necessary to break the mite's life cycle.

Glasshouse Thrips

SYMPTOMS Foliage develops a silvery white discoloration on the upper surface. The leaves are marked with brown excrement spots and narrow bodied insects up to 2mm long may be present. A wide range of indoor plants is attacked, especially palms and citrus plants. Climate change means that this greenhouse pest is now becoming a problem on outdoor plants in sheltered gardens, where it has a particular liking for *Viburnum tinus*.
CAUSE A small sap-sucking insect, *Heliothrips haemorrhoidalis*, which is active all year round on indoor plants. The adult thrips are blackish brown, while the immature nymphs are creamy white.

CONTROL On ornamental plants, spray with a systemic or contact insecticide when signs of damage are seen (*see p.94*); on fruiting citrus plants, use an organic spray (*see p.89*).

Glasshouse Whitefly

SYMPTOMS White-winged insects, 2mm long, readily fly up when plants are disturbed. The adults and the immobile, whitish-green, oval, scale-like nymphs live on the underside of leaves of many houseplants and greenhouse vegetables and ornamentals. The upper leaf surfaces become sticky with honeydew excretion and a black, sooty mould often develops on this substance. Glasshouse whitefly can spread to garden plants during the summer, but will not survive a cold winter.
CAUSE A sap-sucking insect, *Trialeurodes vaporariorum*, which breeds continuously throughout the year on indoor plants.
CONTROL Pesticide resistance is a widespread problem with this pest. Biological control with a tiny parasitic wasp, *Encarsia formosa*, is often the best remedy on indoor plants between mid-spring and mid-autumn when temperatures are high enough to suit the parasite. *Encarsia* must be introduced before a heavy infestation has developed, as it needs time to breed before it can start reducing the whitefly population. Sticky yellow traps hung just above plants will catch adult whiteflies and indicate when they are starting to appear. Organic insecticides will reduce whitefly numbers and have no lasting impact on *Encarsia*. Synthetic insecticides are dangerous to the parasite and may give poor control of whitefly if a resistant strain is present. Several applications at about five-day intervals will be required even if the whitefly is susceptible. Longer control on pot-grown ornamental plants can be achieved by spraying

with or using a systemic insecticide as a compost drench (*see p.94*). This can be used on greenhouse tomatoes, peppers, and aubergines.

Gleditsia Gall Midge

SYMPTOMS Leaves at the shoot tips of honey locust (*Gleditsia triacanthos*) fail to expand into the normal pinnate form. The leaflets are folded over and make pod-like galls (*see p.42*) which contain up to seven whitish-orange maggots. These galls first appear in early summer; by late summer, infestations may be so severe that normal leaf production comes to a complete halt. The cultivar 'Sunburst' seems to be particularly susceptible and produces pale yellow galls, sometimes with a pink flush.
CAUSE A tiny fly, *Dasineura gleditchiae*, that originates from North America but arrived in Europe during the 1970s. Grey, and 2mm long, it has three generations, with female midges laying eggs in developing leaves at the shoot tips in late spring and mid- to late summer.
CONTROL Small trees can be sprayed with a contact insecticide (*see p.94*) when the midges are seen on the shoot tips, particularly against the first generation in late spring and early summer. Tall trees cannot be sprayed properly and this pest often has to be tolerated. Fortunately, the tree is able to make enough normal leaves to maintain growth before heavy infestations develop in mid- to late summer.

Gloeosporum Rot

See BITTER ROT *p.116*.

Golden Root Mealybug

SYMPTOMS Pale yellow, wingless, soft-bodied insects,

2–3mm long, are present amongst the roots of infested plants. Their bodies are covered with a waxy secretion that rubs off and gives a pale yellow coating on soil particles and roots. Affected plants show poor growth if heavily infested. It has been found mainly on the roots of *Meconopsis* and *Primula* species, but can attack a wider range of plants.

CAUSE A sap-sucking insect, *Chryseococcus arecae*, that originates from New Zealand and was first recorded in Britain in 2012. It has been found in Scotland and northern England and is likely to spread further. Other species of root mealybugs in Britain are covered in a white wax and can be a problem in greenhouses on the roots of pot plants (*see p.155*). The golden root mealybug is tolerant of low temperatures and infests plants growing outdoors in gardens.

CONTROL Pesticides currently available for plants growing in open soil are unlikely to give effective control. A systemic insecticide is available as a potting compost drench for treating container-grown plants (*see p.94*).

Gooseberry Dieback

SYMPTOMS Stems of gooseberries or sometimes blackcurrants die back. Occasionally the whole bush is affected, but more commonly only certain areas deteriorate while most of the bush appears to be perfectly healthy. Infection will spread, and if the main stem or base of the plant is affected, is it possible that the whole plant will die back. During damp weather, fuzzy, grey fungal growth may develop around the damaged areas. Foliage on infected stems yellows, turns brown, and then dies back, falling prematurely. Any developing fruits shrivel and die.

CAUSE The fungus *Botrytis cinerea*. The spores are spread on air currents and by rain or water splash.

CONTROL Prune out infected areas promptly, removing any wood that shows staining or discoloration and cutting back to a healthy-looking bud. The infection may also occur on the fruits, so these should also be removed immediately to help minimize the risk of spread of the infection.

Gooseberry Mildew

See AMERICAN GOOSEBERRY MILDEW *p.107*.

Gooseberry Rust

SYMPTOMS In spring, reddish-orange pustules develop on the older, lower leaves and occasionally on the fruits and stems. In summer, the cluster cup stage of the fungus develops on the pustules, causing extensive distortion and curling of the leaves.

CAUSE The fungus *Puccinia caricina*, which may also attack sedges (its alternate host) and which is most damaging to gooseberries following dry weather in early spring.

CONTROL Remove infected areas promptly. Remove sedges growing in the vicinity, as these may harbour the infection.

Gooseberry Sawflies

SYMPTOMS Gooseberry bushes can become rapidly defoliated (*see p.47*), often being reduced to bare stems by the time the berries are ready for picking. Three species of sawfly can attack gooseberry. They have pale green caterpillar-like larvae up to 20mm long, heavily marked with black spots in two of the species. Damage starts in mid- to late spring, but as there can be two or three generations, there may be further damage later in the season. Red and white currants are also attacked in a similar manner.

CAUSE Larvae of the common gooseberry sawfly (*Nematus ribesii*), the lesser gooseberry sawfly (*N. leucotrochus*) and the pale gooseberry sawfly (*Pristiphora appendiculata*).

CONTROL Inspect gooseberry and currant bushes carefully for larvae from mid-spring. Eggs are often laid on the underside of leaves in the centre of bushes, so larvae may go unnoticed until they have eaten upwards and outwards and there are few leaves left. Remove by hand or spray affected plants with a contact insecticide (*see p.94*).

Greenfly

See APHIDS *p.108*.

Green Spruce Aphid

SYMPTOMS The foliage on spruce (*Picea*) develops a whitish-yellow mottling, then turns brown and drops off during the late winter and spring. New growth in the spring is not affected. Dark green aphids with red eyes can be found among the foliage, which is often blackened by sooty mould growing on the aphids' sugary excrement.

CAUSE A sap-sucking aphid, *Elatobium abietinum*. Unlike most greenfly, it is active between autumn and spring and can be particularly damaging in mild winters.

CONTROL Spray small trees with a contact or systemic insecticide (*see p.94*) in the autumn, and again if symptoms are noticed in the winter or spring. Badly defoliated plants will take several years to regain their appearance; encourage better growth by feeding with a general fertilizer in the spring and keep trees watered in dry summers.

Grey Mould

SYMPTOMS Grey or occasionally off white or grey-brown fuzzy fungal growth develops on infected areas (*see p.51*). Grey mould is a very common fungus and, because it can live on most living and dead plant material, its spores are almost always present in the air. Grey mould may attack all above-ground parts of most plants, usually gaining entry via wounds or points of damage. Fruit infections may originate from infection via the open flowers, the fungus remaining dormant until the fruits start to ripen (*see p.16*). Prior to the development of the fungal growth, the infected plant tissue usually discolours, often turning brown and soft. Growth above points of infection may deteriorate, with leaves yellowing and wilting and flowers or fruits dying off (*see p.63*). Grey mould infection on petals or fruits may initially cause a hypersensitive reaction which results in colour changes but no rotting. On flowers, bleached white or pale brown spots (known as ghost spots) may form (*see p.60*). The fungus may also produce numerous small black sclerotia (resting bodies) which can withstand a wide range of weather conditions and then infect plants when conditions are at their most favourable for the pathogen's growth.

CAUSE The fungus *Botrytis cinerea*. Its spores are readily spread by rain or water splash and on air currents. The fungus may also carry over from year to year as sclerotia in the soil or on infected plant debris.

CONTROL Control is difficult, because the fungus is so widespread. Remove all dead and injured parts before they have a chance to become infected. Remove infected areas promptly, cutting right back into healthy growth. Do not leave plant debris lying around.

Grifola frondosa

SYMPTOMS Overlapping, greyish-tan to honey-coloured fungal brackets develop around the base of a tree, on its roots or on tree stumps. The margin of each "frond" is white. The fruiting body may measure up to 30cm across, the individual "fronds" measuring 20–60mm across. If bruised, the fungus has a distinct mouse-like smell. The fungus causes a butt rot of susceptible trees, and infected wood becomes soft and pulpy. White mycelium (a fungal sheet) develops in decaying areas. The fruiting body usually appears in late summer or autumn and disintegrates with the first frosts.

CAUSE The fungus *Grifola frondosa*, easily confused with the much larger *Meripilus giganteus* (p.155).

CONTROL Affected trees may be rendered dangerous and so should be inspected by a reputable tree surgeon and felled if necessary.

Ground Beetles

Ground beetles or carabids are mostly predatory insects that feed on a variety of insects and other invertebrate animals. They vary in size from 2–30mm and are mostly black, brown, or bronze-green. They have long, slender legs that enable them to run rapidly when disturbed. Adult ground beetles live and feed mainly in or on the soil surface but some species will climb up plants at night to prey on aphids, small caterpillars and other insects. The larvae live in the soil and are also predatory. Some adult ground beetles feed on plant seeds and while most are harmless, damage can occur on strawberry fruits.
• *see also* STRAWBERRY SEED BEETLES *p.187.*

Gummosis

See CUCUMBER GUMMOSIS, *p.129.*

Gypsy Moth

SYMPTOMS A wide range of trees and shrubs, including conifers, can be severely defoliated by caterpillars in early summer. The hairy, blackish-yellow caterpillars, up to 70mm long, are variable in colour but distinctive, due to a pair of blue warts on each of the front five segments of their bodies. The other segments have reddish warts on the upper surface.

CAUSE Caterpillars of a moth, *Lymantria dispar*. An Asiatic form of this moth has moved into Europe in recent years and is proving to be a more formidable pest than its European counterpart. It has a wider range of hosts and the females fly more readily, thus dispersing the pest more widely.

CONTROL Look for caterpillars, which hatch from overwintered eggs laid in batches on the bark, from mid-spring onwards. Small trees and shrubs can be treated with a contact insecticide (*see p.94*). Small caterpillars are more susceptible than older larvae.

H

Halo Blight

See BEAN HALO BLIGHT *p.113.*

Hawthorn Gall Midge

SYMPTOMS In early summer, hawthorn shoot tips fail to extend and instead develop a rosette of small leaves. Several orange-white, legless larvae, up to 2mm in length, live within each rosette, which then dries up and turns brown later in the summer.

CAUSE A small fly, *Dasineura crataegi*, that has several generations during summer. Clipped hawthorn hedges are always more heavily infested than uncut hawthorns.

CONTROL This pest causes little real damage, although old, dried-up galls can be unsightly. There is no point in using insecticides; as only clipped hawthorns are particularly susceptible, the damaged shoots will be removed when the hedge or bush is trimmed.

Hebe Downy Mildew

SYMPTOMS Yellow or buff-coloured discolouration develops on the upper leaf surface, corresponding with off-white, fluffy fungal growth beneath. Infected leaves die and drop early, and may leave the plant with sparse foliage. Lower leaves and those in congested parts of the plant are most likely to be attacked.

CAUSE The fungus *Peronospora grisea*, which is encouraged by moist, humid conditions and poor air circulation.

CONTROL Remove infected leaves promptly and destroy any severely infected plants. Avoid overhead watering, especially in the evening, as this encourages the spread of disease.

Hellebore Black Death

SYMPTOMS Black streaks mostly along the leaf veins, sometimes in the form of ringspots. The stems and leaves become distorted and flowers may become infected. Cultivars and hybrids of *Helleborus hybridus* are particularly susceptible. Infected plants may show no symptoms for a year or more so can be a source of infection.

CAUSE A type of carlavirus, most likely *Helleborus net necrosis*. The virus is believed to be spread by aphids feeding on hellebores and acting as vectors.

CONTROL There is no control available. Remove and burn infected plants promptly.

Hellebore Leaf Blotch

SYMPTOMS Slate-grey or greyish-brown lesions develop on the leaves (*see p.34*). They may coalesce, leaving large leaf areas discoloured and dead. Flower stems and occasionally the flowers themselves may also be attacked, causing flower stems to flop over.

CAUSE The fungus *Microsphaeropsis hellebori* (syn. *Coniothyrium hellebori*), which spreads especially rapidly in damp weather.

CONTROL Remove and dispose of affected leaves and other plant parts promptly. Even if the plant is left with only very sparse foliage, replacement leaves should soon be produced. In certain circumstances, *Helleborus argutifolius* and *H. orientalis* appear to show a good degree of resistance to this infection, and so in troublesome sites it may be preferable to concentrate on growing these rather than other species. Avoid growing *H. niger* as this is particularly susceptible.

Hemerocallis Gall Midge

SYMPTOMS The flower buds of daylilies (*Hemerocallis*) become abnormally swollen but fail to open before going brown and drying up (*see p.63*). Petals inside a galled bud are thickened, with a crinkled appearance, and there are many white maggots, 2–3mm long, at the base of the petals.

CAUSE A tiny fly, *Contarinia quinquenotata*, that deposits its eggs in the developing flower buds during late spring and early summer. There is only one generation a year, so buds that develop after the egg-laying period and flower from midsummer onwards escape damage.

CONTROL Pick off and destroy galled buds before the maggots have a chance to escape into the soil to pupate. Pesticide controls are likely to be ineffective since it is difficult to prevent the females from laying eggs, and the grubs are well protected inside the buds. The problem can be largely avoided by growing *Hemerocallis* cultivars that have late flowering periods from midsummer onwards.

Hemispherical Scale

SYMPTOMS Many indoor ornamental plants are susceptible to this pest. When mature, the scales (*see p.29*) have a dark brown, hemispherical, convex shell over their bodies. The shell is up to 4mm across and has an H-shaped ridge on it. The immature scales are flat, oval, and yellowish-brown in colour. They occur on stems and the underside of leaves. Their sugary excrement, called honeydew, falls down and makes upper leaf surfaces sticky. Sooty moulds often grow on this substance.
CAUSE A sap-sucking insect, *Saissetia coffeae*, which breeds continuously throughout the year on houseplants and in heated greenhouses.
CONTROL Heavily infested plants are best disposed of, especially ferns, which generally do not tolerate insecticides. Some scales and honeydew can be removed by wiping the foliage and stems with a damp cloth. Spray with a systemic, contact, or organic insecticide (*see pp.94 and 89*) against the more vulnerable young scale nymphs. Several applications at seven-day intervals may be needed to break the insect's life cycle if using an organic insecticide.

Heuchera Rust

SYMPTOMS The upper leaf surface develops raised and/or sunken areas with corresponding brownish-orange, spore-filled pustules on the lower leaf surface. Occasionally, the pustules also develop on the upper surface. In damp or humid conditions the pustules may turn a greyish-white colour. The petioles may show similar symptoms. If the rust attacks early in the season, infected young foliage becomes distorted. The patches of leaf killed by the fungus may drop out, creating a shothole effect (*see pp.31 and 182*). If the infection is severe, leaves turn brown, shrivel, and die (*see p.30*).
CAUSE The fungus *Puccinia heucherae*, first recorded in Britain in 2004. The fungus produces large quantities of spores, which spread on air currents. Like other rust fungi, heuchera rust tends to be more widespread and damaging during damp weather, or when the foliage is wetted during watering.
CONTROL Always check new purchases carefully to avoid bringing this infection into your garden. Grow heucheras in areas with plenty of air circulation, as damp conditions and high humidity encourage the infection. Avoid wetting the foliage when watering, and try to water in the morning, allowing the foliage to dry off quickly. If the infection does appear, remove infected leaves promptly and burn or dispose of them, as the fungus overwinters on plant debris.

High Temperature Injury

SYMPTOMS Injury may be temporary or result in tissue death and permanent damage. Scorching or scalding of leaves, fruits, and flowers is common, often seen as dry, papery, buff, or brown areas. Wilting may occur if the plant is losing more water than it can take up, due either to excessive transpiration or dry soil conditions. Trees or shrubs which have thin bark may show bark death down the exposed side of their stem or trunk.
CAUSE Excessively high temperatures, or the rapid onset of high temperatures immediately following cool weather.
CONTROL There is rarely anything that can be done once injury has occurred, other than to ensure that further injury is prevented by improving ventilation, provision of shading, damping-down of the greenhouse floor or other means appropriate to the situation in which the plants are growing.

Holly Leaf Blight

SYMPTOMS Circular black spots develop on infected leaves. Premature leaf fall follows. Black lesions form on young shoots and may girdle them, so causing dieback. On holly hedges defoliation occurs in inverted "v" shapes.
CAUSE The fungus-like oomycete *Phytophthora ilicis*, which is spread from tree to tree by rain splash, and far greater distances by flooding and in areas of poor drainage. It is thought that tiny holes created by neighbouring holly leaves are needed for infection to take place. It has been recorded on *Ilex × altaclerensis*, *I. aquifolium* and *I. crenata*. Symptoms are often first seen during the winter and are most severe after periods of cool, wet weather.
CONTROL The fallen leaves may harbour the infection so it is important that these are raked up and disposed of promptly. Avoid bringing infected soil or plants into the garden.

Holly Leaf Miner

SYMPTOMS Leaves develop yellowish-purple blotches (*see p.31*) where the internal tissues have been eaten out by a fly maggot. Leaf mines can be seen throughout the year and may be abundant on hollies trimmed to form a hedge.
CAUSE The larvae of a small fly, *Phytomyza ilicis*, which lays eggs on young foliage in early summer. There is one generation a year.
CONTROL Apart from creating discoloured patches on the foliage, this fly has little impact on the health and vigour of hollies. Mined leaves that turn yellow and drop in early summer are likely to be part of the normal loss of old foliage that occurs at that time. Holly leaf miner generally has to be tolerated, as insecticides do not penetrate the thick, waxy surface of holly leaves.

Hollyhock Rust

SYMPTOMS Bright yellow or orange spots with orangered centres develop on the upper leaf surface. Bracts may also be affected. The lower leaf surface develops numerous raised orange pustules (*see p.30*), each about the size of a pin-head and corresponding with the spots on the upper surface. The spots may join together, killing large areas of leaf tissue and causing it to fall away, leaving the leaves with a ragged and unsightly appearance. The pustules later turn buff-brown in colour.
CAUSE The fungus *Puccinia malvacearum*, which is very common, but most troublesome in damp seasons. The spores are ejected from the pustules and spread on air currents. The fungus overwinters on infected plant debris and on any infected leaves which may remain on the plant during mild winters.
CONTROL Inspect plants regularly and remove any infected leaves. Spray plants at regular intervals from spring and with a suitable fungicide (*see p.94*). Remove potential weed

hosts such as wild mallow (*Malva* species) from the area. Avoid growing other susceptible plants, which include *Abutilon*, *Hibiscus*, *Lavatera*, and *Sidalcea*. There are no resistant hollyhocks available, but the species *Althaea rosea* is thought to be less susceptible.

Holm Oak Leaf Miners

SYMPTOMS Deciduous oaks can be attacked by leaf miners, but damage is most commonly seen on evergreens, such as holm oak (*Quercus ilex*). Damage is particularly noticeable during winter and spring.
CAUSE Caterpillars of a tiny moth, *Phyllonorycter messaniella*, which has three generations a year, create elongate, brownish-white, discoloured blotches on the upper leaf surface. The corresponding areas on the lower surface have a blistered appearance where the lower epidermis has become detached. The caterpillars of another moth, *Ectoedemia heringella*, make narrow, wiggly mines that end in a blotch mine.
CONTROL Treatment is difficult, as insecticides do not readily penetrate into the tough, shiny leaves of holm oak. Many trees may also be too large for effective spraying. Fortunately, oaks can support heavy infestations without adverse effects on their growth. Although the foliage may be unsightly in the winter when the year's accumulated damage is present, the tree's appearance will improve. New growth in spring will cover up the damaged foliage and some of the older leaves will be shed in early summer.

Honey Fungus

SYMPTOMS Affected trees, shrubs, woody climbers, and sometimes woody herbaceous perennials start to die back. Occasionally other plants, like strawberries and potatoes, can become infected. Leaves may discolour and wilt, or fail to leaf up in spring. Death may appear to be quite rapid, or an infected plant may take several years to die. Flowering or fruiting plants may put on a particularly good display or set an unusually productive crop before dying. Resin may exude from around the base of the trunks of infected conifers. The roots and/or stem or trunk bases of infected plants develop a cream mycelium (fungal sheet) on them, sandwiched between the bark and the woody tissues beneath (*see p.50*). The mycelium has a characteristic mushroom-like smell. The fungus also produces tough blackish fungal strands called rhizomorphs. They may show branching and are hollow inside, the blackish outer "rind" being lined with white or faintly pink fungal growth. Rhizomorphs resemble old tree roots and are easily mistaken for them. They vary in length and may be plump or flattened, measuring from under a millimetre to several millimetres in diameter. The rhizomorphs feed on a woody host, or the remains of an old tree stump or dead plant. They may then grow through the soil and latch onto other food sources such as living trees or woody debris. The rhizomorphs may also grow beneath the bark of an infected plant, or be found growing over the surface of a root. Most are found within the top 20cm of the soil. Clumps of honey-coloured toadstools may appear around the base of infected plants in late summer or autumn. The toadstools may also appear around old tree stumps, following the root line of underground roots, or growing from underground rhizomorphs near an infected plant. Toadstools are very variable, but usually grow in clumps and have honey-coloured caps which may be marked with dark brown flecks. The spores are white and may accumulate on the caps of other fungi in the clump. The toadstool's stipe (stem) usually bears a distinct creamy-white ruff or collar.
CAUSE Various species and races of the fungus *Armillaria*. These vary in their ability to infect potential hosts, some being far less pathogenic than others and living only on dead or very stressed plants. It is hard to differentiate between species, except on the basis of the damage they cause. Spread occurs either by rhizomorphs which grow through the soil at the rate of about 1m a year, or by root contact.
CONTROL Remove infected plants promptly, digging out stumps and as much of the root system as possible. Keep plants in good health by regular feeding, mulching, and other relevant care, as this makes them less prone to infection. The spread of disease from the toadstools is minimal, so it is not necessary to clear them away. Plants like birches (*Betula*), cedars (*Cedrus*), most common hedging conifers, cotoneaster, currants (*Ribes*), forsythias, hydrangeas, lilac (*Syringa*), *Malus*, peonies, privet, *Prunus*, rhododendrons, roses, willows (*Salix*), and wisterias, are known to be particularly susceptible and so should not be used to replace plants in gardens with honey fungus. Trees, shrubs, and climbers that show a good degree of resistance include box (*Buxus*), mallows (*Malva*), yews (*Taxus*), bamboos, abutilons, actinidias, carpenterias, celastrus, cercis, catalpas, ceratostigma, chaenomeles, clematis, cotinus, elaeagnus, fothergillas, hebes, kerrias, passifloras, phlomis, pieris, pittosporum, rhus, romneyas, sarcococcas, and tamarix.

Honeydew

This is the sugary excrement produced by some sap-sucking pests such as aphids, whiteflies, suckers, mealybugs, and some scale insects. These pests suck sap from the phloem vessels in leaves and stems. This is rich in sugars, and is only partially digested by the insects, so their excrement is a sticky liquid due to its high sugar content. Although these pests usually live on lower leaf surfaces, it is upper leaf surfaces that become sticky. The insects flick honeydew droplets away from themselves and these then fall down onto leaves and other surfaces below. Under damp conditions non-parasitic fungi called sooty moulds grow on honeydew and form black or greyish-green deposits on the foliage.

Honeysuckle Aphid

SYMPTOMS Greyish-green aphids form dense colonies on the shoot tips and flower buds of honeysuckles in late spring and summer. This can result in the flower buds aborting and shoot tips drying up and browning.
CAUSE A sap-sucking insect, *Hyadaphis passerinii*.
CONTROL Check the new growth and flower buds for the presence of aphids from mid-spring onwards. If necessary spray with a contact, systemic, or organic insecticide (*see pp.94 and 89*). Avoid spraying plants in flower.

Hornbeam Witches' Brooms

SYMPTOMS Extensively branched twigs develop, closely clustered together, to form "witches' brooms" scattered throughout the crown of the tree. An off-white layer of fungal spores may develop on the surface of the leaves from late spring until midsummer. During winter the growths may, at first, be mistaken for birds' nests and can spoil the outline of the tree or hedge, but during the summer they appear as dense, leafy areas and so are less obtrusive.

CAUSE The fungus *Taphrina carpini*, spread by water splash and on air currents.

CONTROL Prune out growths if desired. Spread seems to be very slow, but removal will help to limit the problem. No chemical control is available.

Horse Chestnut Bleeding Canker

SYMPTOMS Drops of reddish-brown to black gummy ooze appear in scattered patches on the bark. Bleeding may then occur on the trunk, main stems, and larger limbs (*see p.54*) and, as the weather warms up, crusty deposits develop. The surrounding bark is killed and the tree itself may die; young trees (10–30 years old) are more severely affected, as they have smaller trunks and are more easily girdled. The wood discolours beneath the bark in affected areas, and secondary wood-rotting fungi may appear from damaged areas. There may also be dieback, thinning of the tree's crown, yellowing of foliage (*see p.54*), and premature leaf drop. Both *Aesculus hippocastanum* and *A. carnea* are affected.

CAUSE The bacterium *Pseudomonas syringae* p.v. *aesculi* is the pathogen generally responsible. The fungus-like *Phytophthora plurivora* and *P. cactorum* are now less commonly involved. Bleeding canker is worst during mild, moist weather in spring and autumn. It is most severe in southern parts of the UK, but does extend into Scotland.

CONTROL There are no chemical control measures available. Remove infected branches promptly. Burn or bury

all infected wood; it must not be kept for later use as firewood, nor should it be chipped on site.

Horse Chestnut Leaf Blotch

SYMPTOMS Chestnut to tan-brown, irregular leaf blotches (*see p.34*), each with a bright yellow margin, develop from midsummer onwards. The damage is often most severe on the leaf margins and leaf tips. Numerous pinprick-sized, black fungal fruiting bodies may develop on the upper leaf surface over the spots. Severely infected leaves turn brown, curl, and fall prematurely.

CAUSE The fungus *Phyllosticta sphaeropsidae* (syn. *Guignardia aesculi*), which, despite causing early leaf fall, has little effect on the tree's overall health because most of the damage occurs quite late in the season.

CONTROL Collect up and dispose of fallen leaves in the autumn to minimize disease carry-over.

Horse Chestnut Leaf-mining Moth

SYMPTOMS Whitish green blotch mines, which later become brown, develop between the leaf veins. These mines resemble the damage caused by the fungal disease horse chestnut leaf blotch. By holding an infested leaf up to the light it should be possible to see the caterpillars, or their circular silk cocoons in which they pupate, within the mines. The caterpillars are up to 5mm in length and the mines may be 40mm long by the time the larvae are fully fed. Heavy infestations can cause early leaf fall and loss of vigour.

CAUSE The larvae of a small leaf-mining moth, *Cameraria ohridella*. This was first detected in Europe in Macedonia in

the 1970s and has since spread rapidly throughout the continent. It was first found in Britain in 2002 and has since become widespread. There are at least three generations between late spring and autumn.

CONTROL Because of the size of most horse chestnuts, there is no effective way of treating them with pesticides. A biological control in the form of a parasite or predator is needed but as yet this pest seems to have no effective natural enemies in Europe. *Aesculus indica* is resistant and could be a replacement for the common horse chestnut, *A. hippocastanum*.

Horse Chestnut Marginal Leaf Scorch

SYMPTOMS Marginal leaf browning or scorching occurs, usually in early to midsummer. Leaves curl slightly and scorched areas turn tan-brown. All or part of the tree may be affected. If the damage is only limited, the rest of the tree usually appears perfectly healthy.

CAUSE The precise cause is unknown, but it is believed that this disease may be due to a xylem-limited bacterial infection (a bacterial infection restricted to the xylem, or water-conducting elements, within the tree).

CONTROL None known, but symptoms may be less apparent in some years than others.

Horse Chestnut Scale

SYMPTOMS Many trees and shrubs, including horse chestnuts (*Aesculus*), limes (*Tilia*), sycamore (*Acer pseudoplatanus*), maples (*Acer*), dogwoods (*Cornus*), magnolias, bay trees (*Laurus nobilis*), elms (*Ulmus*), and *Skimmia japonica*, are attacked. Infestations are

most visible in early to midsummer when the females deposit their eggs under white, waxy threads. These are secreted from the rear edge of the shell-like covering over the insect's body. Mature scales are brown, up to 5mm long and are found on the trunks and larger branches. After egg-laying, the scales die and fall away, leaving circular white egg patches on the bark (*see p.40*). During the summer young scales – flat, oval, yellowish insects – live on the underside of leaves. In autumn they move onto the bark and overwinter as blackish-yellow immature nymphs. Little problem with honeydew and sooty mould occurs with this scale, except on *Skimmia japonica*.

CAUSE A sap-sucking insect, *Pulvinaria regalis*, which has one generation a year. The heaviest infestations occur on patio plants and trees growing in streets and car parks where there is a warmer microclimate due to heat being reflected off concrete, tarmac and brickwork.

CONTROL On large trees, horse chestnut scale has to be tolerated. Fortunately, it causes little real harm apart from the unsightliness of the egg masses. On small trees and shrubs, scrape off the scales and egg masses before the eggs have a chance to hatch in midsummer. If necessary, spray the undersides of leaves with a systemic or contact insecticide (*see p.94*) in midsummer.

Hoverflies

Many of these flies have black-and-yellow striped abdomens and can be seen sitting on flowers or hovering near them. Some have larvae that feed on greenfly and other aphids. They may consume as many as 400 before they are ready to pupate and are helpful in controlling aphids. Other types of hoverfly have larvae that feed in rotting vegetation, in wasp nests, in the mud in pools or in rot holes in trees. The only hoverflies to cause problems in gardens are the narcissus bulb fly (*see p.158*) and sempervivum leaf miner (*see p.182*).

Hydrangea Leaf Spot

SYMPTOMS Grey, brown, or purple-brown spots of varying sizes are usually most apparent on the older foliage. Pinprick-sized fungal fruiting bodies may develop within the damaged areas. A yellow halo appears around the edge of the leaf spot, often indicating that a bacterial infection is responsible. Although spoiling the appearance of the foliage, none of these leaf spots causes long-term damage.
CAUSE Various fungi and bacteria, mostly weak pathogens, only able to attack hydrangeas which are suffering from other problems.
CONTROL Remove severely infected leaves. Improve the vigour of the plant by regular watering, feeding, and mulching.

Hydrangea Powdery Mildew

SYMPTOMS Dense, white powdery deposits form on the upper leaf surfaces, and occasionally on petioles and new stem growth. The affected leaves may become distorted, particularly if infection occurs when they are still young. The leaves usually remain on the plant but may show patchy necrosis. They are rarely killed outright.
CAUSE The fungus *Microsphaera polonica*, which is encouraged by moist or humid air around the foliage and dry soil conditions around roots.
CONTROL Remove any severely affected leaves. Keep the plant

well watered at the base, but avoid overhead irrigation.

Hydrangea Scale

SYMPTOMS This pest is most visible during late spring and early summer when the females deposit eggs among a mass of white, waxy fibres on the stems and foliage of host plants (*see p.28*). Hydrangea is the most frequently attacked plant but *Prunus* and *Acer* are also host plants. The egg masses are oval in shape and up to 5–6mm long. The mature scale is brownish-yellow but drops off when the insect dies, leaving the egg mass as the most obvious sign of infestation. Eggs hatch in midsummer and the flat, oval, pale yellow scale nymphs live on the underside of leaves next to leaf veins during the rest of the summer. They move onto the bark in the autumn and overwinter as immature nymphs.
CAUSE A sap-sucking insect, *Pulvinaria hydrangeae*, which has one generation a year. Heavy infestations reduce the plant's vigour and result in dieback.
CONTROL Spray with a systemic or contact insecticide (*see p.94*) in midsummer against the more vulnerable newly hatched scales. Hydrangeas can be damaged by some pesticides, so avoid treatment when plants are exposed to bright sunlight or high temperatures. Water the plants first if they are dry at the roots.

I

Impatiens Downy Mildew

SYMPTOMS Infected leaves turn yellow and are quickly shed. The flowers may also drop, and a whitish, fluffy growth may develop on the lower leaf surfaces. The leaves rot rapidly.

Leaf drop can be so extensive that the plants may become completely devoid of foliage, except at the shoot tips (*see p.44*),. Badly affected plants will soon die. This infection has so far only been reported on *Impatiens walleriana* and has not extended to the New Guinea impatiens (*I. hawkeri*) or impatiens species found growing in the wild in Britain.
CAUSE The fungus-like organism *Plasmopara obducens*, first recorded in Britain in 2003. It has caused so much damage intermittently that bedding busy Lizzies (*Impatiens walleriana*) have become both less popular and less easily available to purchase. Spores are produced from the lower leaf surfaces and may be spread considerable distances on air currents, or by rain or watering splash. As the pathogen thrives in moist weather, wet summers often lead to more serious outbreaks.
CONTROL Collect infected plants promptly and burn or dispose of them; compost from container-grown plants should also be disposed of. Do not add them to your domestic composting system. It is thought that resting spores may be produced and can remain viable for several years in the soil or compost beneath an infected plant and the area in which it was grown. Wait at least a year before growing bedding busy Lizzies following an outbreak in your garden. Instead, choose the New Guinea hybrids or unrelated bedding plants.

Inonotus hispidus

SYMPTOMS Annual, bracket-shaped, semi-circular fungal fruiting bodies develop on large limbs or the main trunk of trees. Each measure between 10 and 35cm across and is 2–10cm deep. They are covered in dense, hair-like growth and are an orangey-rust colour initially, turning dark brownish-black as

they age. Their presence causes a yellowish-orange discoloration of the wood. Most common on ash, walnut, plane, apple, and elm, but other broadleaved trees may be infected.
CAUSE The fungus *Inonotus hispidus*, which usually enters through wounds or stubs left by poor pruning.
CONTROL Infected trees may be rendered dangerous, their safety depending largely on where the decay has occurred. Employ a reputable tree surgeon to advise as to the best course of action.

Iris Ink Disease

SYMPTOMS Blue-black streaking appears on the outer surface of bulbs, most commonly those of *Iris reticulata* and Dutch irises (*I. x hollandica*). In extreme cases, or if left unchecked, the whole bulb turns black and rots, leaving just the outer shell full of black spores. If infected bulbs produce foliage, it is streaked yellow initially and later develops a red discoloration, particularly towards the tips. Black necrotic spots may develop on the yellowed areas.
CAUSE The fungus *Bipolaris iridis*, which produces spores that are spread on air currents. The fungus can overwinter on infected bulbs, bulb debris, and possibly in soil around infected bulbs.
CONTROL Dispose of any bulbs showing symptoms. Remove any plants which show foliar symptoms, and grow irises on a fresh site.

Iris Leaf Spot

SYMPTOMS Water-soaked spots develop on the foliage, later turning brown with grey centres and yellow margins. The spots coalesce as they enlarge, thereby destroying the foliage. Affected irises are weakened by the loss of functional foliage and so are less vigorous and flower poorly. Bearded irises are most susceptible, but others may be infected too.

CAUSE The fungus *Cladosporium iridis*, which is encouraged by mechanical injury and wet weather. It produces spores which are spread on air currents and by rain and water splash. It overwinters on infected leaves and plant debris.

CONTROL If infection occurs after flowering, cut off badly affected areas. Keep plants well fed and ensure they are planted in a suitable, sunny site; they are less susceptible to infection if growing vigorously.

Iris Rhizome Rot

SYMPTOMS Foliage yellows and withers, starting at the leaf tips or leaf margins. A soft, often foul-smelling rot develops at the base of the affected leaves and rapidly spreads down the rhizome, with the younger parts of the rhizome usually affected first (*see p.21*). As the infection spreads, the foliage symptoms become more extreme, and leaves may yellow completely and topple over. Most common on bearded irises, but others may be affected too.

CAUSE The bacterium *Pectobacterium carotovorum*, which is soil borne and is particularly prevalent in heavy, wet soils and in wet seasons. The bacteria enter the iris via wounds.

CONTROL Plant irises on a well-drained soil. Choose only healthy rhizomes, and plant them shallowly. Avoid injuring the rhizomes and control pests such as slugs which may damage them. Remove and dispose of badly affected irises. It may be possible to save rhizomes which are only slightly affected by cutting off all damaged areas right back to sound, healthy tissue.

Iris Rust

SYMPTOMS Numerous small, raised, lens-shaped orange to yellow pustules, measuring about 1mm in length, develop on the leaves (*see p.27*). These enlarge slightly and erupt to produce masses of yellow-orange spores. The infected leaves discolour, turning yellow at first and later, if the infection is severe, browning and dying off. Later in the season the fungal pustules are dark brown or black. Older leaves are generally the most badly affected.

CAUSE The fungus *Puccinia iridis*, which is encouraged by wet weather and excessively soft leaf growth.

CONTROL Cut off severely affected leaves. Infection usually occurs towards the end of the summer, so it does not cause too much loss of vigour.

Iris Sawfly

SYMPTOMS V-shaped notches are eaten from the leaf margins initially (*see p.45*), but later there is extensive defoliation, particularly on the upper parts of the leaves. Only waterside irises are attacked, especially yellow flag iris, *Iris pseudacorus*.

CAUSE The caterpillar-like larvae of a sawfly, *Rhadinoceraea micans*. It is of local occurrence in Britain, but is numerous where it is found. The greyish-brown larvae, up to 25mm long, are active during early to midsummer. The adult insects are black, with smoky-grey wings, and can be seen resting on waterside irises in late spring and early summer. There is one generation a year.

CONTROL The larvae are susceptible to pesticides (*see p.96*), but they should not be used where there is a risk of the spray getting into the water. Fish, frogs, and other pond wildlife are very vulnerable to insecticides. Removal of the larvae by hand is the best option in most gardens.

Iris Scorch

SYMPTOMS The upper parts of affected leaves wilt and bend over and become reddish-brown. The symptoms may initially be confused with those of iris rhizome rot, but the leaf discoloration is different and there is no deterioration of the rhizome. The roots die back; some rot and others disintegrate internally, leaving just a hollow outer "tube" with the cotton-thread-like remains of the central cylinder inside.

CAUSE The precise cause is not known, but the problem seems to be caused by root injury in winter or early spring. Winter waterlogging may be responsible.

CONTROL Ensure good drainage. It can sometimes be possible to save rhizomes that are only slightly affected by lifting, removing all deteriorating roots and leaves, washing thoroughly, and then replanting on a new, very well-drained site. Foliar feeding, to stimulate the production of replacement roots, can also be beneficial.

Iron Deficiency and Lime-induced Chlorosis

SYMPTOMS Usually seen in combination with manganese deficiency. Leaf yellowing, or chlorosis (*see p.39*), is often combined with the development of brown areas of discoloration which start at the leaf margins and then spread between the veins. Younger growth is usually affected earlier and more severely than older growth.

CAUSE Acid-loving plants and lime-hating plants have roots which are poorly adapted for the absorption of necessary trace elements from an alkaline soil. If conditions are too alkaline, they readily develop deficiency symptoms, in particular of iron and manganese.

CONTROL Avoid growing lime-hating plants on a soil

which is not sufficiently acid. Treat affected shrubs with a chelated compound containing iron, manganese and other trace elements which will be available to the plant because they are in a form which does not become "locked up" in the alkaline soil. Use acidic mulches such as chopped, composted bracken or conifer bark. Incorporate acidic materials such as these into the planting hole at planting. Feed plants with a fertilizer formulated for use on acid-loving (lime-hating) plants. Before planting, consider acidifying soil using sulphur or aluminium sulphate and ferrous sulphate. Sulphur treatments can also be used around existing plants. Follow instructions carefully.

Irregular Watering

SYMPTOMS Fruits are unusually small and may have a particularly thick skin. The surface may be ruptured, with splits caused when the fruit swells in response to the sudden availability of water (*see p.23*). Once the fruit's surface is broken it is open to infection by secondary organisms, such as grey mould, which may then cause further, rapid deterioration. Other fruit symptoms may include blossom end rot (*see p.117*). Leaves fail to grow at a regular rate and so develop an uneven, almost puckered surface (*see p.40*). They may also appear unusually small. Buds may fail to form, or form and later drop, and bark may split (*see p.52*).

CAUSE An irregular or erratic supply of water at the plant roots. Plants growing on a light, sandy soil with a low organic matter content are particularly prone, as are any plants growing in containers.

CONTROL Keep plants adequately watered at all times. Incorporate plenty of bulky

organic matter into light soils before planting. Apply a bulky mulch around the root areas to encourage soil moisture retention.
• see also BUD DROP *p.121;* BUD, DRY *p.121.*

Ivy Leaf Spot

SYMPTOMS Brown or grey, near-circular spots develop on ivy foliage. On pale-leaved or variegated forms, the spots may be surrounded by a narrow band of reddish-purple discoloration. On variegated leaves, leaf-spotting is usually most severe on the pale or white areas of the leaf. Numerous, pinprick-sized, raised fungal fruiting bodies may develop on the spots, usually in concentric rings.
CAUSE Various fungi, including *Mycosphaerella hedericola*, *Colletotrichum trichellum*, and *Phyllosticta hedericola.*
CONTROL Remove severely infected leaves to limit disease spread. Although unsightly, ivy leaf spot infections will not lower the overall vigour of the plant.

Ivy on Tree Trunks

SYMPTOMS Ivy growing up the trunk and, if left unchecked, into the crown of trees. On young trees, growth may be slightly constricted, but this is unusual on larger trees.
CAUSE Ivy, with its climbing, self-clinging habit, moves onto and up suitable surfaces, including tree trunks. It is, however, rare for it to cause damage, but young or poorly growing trees may have their stem or trunk growth constricted. Dense ivy growth on trunks may also mask the development of other problems, such as decay-causing bracket fungi, but it does provide a useful wildlife habitat and nectar source.
CONTROL Rarely justifiable. If desired, the ivy should be removed regularly, before it develops a lot of growth. Established plants can be cut off at ground level and, once dead, their top growth pulled away from the trunk and branches of the tree.

J

Juniper Aphids

SYMPTOMS Leaves and stems become coated with sooty moulds that grow on the sugary excrement of aphids. Heavy attacks cause yellowing and dieback in mid- to late summer.
CAUSE Two species of aphid occur on junipers. *Cinara fresai* is 2–4mm long and brownish-black. Colonies on stems close to the soil often have a protective covering of soil provided by ants, which collect honeydew from the aphids. The other species, *C. juniperi*, measures 2–3mm in length, is pale brown, and occurs mainly on younger growth. Both feed by sucking sap.
CONTROL Look for aphid infestations from mid-spring onwards and, if seen, spray with a contact or systemic insecticide (*see p.94*) on plants small enough to be sprayed.

Juniper Scale

SYMPTOMS Flat, rounded, or elongate and rectangular white objects with a small yellowish patch develop on the foliage of junipers, cypresses, Leyland cypress, and also thuyas. The rounded scales are protective covers for the females and measure up to 1.5mm long. The smaller rectangular scales cover the males. Heavy infestations can cause yellowing (chlorosis) and dieback of green shoots (*see p.53*).
CAUSE A sap-sucking insect, *Carulaspis juniperi*, which has one generation a year.
CONTROL On plants small enough to be sprayed, apply a contact or systemic insecticide (*see p.94*) in early summer when the more vulnerable newly-hatched scale nymphs are present.

Juniper Webber Moth

SYMPTOMS Patches of brown foliage are bound together with silk threads. Dark brown caterpillars with paler stripes running along their bodies, up to 11mm long, live within the webbing (*see p.48*). The leaves damaged by the caterpillars dry up and die in early summer. Compact, upright junipers, such as 'Skyrocket', can be badly damaged.
CAUSE Caterpillars of a moth, *Dichomeris marginella*, which has one generation a year.
CONTROL Infestations can sometimes be dealt with by cutting out webbed shoots in late spring or early summer. Alternatively, spray forcibly with a contact insecticide (*see p.94*) in order to penetrate the webbing.

K

Kabatina Shoot Blight

SYMPTOMS Cankers develop on the shoots of Leyland cypress, certain thuyas, junipers, and cypresses. The cankers usually develop at the base of small branches and cause dieback and, occasionally, death.
CAUSE Fungi of *Kabatina* species. The fungal spores are carried on air currents and by rain splash.
CONTROL Remove affected shoots promptly to minimize spread of the disease.

Kerria Twig and Leaf Blight

SYMPTOMS Small, reddish-brown spots (1–5mm in diameter) develop on both the upper and lower surfaces of leaves (*see p.44*).Each spot has a distinct purple border. In damp or wet weather, clumps of whitish spores may also be visible in the centres of the spots. There may be large numbers of spots on a single leaf. As the spots increase in size, they coalesce; the infected leaves turn yellow, then brown, and drop prematurely. Sunken, elliptical, brownish-purple stem cankers also develop. These are visible on the stems throughout the year. If they enlarge, they may girdle the infected stems and cause extensive dieback.
CAUSE The fungus *Blumeriella kerriae*, which was first recorded in Britain in 2014. It appears to attack only kerrias.
CONTROL Prune out any cankered stems and collect up fallen leaves from beneath the infected plant. All infected material should be burnt or taken to a local authority green-waste collection centre for composting, as the temperatures reached in home-composting are unlikely to be sufficiently high to kill the fungus.

Knopper Galls

See ACORN GALL WASP *p.106.*

L

Laburnum Leaf Miners

SYMPTOMS Whitish-brown, dried-up patches develop on laburnum leaves (*see p.32*) between late spring and early autumn. If the leaf is held up to the light it is often possible to see leaf-mining larvae feeding within the leaf.
CAUSE Two leaf miners commonly occur on laburnum. The larvae of a leaf-mining fly, *Agromyza demeijerei*, cause

narrow linear mines along the leaf margin, which later enlarge into irregular blotch mines extending from the margin. Caterpillars of a leaf-mining moth, *Leucoptera laburnella*, cause roughly circular mines in the centre of leaves, but in heavy attacks the mines join together and lose their distinctive shape.

CONTROL Attacks in late summer may be extensive and unsightly, but they have little effect on the tree's health and vigour and so can be tolerated. None of the insecticides available to home gardeners is likely to give control.

Lacebugs

See PIERIS LACEBUG *p.168*; RHODODENDRON LACEBUG *p.176*.

Lacewings

The larvae of lacewings are beneficial insects that feed on aphids and other small insects. Some lacewing larvae carry the sucked-out remains of earlier meals amongst clumps of bristles on their upper surface. A fully grown larva is up to 10mm long and, apart from a pair of sharp jaws protruding at the front, is hidden under a tangled mass of dead aphids. The harmless adult lacewings have slender, elongate bodies which are pale green, black, or brown. They have two pairs of transparent wings, sometimes with darker markings, which have many veins – hence the name lacewing. The adults have long, thin antennae and feed on pollen and nectar or honeydew. One species of green lacewing, *Chrysoperla carnea*, overwinters as adults and is often found indoors.

Lackey Moth

SYMPTOMS Gregarious, hairy caterpillars occur on a wide

range of trees and shrubs from mid-spring to early summer. They are up to 48mm in length and have slate-grey heads with red, white, black, and blue stripes on their bodies. The caterpillars cover their feeding area with silk webbing and can cause extensive defoliation (*see p.46*).

CAUSE Caterpillars of *Malacosoma neustria*, which hatch in the spring from eggs laid in late summer of the previous year. These form a solid band of several hundred eggs glued together around a twig.

CONTROL Prune out egg bands or the webbed shoots when they are seen. For extensive infestations, use a contact insecticide (*see p.94*) against the young larvae.

Ladybirds

Most of these beneficial insects, including the harlequin ladybird, feed on greenfly and other aphids, both as adult beetles and larvae. Other ladybirds prey on scale insects or red spider mites, while some feed on powdery mildews. Adult ladybirds are familiar insects, especially those with wing cases that are red with black spots (*see p.60*), but some species and colour forms are yellow and black, or brown with white spots. The larval stages are generally slate-grey in colour with orange or white spots on the upper surface. One ladybird, *Cryptolaemus montrouzieri*, is commercially available for controlling mealybugs in greenhouses.

Laetiporus sulphureus (Chicken of the Woods)

SYMPTOMS Overlapping groups of annual, bright yellowish-orange fungal brackets, measuring between 10–40cm in total diameter, develop on the

trunk or large limbs of host trees (*see p.54*). Each bracket is thin and has a wavy margin, with soft flesh resembling cooked chicken, and a yellow undersurface. Hardwoods, including oak, sweet chestnut, cherry (*Prunus*), and occasionally beech, and conifers, including larch, yew, and pine, may become infected. The infected wood discolours, turning yellowish-orange at first and later breaking up into brown, cubical blocks. Off-white to yellow mycelium (fungal sheets) develop between the blocks of decaying wood.

CAUSE The fungus *Laetiporus sulphureus*, which produces numerous spores from its lower surface.

CONTROL Infected trees may become dangerous as the strength of wood is reduced. An arboriculturist should examine the tree and advise on its state of health.

Larch Adelgid

SYMPTOMS Small patches of fluffy white wax appear on the foliage in early summer. Heavy infestations can cause leaf yellowing, and sooty moulds develop on the pests' sugary excrement (*see p.48*).

CAUSE Small, black sap-sucking insects, *Adelges laricis*, that resemble aphids and secrete waxy fibres.

CONTROL Large trees cannot be sprayed effectively but are usually able to tolerate infestations. Small trees, however, can be sprayed with a contact or organic insecticide (*see pp.94 and 89*).

Large Rose Sawflies

SYMPTOMS Rose stems become defoliated by caterpillar-like larvae in early and late summer. The larvae are up to 25mm long and have bodies that are pale green on the sides and yellowish green on the

upper surface. The upper surface has many black spots arranged in rows across the body on each segment. The adults have orange-yellow abdomens, while the head and thorax is wholly or mostly black. Damaged shoots often have an elongate oval split below the damage zone where eggs were inserted by the female sawfly.

CAUSE Two species of sawfly, *Arge ochropus* and *A. pagana*. The latter is the more common and has two or three generations during the summer, while the former usually has just one. Females insert a cluster of eggs into the soft young extension shoots on wild and cultivated roses, and are sometimes seen attached to the stems.

CONTROL Remove the larvae by hand or apply a contact insecticide (*see p.94*).

Lavatera Leaf and Stem Rot

SYMPTOMS Red-brown necrotic spots develop on stems and occasionally on leaves and petioles of lavateras and hollyhocks. The spots enlarge and coalesce and may girdle stems to cause dieback or even death. Tiny, black, bristle-like spore-bearing structures may develop on the surface of the lesion.

CAUSE The fungus *Colletotrichum malvarum*, which can be seed-borne.

CONTROL Remove and dispose of infected plants promptly. Grow replacement plants on a fresh site and do not save seed from those showing infection.

Lavender Shab

SYMPTOMS The plant wilts, discolours, and dies back, usually stem by stem. The whole plant may be killed. Numerous tiny, blackish-brown, raised fungal fruiting bodies the size of pinpricks develop on the dead foliage and stems. The symptoms resemble those caused by wet soil or winter cold injury, but are unique in having the fungal fruiting bodies.

CAUSE The fungus *Phomopsis lavandulae*, which enters the plant through pruning wounds or other points of damage on the bark.

CONTROL Prune out infected areas immediately, cutting well back into healthy wood. It may not be possible, however, to save infected plants, as symptoms develop rapidly.

Leaf and Bud Eelworms

SYMPTOMS Brown islands or wedges develop in leaves (*see p.37*) where the internal spread of the eelworms is restricted by the larger leaf veins. Eventually the entire leaf will turn brown and dry up. Typically, the younger leaves at the shoot tips are free of infestation, but the older leaves lower down the stem show progressively more severe damage.

CAUSE Two microscopic species, *Aphelenchoides ritzemabosi* and *A. fragariae*, attack a wide range of plants in gardens and greenhouses.

CONTROL These eelworms cannot be controlled with the pesticides available to home gardeners. Badly infested plants should be burnt. With indoor plants, avoid wetting the foliage, as eelworms can emerge from leaves and spread rapidly in a surface film of water.

• *see also* CHRYSANTHEMUM EELWORM *p.126*.

Leaf Curl

See PEACH LEAF CURL *p.164*.

Leaf-cutting Bees

SYMPTOMS Circular or elliptical pieces are cut from

the leaf margins (*see p.46*) of roses, wisterias, epimediums, and other plants. The smooth outline and regular shapes of the missing pieces readily distinguish leaf-cutting bee damage from caterpillar or other pest damage. The female bee uses the leaf pieces to make thimble-shaped cells in her nest. This may be in a hollow plant stem or in a tunnel dug in rotten wood or soil. Sometimes a bee makes use of soil in flower pots and the cells, resembling old cigar stubs, come to light when the plant is repotted. When a cell has been built, the bee stocks it with nectar and pollen, lays a single egg, and completes the cell by capping it with circular leaf pieces. A nest may contain up to twenty cells.

CAUSE Leaf-cutting bees, *Megachile* species, which resemble honeybees but have ginger hairs beneath their abdomens. They are solitary bees, each female constructing her own nest during the summer. They frequently visit a plant repeatedly to collect material and by the time their nest is finished they may have removed several hundred pieces.

CONTROL Like other bees, these are useful pollinators and should be tolerated.

Leafhoppers

Leafhoppers are sap-sucking insects, mostly 2–3mm long, which live on the undersides of leaves. The adults readily jump off leaves when disturbed, but their wingless nymphs are less active. Their feeding usually causes a coarse, pale mottling of the upper leaf surface. Garden plants frequently attacked are blackberries, plums, *Phlomis*, salvias, mints, beech, and many others. General control measures for leafhoppers are given in the entry for rose leafhopper (*p.179*).

• *see also* GLASSHOUSE LEAFHOPPER *p.141*;

RHODODENDRON LEAFHOPPER *p.176*.

Leaf Miners

Leaf-mining is a habit adopted by many insects, mainly the larvae of some flies, moths, sawflies, and beetles. The damaged parts of the leaf dry up and become white or brown. The mines may be straight or meandering lines, circular or irregular blotches, and the form of the mine is usually distinctive for any particular leaf-mining insect. If a mined leaf is held up to the light, the larva or pupa can often be seen at the end or edge of the mined area.

• *see also* ALLIUM LEAF MINER *p.107*; AQUILEGIA LEAF MINERS *p.110*; APPLE LEAF MINER *p.109*; ASTRANTIA LEAF MINER *p.112*; BEECH LEAF-MINING WEEVIL *p.114*; BEET LEAF MINER *p.115*; CELERY LEAF MINER *p.125*; CHRYSANTHEMUM LEAF MINER *p.126*; DELPHINIUM LEAF MINER *p.131*; GEUM SAWFLIES *p.140*; HOLLY LEAF MINER *p.144*; HOLM OAK LEAF MINERS *p.145*; HORSE CHESTNUT LEAF-MINING MOTH *p.146*; LABURNUM LEAF MINERS *p.149*; LEEK MOTH *p.152*; LILAC LEAF MINER *p.153*; PYRACANTHA LEAF MINER *p.173*; SEMPERVIVUM LEAF MINER *p.182*.

Leaf Scorch

SYMPTOMS The plant grows well, but its foliage is spoiled by dry, brown scorching, which is usually first seen around the leaf edges. The whole leaf may ultimately be affected, or the central area may retain its normal colour until the end of the season. If the problem is severe, or occurs in several years in succession, plant growth is affected and dieback may occur. On maples and sycamore (*Acer*), stems which deteriorate as a result of scorch often develop a white surface discoloration.

In extreme cases extensive dieback, and occasionally even death, may occur.

CAUSE Various factors may be involved. The most common are an over-wet or excessively dry soil, cold winds, or hot, bright sun. Plants lacking in vigour due to some other cause, or those which are not yet properly established, are particularly prone to scorch.

CONTROL Attempt to determine which adverse condition or conditions are involved, and improve growing conditions accordingly. If poor drainage is responsible, improve this and apply a foliar feed throughout the growing period to encourage replacement root growth. If the soil tends to become too dry, water well and apply a bulky mulch around the root area. Consider erecting windbreaks or planting to provide wind protection. If necessary, and feasible, consider moving the plant to a more suitable site in the autumn.

Leaf Spots (Bacterial)

SYMPTOMS Necrotic patches, usually circular or angular in outline, develop on foliage. Many bacterial leaf spots are surrounded by a bright yellow "halo" (*see pp.26, 44*). There are no minute, raised fungal fruiting bodies (pycnidia) as seen on fungal leaf spots.

CAUSE Various bacteria may be involved, and these are usually spread by rain or watering splash from leaf to leaf or, occasionally, from stem lesions to leaves (for example, bacterial canker may cause leaf spotting on *Prunus*). In most cases the leaf spotting itself is not very damaging, but may indicate that the plant is suffering from another, more serious problem. Examples include *Pseudomonas maculicola* on brassicas, *Pseudomonas* *lachrymans* on cucumber, black blotch on delphiniums, and *Pseudomonas mori* on mulberry. **CONTROL** Remove infected leaves promptly and avoid overhead irrigation.

Leaf Spots (Fungal)

SYMPTOMS Necrotic spots, usually circular and either grey or brown (*see p.25*), develop on the leaves and may coalesce, causing large areas of dead leaf tissue to develop (*see p.25*). Occasionally leaves may be killed, but in most cases leaf spots are not very damaging in their own right, though their presence may indicate that the host plant is suffering from some other, rather more fundamental problem. The spots often show concentric zones of discoloration, and may bear numerous, raised, pinprick-sized, black or brown fungal fruiting bodies. Many common examples have entries elsewhere under their host name; others include *Phyllosticta antirrhini* on antirrhinums; *Discula quercina* on oaks and beeches; *Ramularia beticola* and *Cercospora beticola* on beet crops; *Septoria* on chrysanthemums, pinks, and carnations (*Dianthus*), foxgloves (*Digitalis*), hebes, hawthorn, and cyclamens; *Ascochyta cinerariae* on cinerarias; *Colletotrichum clivae* on clivias; *Marsonina daphnes* on daphnes; *Phyllosticta magnoliae* on magnolias; *Ramularia* on primroses and polyanthus, and *Phyllosticta wistariae* on wisterias. **CAUSE** A wide range of fungi (see above, and also references to specific entries below). **CONTROL** Remove affected leaves if necessary. Rake up and dispose of fallen leaves at the end of the season in order to minimize infection carry-over.
• *see also* ALTERNARIA LEAF SPOT *p.107*; ARBUTUS LEAF SPOT *p.111*; BRASSICA LIGHT LEAF SPOT *p.119*; CELERY LEAF SPOT *p.125*;

HYDRANGEA LEAF SPOT *p.147*; IVY LEAF SPOT *p.149*; PANSY LEAF SPOT *p.162*; PEA LEAF AND POD SPOT *p.163*; RHODODENDRON LEAF SPOT *p.176*; STRAWBERRY LEAF SPOT *p.187*; WATER LILY LEAF SPOT *p.195*.

Leaf Weevils

SYMPTOMS Ragged holes are eaten in the foliage of many deciduous trees, especially *Prunus*, *Malus*, *Sorbus*, birches (*Betula*), oaks (*Quercus*), and beech (*Fagus*), in late spring and early summer. Turquoise-green or golden-brown beetles, 3–7mm in length, may be present in large numbers. **CAUSE** Several *Phyllobius* species occur in gardens and may cause extensive damage to young foliage in mid-spring to early summer. **CONTROL** Small trees can be sprayed with a contact or systemic insecticide (*see p.94*) if leaf weevils are numerous enough to cause problems.

Leafy Gall

SYMPTOMS Numerous small, bunched, often somewhat distorted, thickened, or fasciated leaves develop around stem bases (*see p.55*). The plant may also produce normal growth. **CAUSE** The bacterium *Rhodococcus fascians* (syn. *Corynebacterium fascians*), which is soil-borne and enters wounded plant tissue. **CONTROL** Remove and dispose of affected plants, together with the soil or compost in which they are growing. Avoid spreading the bacteria by washing hands and tools after handling infected plants. Thoroughly scrub pots, trays, and greenhouse staging.

Leatherjackets

SYMPTOMS Legless, greyish-brown tubular larvae, up to 45mm long and without obvious heads, live in the soil (*see p.19*). They are mainly a lawn pest, causing yellowish-brown patches from late winter to midsummer where the grubs have severed the roots (*see p.66*). Leatherjackets can also cause problems in vegetable gardens and flower borders where they kill seedlings and young plants. **CAUSE** Leatherjackets are the larvae of crane flies (daddy-long-legs). Many species occur in gardens, mainly *Tipula* and *Nephrotoma* species. Eggs are laid in the soil mainly in late summer. **CONTROL** There are no suitable insecticides for leatherjacket control. Biological control with the pathogenic nematode *Steinernema feltiae* can be used against the larvae when the soil is moist and warm (at least 14°C/57°F). Leatherjackets can be removed from lawns by covering small areas with black polythene overnight after heavy rain or irrigation. The grubs will come up onto the lawn surface and can be removed when the polythene is lifted.

Leek Moth

SYMPTOMS Onion and leek foliage develops whitish-brown patches (*see p.17*) where caterpillars up to 12mm long have been feeding as leaf miners. The older caterpillars also bore into the stems of leeks and bulbs of onions. There are two generations during the summer, with leaf mining occurring in late spring to midsummer and late summer to mid-autumn. When fully fed, the caterpillars emerge and pupate in silk cocoons on the foliage. **CAUSE** Caterpillars of a small moth, *Acrolepiopsis assectella*, which in Britain is mainly a problem in southern England. **CONTROL** There are no insecticides approved for use against leek moth in gardens. Where leek moth is a regular problem, grow leeks under insect-proof netting to exclude the egg-laying moths. Search for pupae in their net-like cocoons on the leaves, and crush them.
• *see also* ALLIUM LEAF MINER *p.107*.

Leek Rust

SYMPTOMS The outer leaves develop numerous bright orange, lens-shaped, raised fungal pustules, each 1–2mm in length (*see p.30*). These erupt producing conspicuous masses of bright orange spores. Affected leaves yellow and die back. The inner growth is rarely, or only mildly, affected. **CAUSE** The fungus *Puccinia porri* (syn. *P. allii*), which may also affect related plants such as chives, onions, shallots, garlic and, occasionally, ornamental alliums. The disease is encouraged by damp conditions and wet weather. **CONTROL** Remove and dispose of affected leaves at harvest. Dispose of all debris at the end of the season and grow leeks and other susceptible crops on a fresh site. Avoid excessive use of high-nitrogen fertilizers, as these encourage the production of soft growth which is then more readily infected. At planting, apply sulphate of potash at 15–20g per m² to harden growth slightly. Space plants widely to improve air circulation. Where available, choose leek cultivars resistant to this infection.

Leek White Tip

SYMPTOMS Leaf tips turn yellow at the end of summer. As the season progresses, the damage spreads down the leaves; they then either dry out and become pale brown or become bleached and papery in texture, or they rot. **CAUSE** The fungus-like *Phytophthora porri*, which can

persist in plant debris in the soil for several years after infection, even in the absence of any new leek crops during this time.
CONTROL Remove affected leeks promptly. Do not grow either leeks or other susceptible plants, such as onions or ornamental alliums, in the contaminated soil for at least five years.

Leopard Moth

SYMPTOMS Apple trees (*Malus*) are the principal host plant, but *Pyrus*, *Sorbus*, sycamores and maples (*Acer*), birches (*Betula*), hawthorn (*Crataegus*), oaks (*Quercus*), and many others are also attacked. The creamy yellow or white caterpillar, which measures up to 55mm in length, has many black spots on its body. It tunnels in branches or trunks which have a diameter of up to 10cm, and is often detected when the weakened trunk or branch snaps on a windy day. Another sign of infestation is pellets of compacted sawdust (the caterpillar's excrement) coming from a hole in the trunk or branch.
CAUSE Caterpillars of a moth, *Zeuzera pyrina*, which take two years to complete their feeding. Often the infestation consists of a single caterpillar, but this can have serious consequences if it is present in the trunk of a young tree.
CONTROL Preventive measures are not possible since attacks are sporadic, and spraying against the adult moths during summer is unlikely to be effective. If caterpillars are detected, damaged branches can be pruned out, or a piece of wire inserted into the tunnel to skewer the caterpillar.

Lettuce Downy Mildew

SYMPTOMS Angular yellow areas of discoloration develop on the leaves (*see p.37*), the older, outer leaves being affected first. The infected areas turn pale brown and either papery or soft. Off-white, fluffy growth, occasionally with a slightly purple tinge, develops on the lower leaf surface, corresponding with the patches of discoloration. Severely affected leaves die off. Seedlings and mature plants may be affected.
CAUSE The fungus-like *Bremia lactucae*, which is encouraged by humid or damp growing conditions.
CONTROL Remove infected leaves promptly, or the entire plant if the infection is particularly severe. Do not grow lettuces too close together and ensure air can circulate freely. Grow resistant lettuces such as 'Chartwell', 'Leny', and 'Lettony'.

Lettuce Grey Mould

SYMPTOMS Fluffy grey fungal growth develops on lettuce leaves (*see p.43*). If the grey mould infection takes place towards the base of the plant, the stem rots, often together with the internal tissues there, causing a yellowish-orange or brown slimy rot. The whole plant may then die off. Seedlings are also susceptible.
CAUSE The fungus *Botrytis cinerea*, which is encouraged by damp growing conditions and often enters the plant through wounds. It may also be seed-borne in lettuce. The spores are carried on air currents. Black fungal resting bodies (sclerotia) may form on the infected tissues, but will not yet have developed if the plants are removed promptly.
CONTROL Remove all plant debris which may harbour the fungus. Avoid injury to the plant. Control likely sources of injury, such as snails or slugs. Improve ventilation, and water early in the evening to allow time for the foliage to dry off

before nightfall. Remove leaves infected with the fungus.

Lettuce Root Aphid

SYMPTOMS Lettuce plants tend to wilt and make slow growth in sunny weather. When the plants are dug up, a white, powdery wax can be seen on the roots and soil particles. Creamy yellow aphids, measuring up to 2mm long, will be found on the roots.
CAUSE Sap-sucking insects, *Pemphigus bursarius*, which feed on the roots of outdoor lettuce during mid- to late summer.
CONTROL Some lettuce cultivars, such as 'Lakeland', are resistant to lettuce root aphid. Root aphids are more difficult to control than leaf-feeding types. Infestations are unlikely to be controlled by the insecticides currently available to home gardeners. Keep the plants well watered and grow lettuce crops on a new site each year.

Lettuce Viruses

SYMPTOMS Leaves show yellow mottling, mosaic patterning, or vein clearing (yellowing immediately adjacent to the veins). Plants are stunted and the foliage may be distorted.
CAUSE Various viruses, including lettuce big vein, lettuce mosaic, cucumber mosaic, and beet western yellows virus.
CONTROL Remove plants showing symptoms promptly. Control pests such as aphids, which may spread the viruses. Wash hands thoroughly after handling infected plants. Grow lettuces on a fresh site. Choose resistant cultivars, where available, such as those listed in seed catalogues as showing resistance to lettuce mosaic viruses.
• *see also* CUCUMBER MOSAIC VIRUS *p.129*.

Lilac Blight

SYMPTOMS Angular brown spots develop on lilac leaves. Affected leaves die back. Buds blacken and die, and young shoots may be girdled and wilt and die (*see p.51*). Large, elongated cankers may form on older stems. Secondary infection by grey mould is common and may mask the symptoms.
CAUSE The bacterium *Pseudomonas syringae*, which is readily spread on air currents and by rain and water splash.
CONTROL Prune out affected areas of the plant promptly, cutting well back into perfectly healthy tissue.

Lilac Leaf Miner

SYMPTOMS The leaves of lilac, privet, and ash develop brown shrivelled areas (*see p.32*) where the internal tissues have been eaten by leaf-mining caterpillars. When half-grown, the whitish-green caterpillars emerge from their mine and roll up the leaf tip to form a tube held in place by silk threads. There are two generations, with damaged leaves being seen from late spring to early summer and from late summer to early autumn.
CAUSE Caterpillars of a small moth, *Gracillaria syringella*.
CONTROL Pick off affected leaves; clipping privet hedges will remove many of the mined or rolled leaves.

Lily Beetle

SYMPTOMS Bright red beetles, measuring up to 6–7mm long, and with black heads, appear on lilies (*Lilium*) and fritillaries from early spring onwards. In midsummer the larval stage also occurs on the foliage. The grubs have rotund bodies, up to 8–9mm long, and are orange-red with black heads. They cover themselves with their own wet, black excrement. Both the grubs and the adults eat leaves, flowers and seed pods, and can ruin plants.

CAUSE A leaf beetle, *Lilioceris lilii*, that is active as an adult from spring to autumn. This pest occurs throughout England and in parts of Scotland, Wales, and Ireland.

CONTROL The long period during which the adults and larvae are active makes this a difficult pest to control. With small numbers of lilies and fritillaries, it is possible to protect them by picking the adults and grubs off by hand. If necessary, spray with a systemic or contact insecticide *(see p.94)*; the grubs are more vulnerable than adult beetles.

Lily Disease

SYMPTOMS Dark green, water-soaked spots develop on the foliage. Affected leaves soon turn brown or bleached and die off. Leaves wither from the base of the stem upwards, wilt, and remain hanging on the stem. Flower buds are killed or damaged, so that they become very distorted, and plants fail to flower. If infection spreads into the stem, the plant topples over. The bulb usually remains healthy but may start to rot.

CAUSE The fungus *Botrytis elliptica*, which produces numerous spores that are spread both on air currents and by rain and water splash. The fungus may persist from year to year as black sclerotia (fungal resting bodies) that develop on infected plant debris. Wet weather encourages both the development and spread of the disease.

CONTROL Remove and dispose of infected leaves, or the whole of the top growth if necessary. Bulbs from infected plants can be kept, provided they show no rotting. Remove all debris at the end of the season. Some species, notably

Lilium hansonii, L. martagon, L. pyrenaicum, and *L. davidii* var. *wilmottiae,* seem to show a degree of resistance.

Lily Viruses

SYMPTOMS Foliage shows yellow mottling, streaking, and mosaicing. Plants are usually stunted, and leaves show twisting or other deformity. Flower buds may not develop, or turn dry and brown and die off.

CAUSE Various viruses, including tulip-breaking virus, lily mosaic, and lily mottle. Tulip-breaking virus also causes flower-breaking symptoms on tulips and so lilies and tulips should not be grown in close proximity.

CONTROL None available. Remove infected plants, and wash hands thoroughly after handling them. Spray against aphids, as these may be vectors of the viruses.

Lime-induced Chlorosis

See IRON DEFICIENCY AND LIME-INDUCED CHLOROSIS *p.148.*

Lime Nail Gall Mite

SYMPTOMS From late spring onwards, slender, tubular, red or yellowish-green projections, up to 5mm long, develop on the upper leaf surface of lime trees, *Tilia* species *(see p.27).*

CAUSE Microscopic gall mites, *Eriophyes tiliae*, which induce the galls' growth by secreting chemicals as they feed on the underside of the leaves in early summer. The galls are hollow and the mites live within them during the summer months.

CONTROL Apart from creating the galls, the mites have no harmful effects on the tree and can be tolerated.

Lupin Anthracnose

SYMPTOMS Necrotic brown patches develop on the stems, foliage, flower stalks, and seed pods. Bright orange spores may also appear on these areas, and the plants may die back. The petioles also develop these symptoms and distort in a corkscrew fashion. Although this infection rarely kills lupins, it leaves them very unattractive *(see p.41).*

CAUSE A species of *Colletotrichum* fungus, which first became a problem on ornamental lupins in the 1980s, and is now considered to be the most damaging lupin disease. The infection is encouraged by wet conditions and is commonly seen in spring and summer, especially during warm, wet weather.

CONTROL If only a few lesions are seen, remove and dispose of infected leaves promptly. If the plant is severely infected it should be dug up and binned or burned. Collect any lupin debris left on the soil surface, which can harbour infection. As the spores are spread by rain and water splash, and the infection is encouraged by damp foliage, avoid overhead watering and ensure that plants are well spaced, allowing good air circulation. The fungus may be found on the stem at or just below soil level, and may also occur on the roots, so cutting back hard an infected plant is unlikely to solve the problem. As the fungus may be seed-borne, do not save seed from infected plants.

Lupin Aphid

SYMPTOMS Dense colonies of whitish-grey aphids, which can be up to 4.5mm long, occur on herbaceous and tree lupins during late spring and summer *(see p.59).* They live on the underside of leaves and on flower spikes, which become sticky with honeydew excreted

by the pests. Infestations can be so heavy that plants wilt and die.

CAUSE A sap-sucking insect, *Macrosiphum albifrons*, which is a North American species that arrived in Britain in 1981 and is now widespread in Europe.

CONTROL Look for aphids on leaves from mid-spring; spray with a systemic, contact, or organic insecticide *(see pp.94 and 89)* before a damaging infestation develops. Avoid spraying plants in flower.

M

Magnesium Deficiency

SYMPTOMS Yellowing develops between leaf veins (interveinal chlorosis) and also around the leaf margins, leaving clear green bands immediately adjacent to the veins *(see p.37).* As the green pigment deteriorates, other pigments may become visible, so instead of yellowing, red, purple, or brown discoloration develops between the veins. Camellias affected by this deficiency often develop distinct, raised, brown interveinal patches on the lower leaf surface. When magnesium is in short supply, it is transported out of the older leaves into new, developing leaves; older, lower leaves are therefore the ones most readily affected.

CAUSE A deficiency of magnesium, which often occurs in plants growing in very acid soils or composts, or after periods of heavy rain or watering. Magnesium is readily leached through the soil, particularly when soil water levels are high. The use of high potash fertilizers or sulphate of potash may also increase the occurrence of this deficiency, as the magnesium may be rendered unavailable because of the high levels of potassium in the soil.

CONTROL For a rapid response, apply magnesium as a foliar spray. Use Epsom salts at the rate of 200g in 10 litres (8oz in 2½ gallons) of water to which is added a wetting agent such as a few drops of mild liquid detergent or soft soap. Several applications are usually needed, at weekly to fortnightly intervals. Alternatively, apply Epsom salts to the soil at 40g per m².

Mahonia Rust

SYMPTOMS Bright orange spots develop on upper leaf surfaces (see **p.39**), with corresponding dark brown spore-filled fungal pustules beneath (see **p.25**). The leaf develops bright, often red or orange discoloration, and falls prematurely.
CAUSE The fungus *Cumminsiella mirabilissima*, which is encouraged by moist or humid air.
CONTROL Clear up and dispose of fallen leaves to minimize overwintering of the fungus. If feasible, improve air circulation around the plant. Infection is rarely damaging enough to merit treatment, but if necessary, spray with a suitable fungicide (see p.94).

Manganese Deficiency

See IRON DEFICIENCY AND LIME-INDUCED CHLOROSIS *p.148*.

Massaria Disease

SYMPTOMS Lesions in the form of a band of dead bark develop on the upper surfaces of larger branches of London plane trees (*Platanus* x *hispanica*), usually starting at the branch collar and tapering to a distinct pointed tip. The infection appears most commonly in limbs thought to be pre-disposed to it by drought, damage, or other stresses. Infection may also cause entire branches to drop. There is a soft rotting of the wood of affected branches (see **p.55**). The older, shaded branches towards the base of the tree seem to be most commonly affected. Trees under the age of 40 are rarely attacked.
CAUSE The fungus *Splanchnonema platani* (syn. *Massaria platani*), which was first recorded in Britain in 2009 and only attacks London plane trees. While this fungus is thought to be the major factor behind the infection, more research is still needed.
CONTROL At present there is no control method available, although plane trees that are regularly maintained are rarely affected. It would be prudent to have an infected tree inspected by an experienced tree surgeon or arborist.

Mealy Cabbage Aphid

SYMPTOMS All brassicas are susceptible to this pest between mid-spring and mid-autumn. Yellow patches appear on the foliage, and the underside of these areas will have dense colonies of whitish-grey aphids (see **p.48**). During early summer the aphids infest leaves developing at the shoot tips of young plants. These leaves become distorted and have a pale, mottled appearance. Heavy infestations can kill shoot tips, causing secondary buds to develop as side-shoots, resulting in multi-headed plants.
CAUSE Sap-sucking insects, *Brevicoryne brassicae*, which can also attack swede and radish.
CONTROL Spray with a contact insecticide labelled for use on the particular crop (see p.94), especially if young plants are under attack.

Mealy Plum Aphid

SYMPTOMS The shoot tips and undersides of leaves on plums have dense colonies of whitish-green aphids on them during late spring to mid- or late summer. The foliage and fruits become sticky with honeydew excreted by the aphids. A black, sooty mould often grows on this substance.
CAUSE A sap-sucking insect, *Hyalopterus pruni*.
CONTROL Look for the aphid from late spring onwards, and if necessary spray with a contact or systemic insecticide (see p.94). Sooty mould can be wiped off fruits with a damp cloth. A plant oil winter tree wash applied in early winter to control overwintering eggs can reduce the need to spray in spring.

Mealybugs

SYMPTOMS Cacti and succulents are the main host plants, but many other house plants and those grown in greenhouses are attacked. Mealybugs are soft-bodied, greyish-white or pink insects up to 4mm long. They frequently infest leaf axils (see **p.48**) and other inaccessible places, such as between twining stems (see **p.58**). A white, fluffy wax is secreted from the insects' bodies and this also conceals the eggs. Infested plants become sticky with honeydew excreted by the pest. Sooty moulds may grow on the honeydew.
CAUSE Several sap-sucking mealybugs (*Pseudococcus* and *Planococcus* species) occur in greenhouses and on house plants.
CONTROL During midsummer, when temperatures are high, biological control with a ladybird predator, *Cryptolaemus montrouzieri*, can reduce infestations. At other times spray thoroughly with a systemic, contact, or organic insecticide (see pp.94 and 89).
• see also GOLDEN ROOT MEALYBUG *p.141*; PHORMIUM MEALYBUG *p.167*; ROOT MEALYBUGS *p.178*.

Melon Cotton Aphid

SYMPTOMS Small, dark green or black aphids form dense colonies on the foliage and flowers of chrysanthemums (see **p.61**). Other plants likely to be attacked include cucumbers, melons, begonias, dahlias, and cinerarias. Heavy infestations cause stunted growth and the early death of blooms. Plants are soiled, with cast aphid skins, a sticky excrement called honeydew and sooty moulds.
CAUSE A sap-sucking insect, *Aphis gossypii*, which can also damage plants by transmitting tomato spotted wilt virus. It is mainly a greenhouse problem, where it can remain active throughout the year.
CONTROL This aphid has a high degree of pesticide resistance, so most of the usual chemicals available to home gardeners fail to work. Biological control with parasitic wasps, *Aphidius colemani*, or predatory midge larvae, *Aphidoletes aphidomyza*, can help during the warmer summer months. A systemic spray (see p.94) is usually effective.

Meripilus giganteus (Giant Polypore)

SYMPTOMS Large, fleshy, golden-brown fungal brackets develop around the base of affected trees or in close proximity to the trunk or roots (see **p.55**). The lower surface is creamy white (but bruises dark when handled). The fruiting bodies are conspicuous for their large size (up to 30cm across) and the number that are produced (they may total up to about 12kg in weight). They usually appear in late summer or autumn, and deteriorate into a smelly, black mass when frosted. The fungus causes root decay, often attacking the lower roots first, so the damage may not be apparent on initial investigation. The roots become hollow and so make the tree unsafe. Infected trees often show dieback of branches; they produce small, sparse leaves, and may be blown over in heavy winds.

CAUSE The fungus *Meripilus giganteus*, also known as the giant polypore. Mature or over-mature trees of beech (*Fagus*), oak (*Quercus*), and occasionally robinia may be attacked.

CONTROL There is no control available, and prompt inspection by a professional arboriculturist is advisable, as trees may be rendered dangerous. Removal of the fungal growth will not prevent further damage to the tree but will prevent spores being spread to other potential hosts.

• see also GRIFOLA FRONDOSA p.143.

Mice

Several species of mice, such as *Apodemus* and *Mus* species, can cause problems in gardens. They dig holes where peas, beans, and sweetcorn have been sown, or where crocus corms have been planted. Green shoots are left lying on the soil surface but the corms and seeds are eaten. They sometimes also bite off strawberries and other fruits before they are ripe, leaving the fruits among the plants. Mice also cause damage in the autumn to spring period, when they come indoors and feed on stored fruits, vegetables, and seeds.

CONTROL Set mouse traps in places where mice are causing damage. In the garden, put traps under the cover of logs or bricks, away from birds and pets. After planting crocuses, press the ground down firmly to make

it harder for mice to locate the corms.

Michaelmas Daisy Mite

SYMPTOMS Infested plants are stunted and have a brownish scarring on the stems. The flowers are reduced in size and number, with many of them being converted into rosettes of small green leaves (*see p.63*). *Symphyotrichum novi-belgii* cultivars are particularly susceptible.

CAUSE Microscopic sap-sucking mites, *Phytonemus pallidus*, which live at the tips of shoots and in the flower buds.

CONTROL None of the pesticides available to home gardeners is effective. Infested plants should be scrapped, as their condition will not improve. Other forms of Michaelmas daisy, such as *Symphyotrichum novae-angliae* and *Aster amellus*, are less susceptible and should flower satisfactorily.

Michaelmas Daisy Wilt

SYMPTOMS The foliage wilts and later withers. The plant is usually affected in sections. Dark brown discoloration of the vascular tissue (conducting elements) within the stem may be apparent if the bark is peeled back. The whole plant is usually killed.

CAUSE The fungus *Phialophora asteris*, which is soil-borne. It may enter via damaged roots, or via pruning tools that have been used on an infected plant and then on a healthy plant without disinfecting them.

CONTROL No control is available. Remove infected plants and the soil in the vicinity of their roots. Do not plant Michaelmas daisies on the same site.

Midges

See CHIRONOMID MIDGES *p.126*; GALL MIDGES *p.139*.

Millipedes

SYMPTOMS Millipedes feed mainly on rotting organic matter, but will sometimes damage seedlings and other soft plant tissues, such as strawberry fruits. They can also extend damage initiated by other pests, such as slugs on potato tubers. Millipedes have long, segmented bodies, up to 40mm long, which may be cylindrical or flat. Each body segment has two pairs of legs, unlike the superficially similar predatory animals called centipedes (*see p.125*), which have one pair of legs per segment.

CAUSE Many species can be found in garden soils. The one most often associated with plant damage is the spotted snake millipede, *Blaniulus guttulatus* (*see p.20*). It is cream coloured, with a row of red dots on either side of a cylindrical body up to 25mm long.

CONTROL Protect strawberry fruits by using straw to lift the berries clear of the soil. Damage to potato tubers is generally initiated by slugs, so it is necessary to control these instead (*see p.183*). Millipedes are encouraged by organic matter in the soil, so avoid adding more compost if millipedes are numerous enough to cause problems. They are not controlled by any of the pesticides available to home gardeners.

Mint Beetles

SYMPTOMS Holes are eaten in the foliage of wild and cultivated mints in early to late summer. Emerald green or dark blue beetles, 7–9mm long, and their matt-black grubs are present on the plants.

CAUSE The green mint beetle, *Chrysolina herbacea* (*see p.47*) and the blue mint beetle, *C. coerulans*, which has become established in Britain since 2011. Both have one generation a year and overwinter in leaf litter as adult beetles.

CONTROL Adult beetles are easy to see and it is often enough simply to pick them off by hand. If plants are heavily infested, spray with an organic or contact insecticide approved for use on edible herbs (*see p.95*). Do not spray watermint due to the danger to aquatic life. The larval stage is more susceptible to pesticides than the adult beetles.

Mint Rust

SYMPTOMS Leaves and stems become flecked with yellow and distorted. The stem may be conspicuously curled, often shortly after growth has begun in spring. Numerous orange-grey, cup-like fungal bodies develop on the lower leaf surface and the affected areas of stem. These become yellowish-orange and later blackish-brown. Mint is the most common host, but several other herbs, including marjoram and savory, may also be affected.

CAUSE The fungus *Puccinia menthae*, which survives the winter on the remains of the plants or as resistant, resting spores in the soil.

CONTROL It may be possible to kill the fungus by burning off all of the top growth of mint plants and scorching the soil surface at the end of the season. This is, however, difficult to do safely and reliably; it may be more advisable to dispose of all infected plants, replanting fresh stock on a new site.

Mites

Mites are allied to spiders and most have four pairs of legs, although gall mites have only two pairs. All the plant pest species of mites are less than 1mm in length. They feed by sucking sap from their hosts. Gall mites secrete chemicals as they feed, causing the plant to produce abnormal growths that enclose the mites. Most mites capable of damaging plants belong to the red spider mite group. These cause fine, pale mottling of foliage and this can lead to defoliation. Heavily infested plants may become festooned with silk webbing. Tarsonemid mites infest shoot tips and flower buds, causing stunted and distorted growth. Many mites are harmless, since they feed on fungal growth on rotting plants or on algae (*see* oribatid mites, *p.162*). Some mites are predators of pest insects or of other mites, and can be used as biological controls in greenhouses. Examples are *Phytoseiulus persimilis* for the control of glasshouse red spider mite, *Hypoaspis miles* for fungus gnats, and *Amblyseius* species for western flower thrips.

• *see also* **Gall mites** – ACER GALL MITES *p.106*; BIG BUD MITES *p.115*; BROOM GALL MITE *p.120*; ELM GALL MITE *p.134*; FELT GALL MITES *p.135*; FUCHSIA GALL MITE *p.138*; LIME NAIL GALL MITE *p.154*; MOUNTAIN ASH BLISTER MITE *right*, PEAR LEAF BLISTER MITE *p.164*; PLUM LEAF GALL MITE *p.169*; RASPBERRY LEAF AND BUD MITE *p.175*; RED BERRY MITE *p.176*; VINE ERINOSE MITE *p.193*; WALNUT BLISTER MITE *p.194*.
Red spider mites – BEECH RED SPIDER MITE *p.114*; BOX RED SPIDER MITE *p.118* BRYOBIA MITES *p.121*; CONIFER RED SPIDER MITE *p.128*; FRUIT TREE RED SPIDER MITE *p.138*; GLASSHOUSE RED SPIDER MITE *p.141*.
Tarsonemid mites – BULB SCALE MITE *p.121*; MICHAELMAS DAISY MITE *p.156*; STRAWBERRY MITE *p.187*; TARSONEMID MITES *p.188*.

Moles

Moles, *Talpa europaea*, are infrequently seen, as they spend most of their lives below soil level. They create a system of underground tunnels and chambers in which they live and feed on earthworms and other soil-dwelling invertebrates. The soil excavated during tunnel construction is deposited in heaps on the surface of lawns and flowerbeds (*see* **p.67**) All of the mole activity in a small garden could be the work of a single animal. Moles also make shallow tunnels above which the soil is pushed up as a ridge.
CONTROL A variety of controls and deterrents is available. The cheapest and often most effective method is to use a mole trap. The position of a mole's tunnel is located by probing the soil around a recent molehill with a stick. When found, open up the tunnel with a trowel at least 30cm away from the molehill. Insert the trap, taking care to align it with the direction and depth of the tunnel. Cover the trap with turf or an inverted bucket to stop light entering the tunnel. If the mole keeps pushing soil into the trap, reset it elsewhere in the tunnel system. An alternative to trapping is an electronic device. These emit noises said to drive moles away, but they give variable results.

Moths

Moths, like butterflies, are characterized by having wings that are covered by many small, overlapping scales. These scales are coloured, and their arrangement on the wings gives the distinctive colour patterns by which the different moth species can be recognized. Moths vary considerably in size; some of the smallest, like some of those with leaf-mining larvae, have a wing-span of about 5mm, while the biggest hawk moths have a wing-span of up to 80mm. In all cases it is the caterpillars that cause damage to plants, rather than the adults, which take nectar from flowers. Most moths are not pests, as their caterpillars do not feed on cultivated plants, but those listed below can cause damage to garden plants.

• *see also* APPLE LEAF MINER *p.109*; BOX TREE MOTH *p.118* BROWN-TAIL MOTH *p.120*; BUFF-TIP MOTH *p.121*; CABBAGE MOTH *p.122*; CARNATION TORTRIX MOTH *p.123*; CHINA MARK MOTH *p.126*; CODLING MOTH *p.127*; COTONEASTER WEBBER MOTH *p.128*; CURRANT CLEARWING MOTH *p.130*; DELPHINIUM MOTH *p.131*; ELEPHANT HAWK MOTH *p.134*; GYPSY MOTH *p.143*; HORSE CHESTNUT LEAF-MINING MOTH *p.146*; JUNIPER WEBBER MOTH *p.149*; LACKEY MOTH *p.150*; LEEK MOTH *p.152*; LEOPARD MOTH *p.153*; LILAC LEAF MINER *p.153*; MULLEIN MOTH *p.157*; OAK PROCESSIONARY MOTH *p.160*; PEA MOTH *p.163*; PLUM MOTH *p.169*; PYRACANTHA LEAF MINER *p.173*; SMALL ERMINE MOTHS *p.184*; SWIFT MOTHS *p.188*; TOMATO MOTH *p.190*; TORTRIX MOTHS *p.190*; VAPOURER MOTH *p.192*; WINTER MOTH *p.198*; YELLOW-TAIL MOTH *p.199*.

Mountain Ash Blister Mite

SYMPTOMS The foliage of rowan or mountain ash and other *Sorbus* species develop a whitish-green blistered appearance during late spring or early summer. Later in the summer, the pale areas darken and become brown. Sometimes nearly every leaf is affected, and there may be some early leaf fall.
CAUSE Microscopic gall mites, *Eriophyes pyri* on rowan and *E. arianus* on whitebeam, which live inside the blistered areas of the leaves.
CONTROL None of the pesticides available to home gardeners gives control. Light infestations on small trees can be dealt with by removing affected leaves. Heavier infestations should be tolerated, since despite the unhealthy appearance of the foliage, the mites have little effect on the tree's vigour. The degree of infestation can vary greatly from year to year.

Mulberry Bacterial Blight

SYMPTOMS Small, black, angular spots, each with a yellow "halo", develop all over the leaf surface. As the infection progresses, the entire leaf may become yellowed and, if infection is severe, the tree's overall vigour may be reduced. Roughened areas may develop on the bark of shoots and can result in dieback.
CAUSE The bacterium *Pseudomonas mori*, which overwinters in the stem lesions.
CONTROL No control measures are available. Remove affected leaves and prune out any stems showing lesions.

Mulberry Canker

SYMPTOMS Shoots die back due to the presence of small cankers, which girdle the affected stems (*see* **p.54**). Tiny, reddish-brown pustules develop around the cankered area. Symptoms are usually first noticed in the summer.
CAUSE The fungus *Gibberella baccata*.
CONTROL Prune out cankered shoots and areas of dieback promptly in an attempt to limit spread.
• *see also* NECTRIA CANKER *p.159*.

Mullein Moth

SYMPTOMS Caterpillars up to 50mm long, greyish-white with black and yellow markings, are found during summer on verbascums, figwort (*Scrophularia*), and *Buddleja*. They feed on the foliage and flower spikes.

CAUSE Caterpillars of a moth, *Cucullia verbasci*.

CONTROL The caterpillars feed in exposed positions and can be removed by hand. Where this isn't feasible, spray with a contact insecticide (*see p.94*) before extensive damage is caused.

Mussel Scale

SYMPTOMS This pest lives on the older stems of woody trees and shrubs, including apple (*Malus*), cotoneaster, box (*Buxus*), dogwoods (*Cornus*), heathers, and ceanothus. The scales are greyish-brown, 2–3mm long, and in shape and appearance resemble the shellfish of the same name. Heavy infestations can occur to the extent that the bark is entirely covered (*see p.56*), and scales may spread onto fruits and berries. Such plants lack vigour and shoots may die back.

CAUSE Sap-sucking insects, *Lepidosaphes ulmi*, which have one generation a year.

CONTROL On apple, ornamental trees and shrubs, spray with a systemic or contact insecticide (*see p.94*) in early summer to kill the more vulnerable, newly-hatched scale nymphs. Avoid spraying plants in flower.

Mycoplasma Infection

SYMPTOMS Symptoms vary from host to host and with the precise identity of the organism. They may resemble those produced by viruses, and may include greening of flowers and yellow markings on leaves, often combined with distortion and stunting.

CAUSE Various organisms, usually described as "virus-like", thought to be similar in structure and activity to both bacteria and viruses.

CONTROL None is available. Remove any affected plants promptly.

N

Narcissus Basal Rot

SYMPTOMS Infection occurs at the basal plate (the flattened area at the base of the bulb), causing it to discolour and rot. Discoloration and rotting spread upwards through the inner scales, turning them chocolate-brown. A pale pink fungal growth may develop on the basal plate and among the affected scales. Bulbs in store turn brown all over, dry out, and become mummified. Those in the ground usually rot off completely, and by the end of the season there may be no trace of them. Leaf symptoms on growing bulbs include yellowing, stunting, and sparse foliage.

CAUSE The fungus *Fusarium oxysporum* f. sp. *narcissi*, which is soil-borne. It persists in soil on infected bulbs that are not sufficiently affected for the symptoms to have been noticed. The disease is most damaging during or following very hot summer weather.

CONTROL None available. Early planting in autumn, may reduce the risk of infection developing.

Narcissus Bulb Flies

SYMPTOMS Bulbs fail to grow or produce only a few thin leaves in the spring. If the bulb is cut in half vertically, it can be seen that the centre has been eaten away and filled with a fly larva's muddy excrement (*see p.20*). The plump maggot is a dirty cream colour and is up to 16mm long. There is usually only a single maggot in each bulb. Several maggots, each up to 8mm long, indicate that the bulb has been colonized by small bulb fly larvae (*Eumerus* species). These flies lay eggs on bulbs that are already damaged

by other pests or diseases, whereas narcissus bulb fly attacks sound bulbs.

CAUSE Narcissus bulb fly, a hoverfly (*Merodon equestris*), resembles a small bumblebee and lays eggs on the necks of bulbs in early summer, when the foliage is dying down. It also attacks some other bulbs, including hippeastrum and snowdrops.

CONTROL Pesticide controls are not effective, but cultivation techniques can reduce losses. The adult flies prefer warm, sunny places, so plant narcissus bulbs in shady or exposed sites. Mound up soil around the stem bases when plants start to die down, to make them less attractive as egg-laying sites. Valuable bulbs can be protected by growing them under insect-proof netting. This needs to be in position from late spring to midsummer. When small bulb fly larvae are found, investigate whether the bulbs are suffering from other pests such as eelworm or narcissus fly, or from fungal diseases.

Narcissus Eelworm

SYMPTOMS Plants produce stunted and distorted foliage and flower stems. Some viruses cause similar symptoms, so the presence of eelworms needs to be confirmed by cutting the bulb in half transversely. Eelworm activity in the bulb causes concentric brown rings or arcs where the pest has damaged the tissues (*see p.20*). In large plantings of daffodils, the pest gradually spreads, so that each year the area of killed and damaged plants gets larger. New bulbs planted as replacements in infested soil quickly succumb to eelworm.

CAUSE Microscopic nematodes (*Ditylenchus dipsaci*) that live within the bulb and foliage.

CONTROL There are no effective pesticides available

to home gardeners. Dig up and burn daffodils showing symptoms; others growing within a radius of 1m may be infested and should also be removed. Avoid introducing narcissus eelworm into a garden by planting sound bulbs purchased from a reputable supplier. Hot water treatment of bulbs can be attempted, although it is difficult to do this properly without access to equipment that will keep the temperature constant. In late summer, dormant bulbs must be immersed for three hours in water heated to 44.5°C (112°F), after which they are cooled and planted in fresh soil. Too much heat will damage the bulbs, while too little allows eelworms to survive.

Narcissus Leaf Scorch

SYMPTOMS As the leaves emerge, they become red-brown and scorched, and the symptoms then spread down the leaves (*see p.44*). Brown spots appear on the foliage and numerous tiny, pinprick-sized, raised fungal bodies may develop on damaged areas. Occasionally flower stems and spathes may also be attacked. The bulbs do not show rotting or discoloration. In addition to daffodils and narcissi, amaryllis, crinums, hippeastrums, nerines, snowdrops, and sprekelias may be infected.

CAUSE The fungus *Didymella curtisii* (syn. *Stagonospora curtisii*), which may survive in the dry, papery outer scales of bulbs.

CONTROL Remove and dispose of infected leaf tips.

Narcissus Smoulder

SYMPTOMS Infected bulbs produce deformed, chlorotic shoots with blackish-brown, withered leaf tips. Only a few leaves may be affected. In wet weather conditions they may become covered with masses of grey spores. Infected leaves may rot off, and the deterioration then spreads down into the centre of the bulb. Numerous tiny, black sclerotia (fungal resting bodies) develop on the outer scales of the bulb and on infected leaves and debris.

CAUSE The fungus *Sclerotinia narcissicola*, which may persist in the soil as sclerotia on infected bulbs and debris.

CONTROL As soon as symptoms appear on leaves, cut back affected areas. Inspect bulbs carefully and plant only those which appear perfectly healthy and free from sclerotia.

Narcissus Viruses

SYMPTOMS Many different viruses may infect daffodils and narcissi, alone or in combination. The symptoms depend on the cultivar, the virus, and growing conditions. Faint yellow mottling may occur in mild cases, particularly towards the leaf base. Flowers may show bleached streaks (see *p.38*), and occasionally yellow streaking on the flower stem. Certain viruses, in particular narcissus yellow stripe virus, cause bright yellow leaf stripes, often combined with distortion and stunting, and the plant may fail to flower.

CAUSE Various viruses, transmitted by aphids, soil-living eelworms or handling. The vector varies depending on the virus.

CONTROL Mild symptoms are quite common and may never develop seriously. All infected plants are, however, a source of infection, so disposal is usually the best option, particularly for those showing symptoms of narcissus yellow stripe virus.

Nectar Robbing

Many flowers produce nectar in order to attract bees and other pollinating insects. As these insects push their mouthparts into the nectaries at the base of a flower, their bodies become dusted with pollen from the anthers. When they visit other flowers of the same type, some of this pollen rubs off onto the stigma and pollination is achieved. Sometimes, however, bees cheat by biting a hole in the base of the flower (see *p.65*), enabling them to reach the nectar without touching the anthers and stigma. It is mostly short-tongued bumblebees that do this when they cannot reach the nectaries from the front of the flower. Once a short-cut to the nectar exists, other insects will take advantage. Nectar robbing is frequently encountered on runner beans and may contribute to poor flower set. It is not possible to prevent nectar robbing, but flower set is usually adequate if beans are kept well-watered and do not suffer from high temperatures.

Nectria Canker

SYMPTOMS Woody plants show roughened, flaked areas of bark on their branches, twigs, and stems. The cankers enlarge and may girdle the stem, causing dieback. During the summer, white fungal pustules may develop on cankered areas; then in winter, these are replaced by numerous small, red fruiting bodies.

CAUSE Various species of the fungus *Nectria*.

CONTROL Prune out any affected areas immediately after symptoms develop.

Needle Blights and Casts of Pines

SYMPTOMS On young pines, the needles discolour, often showing diffuse brown or black lines. Infected needles may be shed prematurely. Damage to mature trees is insignificant, but young pines may be weakened. Another blight may cause the development of distinct bands of discoloration on the needles.

CAUSE *Lophodermium* species and *Dorthistroma pini.*

CONTROL None available. Rake up and dispose of infected needles as they fall.

Needle Rusts of Conifers

SYMPTOMS On spruce (*Picea*), needles become discoloured. If the fungus *Chrysomyxa abietis* has attacked them, they may develop orange fungal pustules but will remain attached to the tree through winter; or, if attacked by *Chrysomyxa rhododendri*, they may become covered with orange pustules, followed by white spore tendrils, and fall from the tree in autumn. This latter fungus is less common and spends part of its life cycle on rhododendrons. Norway and Sitka spruce are most frequently affected, but most spruces can be attacked. On pines another fungus causes small yellowish spots on the needles, and resin-like exudations may occur. Later, off-white, column-like growths develop on the needles and erupt to produce bright orange spores. This fungus may also attack various herbaceous hosts, including campanulas and some weeds, in particular sow thistle and groundsel.

CAUSE The fungi *Chrysomyxa abietis* and *C. rhododendri* on spruce; the fungus *Coleosporium tussilaginis* on pine and other plants.

CONTROL None available. Dispose of infected needles.

Nematodes

See EELWORMS *p.134.*

Nitrogen Deficiency

SYMPTOMS Leaves are small and chlorotic, and general growth is reduced. On certain plants, red or purple leaf discoloration may develop as the green pigment levels drop. The lower, older leaves are affected first, but if the deficiency progresses, all parts of the plant may be similarly affected. Flowering, fruiting, and tuber or root formation are also reduced.

CAUSE A deficiency of nitrogen, which can develop in any soil, particularly if it is light and has a low organic matter content, or if it has been heavily cropped. The use of large quantities of wood chippings may also result in nitrogen deficiency, as nitrogen is removed from the soil as the lignin in the wood is broken down.

CONTROL Apply nitrogen-rich fertilizers, manures, composts, and mulches to the soil regularly. Grow legumes, which are capable of fixing nitrogen (see *p.77*) using bacteria in their root nodules.

Nut Weevil

SYMPTOMS In late summer, hazel and cob nuts have a round hole, l–2mm in diameter, in their shells (see *p.18*) where beetle grubs have bored their way out before pupating in the soil. The kernel inside the nut shell is eaten by a white grub with a pale brown head.

CAUSE A brown weevil, *Curculio nucum*, which is about 10mm long. Almost half the body length is due to a slender snout or rostrum on the front of the head.

CONTROL Usually only a small proportion of the crop is attacked, so routine control measures are not required. This is fortunate as hazel is often too tall to be sprayed effectively.

O

Oak Gall Wasps

SYMPTOMS English oaks (*Quercus robur* and *Q. petraea*) are the host plant of more than 30 species of gall wasps. Many of these have complex life cycles, involving alternating sexual and asexual (female only) generations, which create different forms of gall (*see p.30*) on different parts of the tree. Some examples are the oak apple gall wasp and the common spangle gall wasp. The former lays eggs in buds in the winter and this causes them to develop into pithy, creamy pink, bun-shaped objects in the spring. The adults that emerge in midsummer go into the soil and lay eggs on the roots, giving rise to spherical swellings that contain the wasp's grubs. When adults emerge they crawl up the tree and complete the two-generation cycle by laying eggs in the buds. The common spangle gall wasp creates flat, yellowish-brown, disc-like galls (*see p.30*) on the underside of oak leaves in late summer. The next generation causes spherical yellowish-red galls on the male catkins in the spring, with adults emerging to lay eggs in the foliage in midsummer. With all gall wasps it is the larval stage that secretes chemicals that induce the plant to produce gall tissues. **CAUSE** Oak apple gall wasp, *Biorhiza pallida*, and common spangle gall wasp, *Neuroterus quercusbaccarum*. Other gall wasps frequently encountered on oaks are the marble gall wasp, *Andricus kollari*, which makes woody spherical galls that measure up to 25mm across on the stems, and the artichoke gall wasp, *Andricus fecundator*, which makes greatly enlarged buds at the shoot

tips. The silk button gall wasp, *Neuroterus numismalis*, has galls shaped like ring doughnuts, 3mm across, on the underside of leaves in late summer (*see p.30*). The cherry gall wasp, *Cynips quercusfolii*, produces reddish-yellow spherical galls measuring up to 18mm across under leaves in late summer.
CONTROL Control measures are unnecessary. None of the gall wasps has any serious effect on the growth of oak trees, even when numerous.
• see also ACORN GALL WASP *p.106*.

Oak Phylloxera

SYMPTOMS In late spring or early summer, yellow spots develop on the upper surface of oak (*Quercus*) leaves. Underneath these spots on the lower leaf surface are orange aphid-like insects that measure up to 1mm in length. The females are often surrounded by a ring of yellow eggs. Later in the summer, the spots on the leaves tend to join together and damaged parts of the leaf dry up and become whitish-brown.
CAUSE A sap-sucking insect, *Phylloxera glabra*, which is a specific oak pest and over-winters on the tree as eggs. The heaviest infestations tend to be on saplings rather than mature trees.
CONTROL Encourage good growth by watering young trees in dry summers. On small trees, phylloxera can be controlled by spraying with a systemic or contact insecticide (*see p.106*) in early summer if damage is seen.

Oak Powdery Mildew

SYMPTOMS Tiny, almost imperceptible cinnamon-coloured spots develop on young leaves (*see p.39*) from late spring onwards. These are soon followed by a layer of white fungal spores and mycelium.

Severely affected leaves may be killed, but the overall vigour of the tree is rarely affected.
CAUSE The fungus *Microsphaera alphitoides*, which overwinters in buds.
CONTROL Only really necessary on very young trees, which are more likely to suffer serious effects. Keep the plants well watered and mulched, avoiding overhead irrigation.

Oak Processionary Moth

SYMPTOMS Oaks, *Quercus* species, can be defoliated during early summer by large numbers of gregarious caterpillars. Young caterpillars are brown but they become grey as they mature. Their bodies are covered in long white hairs. When not feeding, the caterpillars rest in dense clusters under silk webbing on the trunk or branches.
CAUSE Caterpillars of a moth, *Thaumetopoea processionea*, which is a native of central and southern Europe but has spread more widely in recent years. It became established in the London area in 2006. Efforts are being made to prevent it spreading further. In addition to devouring the leaves of oaks, the caterpillars are a health risk. Hairs on the caterpillars' bodies readily break off and have an irritating effect on skin, eyes, and mouths. The adult moths lay their eggs in late summer to early autumn and they hatch the following spring. The caterpillars are present during late spring to early summer, when they pupate under silk webbing on the bark.
CONTROL At the present time in Britain, infestations of this pest outside of west London should be reported to the Forestry Commission, which will advise on the necessary control measures. Avoid close contact with the caterpillars because of their irritating hairs.

Oedema

SYMPTOMS Raised, wart-like outgrowths develop, most frequently on the lower leaf surfaces (*see p.33*). These are the same colour as the leaf initially but may later become corky and brown. Affected leaves may sometimes be distorted but do not always die.
CAUSE The plant takes up more water than it can lose through the leaves by the process of transpiration. As a result, small groups of leaf cells swell up as they become over-filled with water, producing pale green warty growths. If the growing conditions do not improve, cells rupture and die off and so turn brown and corky. Oedema usually develops when humidity is high and water levels around the roots are excessive.
CONTROL Do not remove leaves showing the symptoms. This disorder is not infectious and removal of leaves further reduces the rate of moisture loss and so worsens the problem. Reduce watering and/or improve drainage. Increase air circulation around the plants by spacing them further apart and improving ventilation, as appropriate.

Oleander Scale

SYMPTOMS Flat, round, whitish-brown objects up to 2mm across develop on the stems and foliage of many greenhouse plants. These include oleander, jasmine, asparagus ferns, azalea, palms, and ivies; it also occasionally occurs on spotted laurel (*Aucuba japonica*) growing in sheltered places in gardens. Heavy attacks cause loss of vigour and dieback.
CAUSE A sap-sucking insect, *Aspidiotus nerii*, which breeds continuously throughout the year in greenhouses. Eggs are laid under the protection of the females' bodies and shell-like

scales. No honeydew is excreted, so there is no sooty mould on the foliage.
CONTROL Prune out heavily infested shoots or leaves. Spray with a systemic or contact insecticide (*see p.94*), particularly on the lower leaf surface. Two or three applications at fortnightly intervals may be needed to control the young nymphs as they hatch from the eggs.

Onion Bolting

SYMPTOMS Onions produce flower stems before reaching the normal stage of maturity, so the bulb size is smaller than it should be.
CAUSE Exposure to cold temperatures earlier in the year promotes early flowering. This is usually brought about by a late, cold spring and is common on early cultivars. Very hot, dry conditions and day length may also induce bolting.
CONTROL Little can be done, but the incidence will be reduced by use of late cultivars, which are less likely to be subjected to early cold weather.

Onion Downy Mildew

SYMPTOMS Affected leaves turn grey, wither, and collapse, often with a "boiled" appearance. Close inspection may reveal dense, slightly fluffy, off-white fungus-like growth. However, secondary infections, particularly of the fungus *Alternaria*, may rapidly mask this. If left on the plant, the rotting of the leaves may spread down into the bulb and so cause further losses. The earlier in the season that the disease appears, the greater the reduction in bulb size.
CAUSE The fungus-like *Peronospora destructor*, which is particularly common in mild, wet weather and is encouraged by humid growing conditions.

CONTROL Remove infected leaves promptly. Improve air circulation around the plants by increased spacing and regular weeding. Clear up and dispose of all debris at the end of the season. Avoid overhead watering. Consider growing onions showing some resistance, such as 'Hylander' and 'Santero'.

Onion Eelworm

SYMPTOMS Young onion plants are stunted and abnormally swollen (*see p.18*). The tissues have a soft mealy texture and readily succumb to secondary rots, so the plants usually die before reaching maturity. Onions attacked later in the summer can produce bulbs, but these will also be infested and will rot in store.
CAUSE Microscopic worm-like eelworms or nematodes, *Ditylenchus dipsaci*, that infest the foliage, bulbs, and seeds of onion and allied plants. It can also attack some other vegetables and common weeds, without necessarily causing obvious symptoms, but these alternative host plants allow onion eelworm to survive when onions are not being grown.
CONTROL There are no effective pesticides available to home gardeners, so remove infested plants as soon as they are spotted. Buy onion sets and seeds from reputable suppliers in order to reduce the risk of introducing the pest into a garden. When onion eelworm has occurred, crop rotation can be effective in eliminating it. Good weed control and growing vegetables not affected by onion eelworm, such as lettuce, turnip, swede, and any of the brassicas, for one or two years will clear the soil.

Onion Fly

SYMPTOMS Onion is the most susceptible plant but leeks,

shallots, and garlic can also be attacked. In early summer young plants may collapse as the roots are eaten by white maggots up to 8mm long. Larvae of the second generation in late summer feed on roots and also bore into onion bulbs, allowing secondary rots to gain entry.
CAUSE Maggots of a fly, *Delia antiqua*, which is similar to a housefly in appearance.
CONTROL Onions grown from sets are less susceptible to first generation maggots than seedlings. There are no insecticides approved for use against onion fly. Infested plants should be lifted carefully and destroyed before the maggots go into the soil to pupate. Growing onions and other host plants under insect-proof netting will stop female flies laying eggs near the plants.

Onion Neck Rot

SYMPTOMS Scales of infected onions become soft, pale brown, and semi-transparent (*see p.18*). A dense, grey fungal growth develops over the affected areas and the tissues start to dry out, so that the onion becomes mummified. Sclerotia (resilient black fungal resting bodies), which may be several millimetres in length, develop on the affected areas, particularly around the neck end of the bulb.
CAUSE The fungus *Botrytis allii*, which can persist in the form of sclerotia on onion debris or, having fallen from the bulb, loose in the soil. Infections usually spread by spores or by sclerotia. The symptoms are usually first noticed on onions and shallots in store.
CONTROL Buy seed and sets only from reputable sources, as the fungus can easily be brought in on infected material. Do not grow onions on the same site for more than two years in succession. Improve cultural conditions so that hard, well-

ripened bulbs are produced. Do not apply fertilizers after midsummer and avoid excessive use of high nitrogen fertilizers. Keep crops evenly and adequately watered. Grow red or yellow bulbs instead of white, as these are generally less susceptible. Keep onions dry as they are drying off.

Onion Thrips

SYMPTOMS The foliage of onions and leeks develops a fine, white mottling during the summer. Narrow-bodied insects, up to 2mm long, and black or pale yellow in colour, occur on the leaves. Thrips are more troublesome in hot, dry summers and heavy infestations will check the plant's growth.
CAUSE The adults and nymphs of a sap-sucking insect, *Thrips tabaci*, which can also feed on many ornamental plants.
CONTROL Light infestations can be tolerated, but if much of the green colour is being lost before late summer, spray with an organic insecticide (*see p.89*), but this pest is not easily controlled.

Onion White Rot

SYMPTOMS Dense, white, slightly fluffy fungal growth that develops around the base of the bulb and the roots of members of the *Allium* family, particularly onions, shallots, garlic, and leeks (*see p.18*). Numerous small, black sclerotia subsequently develop in among this growth. The sclerotia may fall off and into the soil where they may remain viable for more than seven years and will be stimulated to germinate when alliums are planted on the site again. The foliage on affected bulbs becomes chlorotic and wilts.
CAUSE The fungus *Stromatinia cepivora* (syn. *Sclerotium cepivorum*).

CONTROL There is no chemical control available. Remove and burn infected plants as soon as they are noticed and do not grow onions or related plants on the same site for at least eight years.

Orchid Viruses

SYMPTOMS Various symptoms, depending on the orchid cultivar and the virus or viruses involved. Yellow or very dark brownish-black streaking of the foliage may appear (see *p.24*). Yellow or brownish-black spots or rings may also develop. Flowering may be affected and the plant's overall vigour declines.
CAUSE Various viruses, which may infect alone or in combination. The mode of transmission varies but handling, pruning tools, and sap-feeding insect pests may all be involved.
CONTROL None is available. Remove and dispose of infected plants promptly and control potential means of spread.

Oribatid Mites

Oribatid mites, such as *Humerobates* species, also known as beetle mites, commonly occur on the bark of a variety of trees and shrubs, and also sometimes on fences. They have shiny black or reddish-brown bodies, about 1mm long, and during the day the mites tend to cluster together in crevices where they can be mistaken for eggs. They are often noticed on dead or dying plants, but are not the cause of the plant's decline. Oribatid mites feed on algae on the bark and are mainly active after dark.
CONTROL No control measures are required, as these mites do not harm plants.

Oriental Chestnut Gall Wasp

SYMPTOMS Pale green, hard, swollen galls with a knobbly appearance develop on the leaves, leaf stalks, and buds of sweet chestnut (*Castanea sativa*) in late spring to early summer (see *p.42*). As the galls mature, they may become pinkish-red. Galls vary in size, but are generally 5–20mm long. Each gall contains several gall wasp grubs. Affected leaves are distorted and undersized.
CAUSE A tiny gall wasp, *Dryocosmus kuriphilus*, originating from China, which is now widespread in Europe and was first found in Britain in 2015. It is a specific pest of sweet chestnut and does not affect horse chestnut trees (*Aesculus hippocastanum*). Eggs are laid in midsummer in buds and insects overwinter as small larvae inside the buds, with galls forming around the larvae as the leaves develop in the spring.
CONTROL Sweet chestnut trees are often too tall to be sprayed effectively, so gall wasp infestations have to be tolerated. Heavy attacks may restrict growth, but the tree will survive.

P

Pansy Downy Mildew

SYMPTOMS Purple-brown spots appear on the upper surface of the leaves, often combined with a yellow "halo". These may be confused with the symptoms caused by pansy leaf spot (see below). The lower leaf surface develops small, grey or off-white patches of fuzzy fungus-like growth that correspond with the spots. Affected leaves become chlorotic and may wither and die off. In severe cases the whole plant may become so weakened that it dies.
CAUSE The fungus-like *Peronospora violae*, which is encouraged by humidity and moist growing conditions.
CONTROL Remove affected leaves.

Pansy Leaf Spot

SYMPTOMS Brown, purple-brown, or very pale buff-coloured spots develop on the foliage (see *p.26*). The older leaves are affected first and most severely. If the affected leaf tissue is killed, the whole area may fall away, producing shothole symptoms.
CAUSE Various fungi, including species of *Ramularia*, *Cercospora* and *Phyllosticta violae*.
CONTROL Rarely necessary; remove any affected leaves.

Pansy Powdery Mildew

SYMPTOMS Powdery, white fungal growth develops on the upper leaf surface. This may cause distortion and leaf yellowing (chlorosis), occasionally followed by leaf death.
CAUSE *Oidium* species, which may be more troublesome if soil conditions are dry and the plants are growing in overcrowded conditions.
CONTROL Ensure that the soil or compost is adequately watered during dry weather, but avoid overhead watering. Pick off severely affected leaves.

Pansy Sickness

SYMPTOMS The whole plant wilts and does not recover, even if watered. The foliage yellows and dies off and the plant is killed. The symptoms may take several weeks to develop fully.
CAUSE The soil-borne *Pythium violae*, and, occasionally, other organisms associated with root and foot rots. *Pythium* builds up in the soil, particularly when pansies or violas are grown in the same site for several years in succession. The fungus seems to be able to survive in soil for several years, even in the absence of suitable host plants.
CONTROL Remove and dispose of affected plants. Do not incorporate them into the compost heap. Plant new stocks of pansies or violas on a fresh site.

Papery Bark

SYMPTOMS The bark peels off as a thin, papery, brown sheet (see *p.56*). All woody parts of the tree may be affected but small- to medium-sized shoots more frequently show these symptoms than do large branches or tree trunks. Shoots may also die back.
CAUSE Problems associated with inadequate functioning of the root system of the tree, most commonly waterlogging.
CONTROL Improve growing conditions, in particular soil drainage. Prune off all dead stems and those showing the symptoms. Prune back to a healthy bud growing from perfectly healthy-looking wood. The removal of the dead wood is necessary not because the problem is infectious, but because once it is dead or injured, the wood is prone to invasion by other organisms which may subsequently cause further dieback.

Parsnip Canker

SYMPTOMS Roughened cankered areas develop on the root, usually reddish-brown, orange-brown or black in colour (see *p.23*). The shoulder of the root is the area most commonly affected, but lesions may occur anywhere.
CAUSE The fungus *Itersonilia pastinacae* or, less commonly, *Mycocentrospora aceria*. The fungus may be spread into the soil from leaf spots and

usually infects via injured fine roots. Other sources of injury, such as that caused by carrot fly, may also provide the initial point of entry.
CONTROL No control is available. Reduce the risk of infection by growing resistant cultivars such as 'Archer', 'Avonresister', and 'Cobham Improved Marrow'. Improve soil drainage and avoid or control any source of injury to the roots. Sowing late (during late spring), in closely spaced rows produces parsnips with smaller roots which seem to be less susceptible than larger-rooted ones.

Parsnip Viruses

SYMPTOMS Yellow markings on the leaves, usually as bands adjacent to the leaves or mosaic-like patterns or flecking. The foliage may be slightly stunted and distorted and overall growth retarded.
CAUSE Various viruses, in particular parsnip yellow fleck virus, which is spread by aphids.
CONTROL Remove infected plants promptly to prevent further spread of the infection. Try to control aphids to decrease the risk of infection.

Pea and Bean Weevil

SYMPTOMS Peas and broad beans have U-shaped notches eaten in their leaf margins (see *p.47*) by greyish-brown beetles that are 3–4mm long.
CAUSE Adult beetles, *Sitona lineatus;* the larval stage lives in the soil and feeds on the nitrogen-fixing nodules on the roots of peas and beans.
CONTROL Normally only a small proportion of the foliage is eaten, and plants can usually tolerate the damage. Control measures are only required where small plants are being

heavily attacked. In such cases, spray plants with a contact insecticide (see *p.94*), but protect pollinators by not spraying while broad beans are in flower.

Pea Downy Mildew

SYMPTOMS The foliage develops yellow blotches of discoloration, each of which corresponds with patches of fluffy, off-white fungal-like growth on the lower surface. Occasionally, yellowish patches of discoloration may also develop on the pods, but this symptom is less common. Infection early in the season may cause serious stunting and little, if any, pod set.
CAUSE The fungus-like *Peronospora viciae*, which is only likely to cause serious damage in wet seasons.
CONTROL Remove any signs of early infection. The pathogen forms spore stages that can persist in the soil, so it is important to remove all debris at the end of the season and grow peas on a fresh site. Grow resistant cultivars such as 'Hurst Green Shaft' and 'Kelvedon Wonder'.

Pea Leaf and Pod Spot

SYMPTOMS Brown or yellow, often sunken spots develop on the leaves (see *p.26*), stems, pods (see *p.17*), and flower stems. Numerous tiny, pinprick-sized, raised fungal fruiting bodies may develop on the lesions.
CAUSE Various fungi, including *Ascochyta pisi* and *A. pinodes*, which most frequently attack fully grown peas. They may also occasionally attack seedlings, which may be killed. The fungus is persistent and carries over from year to year in infected plant debris and, if seeds are collected from infected pods, the resultant seedlings will succumb, too.

CONTROL Remove diseased plant material at the end of the season. Plant fresh seed on a new site. Do not save seed from infected plants.

Pea Marsh Spot

SYMPTOMS Circular brown areas of discoloration develop within the seed. A brown cavity later develops. This spoils the eating quality of the crop and also renders them useless for saving as seed.
CAUSE A deficiency of manganese. This is most likely to develop on poorly drained soils, particularly those which are alkaline and have a high organic matter content.
CONTROL Improve soil conditions and apply sulphur to lower the pH.

Pea Moth

SYMPTOMS Caterpillars, measuring up to 6mm in length, live inside pea pods and feed on the developing seeds (see *p.17*). The caterpillars have black heads and small dark spots on creamy white bodies.
CAUSE Caterpillars of a moth, *Cydia nigricana*. Eggs are laid on peas that are in flower, so early or late sown peas that flower outside the moth's flight period from early to late summer will escape damage.
CONTROL Use an insect-proof mesh to avoid the need for pesticide control. If a contact insecticide is required, apply it when most flowers have gone over and pods are starting to form, using a spray suitable for use on peas (see *p.95*).

Pea Powdery Mildew

SYMPTOMS Powdery white fungal growth develops on the leaves and occasionally the leaf stalks and the stems. Leaves may become slightly distorted and yellowed.

CAUSE The fungus *Erysiphe pisi*, which is encouraged by dry conditions at the plant roots, and by poor air circulation around the foliage.
CONTROL Control is only necessary if the infection occurs early in the season. More usually it develops towards the end of summer and so has little if any effect on productivity. Grow resistant cultivars including 'Kelvedon Wonder' and 'Oregon Sugar Pod'.

Pea Thrips

SYMPTOMS Pea pods are discoloured with a silvery or brownish scarring and the pods may contain only a few seeds at the stalk end. The foliage is also discoloured. Adult thrips are black and have narrow elongate bodies 2mm long. The immature nymphs have a similar appearance but are pale yellow.
CAUSE Sap-sucking insects, *Kakothrips pisivorus*, which can be troublesome in hot, dry summers.
CONTROL Watch for signs of thrips' activity as the pods develop and, if necessary, spray with a contact insecticide approved for peas (see *p.95*).

Pea Viruses

SYMPTOMS Yellow mosaicing, streaking, spotting, or flecking occurs, mainly on the leaves; occasionally symptoms may also be seen on the stems. Plants are stunted and distorted and may fail to flower or may flower but fail to set much of a crop. Pods formed may be discoloured and distorted.
CAUSE Various viruses, including pea enation virus, pea mosaic virus, and bean leaf roll virus, the first being spread by several different species of aphid.
CONTROL Remove and burn affected plants promptly in

order to minimize the risk of the infection spreading. Keep pests such as aphids under control. Avoid handling healthy-looking plants after handling those that are showing symptoms. If necessary, scrub your hands thoroughly. Do not save seed. Grow the cultivars 'Ambassador' and 'Sugar Pod', which show resistance to pea enation virus.

Pea Wilt

SYMPTOMS Rapid wilting of the plant, usually in association with yellowing and dieback of the leaves (*see p.43*). Infection can occur on plants of any age, but is most common towards the middle of the season. Dark brown discoloration of the vascular tissue (the conducting elements) in the stem develops but is only visible if the outer parts of the stem are gently stripped away.
CAUSE The fungus *Fusarium oxysporum* f. sp. *pisi*, which is soil-borne and so particularly troublesome if peas or related plants are grown on the same site for several years in succession.
CONTROL No control is available. Remove infected plants and grow any replacements or future crops on a fresh site. Grow cultivars showing a degree of resistance, such as 'Greenshaft', 'Kelvedon Wonder', 'Sugar Snap', 'Sugarbon', and 'Onward'.

Peach Aphids

SYMPTOMS Peach, nectarine, and almond foliage becomes crinkled and curled, with a yellowish-green colour, during late spring to early summer. Green or black aphids suck sap from the underside of the leaves, although they may die out later in the summer, leaving only whitish cast skins.
CAUSE Several aphids occur on peach and related plants. The more important ones are the peach potato aphid, *Myzus persicae*, which is yellowish-green in colour, and a black species, the peach aphid, *Brachycaudus schwartzi*.
CONTROL Look for aphids on the foliage from mid-spring onwards and spray with a contact insecticide approved for use on the crop (*see p.95*), if they are seen.

Peach Leaf Curl

SYMPTOMS Leaves may become infected before or as soon as they unfurl in the spring. They develop symptoms of puckering and blistering (*see p.40*), being pale green at first and soon turning bright red or purple. A white, powdery spore layer then develops over the leaf surface. Affected leaves drop prematurely. A second flush of foliage is usually produced later in the season and this almost invariably remains healthy. The overall vigour and cropping of the tree is only likely to be affected if the infection occurs several years in succession. On nectarines, but not on peaches, the smooth skin on the fruit may develop slightly raised, rough patches as a result of infection spreading to the fruits. A very similar disease may affect poplars.
CAUSE The fungus *Taphrina deformans*, which produces spores that are carried by wind and rain, and lodge in cracks and crevices both in the bark and in bud scales. This overwintering allows them to cause such early infections. Nectarines, almonds, and apricots may also be affected.
CONTROL Remove infected leaves promptly, preferably before the spore layer develops. Keep the tree well watered and fed to encourage the development of plenty of new replacement growth. Peaches and nectarines that grow under glass are rarely attacked because the glass acts as a shelter, preventing the spores from landing. Temporary protection may work well – this can be provided by the erection of an open-sided polythene shelter over fantrained trees. To be effective, the shelter must be in position from mid-winter until late spring.

Pear and Apple Canker

See APPLE AND PEAR CANKER *p.108.*

Pear and Cherry Slugworm

SYMPTOMS Pale yellow, caterpillar-like larvae, which secrete a black, slimy mucilage over their bodies, giving them a slug-like appearance. They feed on pear and cherry foliage between late spring and mid-autumn. Other plants attacked include hawthorn, Japanese quince, mountain ash, and plums. The larvae can be up to 10mm long and are swollen at the front end, making them club-shaped. They feed by grazing away the upper leaf surface, causing the remaining damaged tissues to dry up and turn brown (*see p.32*).
CAUSE Larvae of a sawfly, *Caliroa cerasi*, which has two or three generations during the summer.
CONTROL On trees small enough to be sprayed, slugworms are easy to control with most contact insecticides (*see p.94*).

Pear Bedstraw Aphid

SYMPTOMS Leaves at the shoot tips on pear become yellowish-green and curled during the late spring. Whitish-grey aphids infest the undersides of the leaves and are sometimes present in such large numbers in late spring to early summer that they swarm over the branches and trunk. The aphids excrete a sugary honeydew that can allow sooty moulds to grow on the foliage.
CAUSE A sap-sucking insect, *Dysaphis pyri*, which in midsummer migrates to wild flowers known as bedstraws (*Galium* species).
CONTROL The aphid overwinters on pears as eggs laid near the buds. In the spring, inspect the foliage and spray where necessary with a systemic or contact insecticide (*see p.94*) if aphids are seen. Poor control is likely once the foliage has become curled. A plant oil winter tree wash applied in early winter to control overwintering eggs can reduce the need to spray in spring.

Pear Boron Deficiency

SYMPTOMS Distorted fruits are marked internally with numerous small, brown spots. Most fruits are affected. The fruit's texture is unpleasant because of the patches of dead, brown cells. Some stem dieback may occur and overall vigour may be reduced. The symptoms may be confused with those of the far more serious pear stony pit virus (*see p.165*).
CAUSE A deficiency of boron.
CONTROL *See* BORON DEFICIENCY, *p.117.*

Pear Leaf Blister Mite

SYMPTOMS In mid- to late spring, the foliage of pears develops blisters or raised blotches that are yellowish-green or pink. On expanded leaves these form broad bands on either side of the midrib (*see p.29*). By midsummer the blisters have darkened and become black. Less than 3mm

in diameter, they should not be confused with the larger black lesions caused by pear scab disease (*see p.165*).
CAUSE Microscopic gall mites, *Eriophyes pyri*, which live within the leaves. The blistered appearance of the foliage is induced by chemicals secreted by the mites as they feed.
CONTROL None of the pesticides currently available to home gardeners gives good control. Fortunately, although blister mite gives a tree an unhealthy and unsightly appearance, it has little impact on the tree's ability to produce fruit. On small trees, affected leaves can be picked off if the infestation is light, otherwise this pest has to be tolerated.

Pear Midge

SYMPTOMS Infested pear fruitlets make rapid initial growth but, a few weeks after petal fall, start to turn black at the eye end of the fruits (*see p.13*). Inside the fruitlets are whitish-orange maggots, which are up to 2mm long. Affected fruitlets become extensively blackened and fall from the tree in late spring and early summer. Sometimes nearly all the potential fruit is lost.
CAUSE Larvae of a small fly or gall midge, *Contarinia pyrivora*, which go into the soil when the fruitlets drop. They pupate inside silk cocoons and emerge as adults the following spring.
CONTROL On small trees, pick off and destroy infested fruitlets before the maggots complete their feeding. Prevent the adult flies from laying eggs by spraying with a contact insecticide (*see p.94*) when the blossom is at the white bud stage – when the petal colour can be seen, but before the flowers open. Do not spray open blossom.

Pear Rust

SYMPTOMS In the spring and summer leaves develop red spots on the upper surface. Each spot is 1–3cm in diameter and may have a bright margin, with the spot darkening with age. Occasionally cankered areas may develop on pear stems. Cankers develop on branches of junipers and in spring masses of orange gelatinous spores may be seen on these cankers.
CAUSE *Gymnosporangium sabinae*, a fungus whose alternate host is *Juniperus sabina*.
CONTROL The only control is to remove any juniper bushes in the vicinity.

Pear Scab

SYMPTOMS Blackish-brown scabby patches develop on the fruits (*see p.13*), and similar but more greenish-grey spots develop on the leaves. In the most severe cases, fruit may be almost entirely covered with these scabby patches and is small, and misshapen. Infected fruits may also crack or split and so become infected with secondary organisms such as brown rot. Affected leaves yellow and fall early.
CAUSE The fungus *Venturia pirina*, which overwinters on scabby patches on the young stems and also on fallen infected leaves. The disease tends to be most prevalent on trees that have crowded branches, and during damp seasons.
CONTROL Rake up and dispose of affected leaves as they fall. Prune out cracked and scabby shoots to limit the overwintering of the fungus. Grow cultivars showing some resistance, including 'Beurre Hardy', 'Jargonelle', and 'Catillac'.

Pear Stony Pit Virus

SYMPTOMS The fruits become knobbly and pitted (*see p.13*).

Within the flesh are found many dead sclerenchyma (stone cells), which make the fruits unpleasant to eat. The symptoms of this virus may be confused with boron deficiency. Affected fruits rarely appear all over the tree, but are frequently borne on just one branch.
CAUSE Stony pit virus, which is most common on old trees. The virus particles are only known to be transmitted by grafting.
CONTROL There is no known cure. Avoid cultivars known to be susceptible, including 'Doyenné du Comice', 'Laxton's Superb', 'Winter Nelis', and 'Anjou'.

Pear Sucker

SYMPTOMS Pear leaves become very sticky with honeydew excreted by blackish-green insects up to 2mm long. The immature nymphs have flattened bodies, while the adults resemble winged aphids. Sooty moulds frequently develop where honeydew has accumulated on the upper surface of leaves.
CAUSE A sap-sucking insect or psyllid, *Cacopsylla pyricola*, which has several generations during summer and overwinters as an adult.
CONTROL Pear sucker is less frequent on pears in gardens than in commercial orchards. Routine control measures are not needed, but look for the pest during late spring to late summer and, if necessary, spray with a systemic or contact insecticide (*see p.94*).

Pedicel Necrosis

SYMPTOMS Flower buds develop normally, but before or just after opening, the pedicel (area of flower stem immediately behind the bud) discolours, shrinks inwards, and flops over. The flower bud then

turns brown and fails to open, or opens and dies off.
CAUSE Extremely rapid or soft growth, such as that produced by excessive use of high-nitrogen fertilizers. A sudden check in growth caused, for example, by drought, inadequate fertilizer use, or the wrong amount of potassium, may also be involved. Plants with very large or double flowerheads may be more prone to this than single varieties.
CONTROL Remove damaged flowerheads. Feed the plant with sulphate of potash at about 20g per m² in spring and autumn. Avoid excessive use of high nitrogen fertilizers. Mulch the soil if necessary to improve its ability to retain moisture.

Pelargonium Rust

SYMPTOMS Rings of dark brown fungal pustules develop on the lower leaf surface, often in concentric rings (*see p.28*). Corresponding yellow blotches develop on the upper surface. Occasionally, the pustules appear on the upper surface, too. The infection spreads rapidly and infected leaves become extensively discoloured, wither and die. If the disease spreads unchecked, the plant may be killed.
CAUSE The fungus *Puccinia pelargonii-zonalis*, which produces spores that are spread on air currents and by water splash. Spread of the fungus is encouraged by humid atmospheres, and so is usually most severe during the winter months and on plants growing in crowded conditions under glass.
CONTROL Remove infected leaves. Spray with a suitable fungicide (*see p.94*). Improve air circulation to reduce humidity. Space plants further apart, or if under cover, open vents, windows, and doors, as appropriate.

Pelargonium Viruses

SYMPTOMS Yellow markings, such as flecks, ring spots, and streaks (*see p.35*), on the foliage. Leaves may, occasionally, be distorted and flowers may show colour-breaking symptoms (pale streaks on the petals). The overall vigour of the plant is rarely affected, but the appearance of foliage and flowers may be spoiled.
CAUSE Various viruses, either alone or in combination, some of which may be transmitted by aphids, others by handling, or by eelworms in the soil.
CONTROL There is no control available, but it is always advisable to inspect plants carefully before purchase and to dispose of any showing severe symptoms.

Peony Wilt

SYMPTOMS Shoots of herbaceous peonies wilt and turn brown at the stem base. In damp conditions, fuzzy grey fungal growth may develop at the base of the stem. The affected stem withers and dies, shortly followed by other stems on the plant. Sclerotia (shiny black fungal resting bodies) form on and within the infected stem. Occasionally the fungus attacks higher up the stem, causing similar symptoms (*see p.62*). Wilt is most common on herbaceous peonies, but tree species may also be affected.
CAUSE The fungus *Botrytis paeoniae*, which produces numerous spores that are carried on air currents and by water splash. The sclerotia are usually responsible for wilt infections, which appear early in the season; subsequent infections may be due to spore spread.
CONTROL Cut out infected stems promptly, cutting right back into healthy growth and below soil level if necessary. If the stems are not removed,

the sclerotia fall into the soil where they can persist for several years. Remove all debris at the end of the season, as this may harbour sclerotia. If a plant has been severely damaged by the infections, scrape off the uppermost few millimetres of soil in the autumn in case it contains sclerotia – this must be done with care to avoid injuring the roots.

Periwinkle Rust

SYMPTOMS Dark brown spore pustules develop beneath the leaves (*see p.27*). The foliage becomes pitted and distorted and the whole plant also appears distorted and may fail to flower.
CAUSE The fungus *Puccinia vincae*, which produces a fungal mycelium that invades all parts of the plant, including the roots.
CONTROL As the fungus is present throughout the plant, there is little point in attempting control, and all infected or obviously distorted plants should be removed and disposed of promptly.

Pestalotiopsis

SYMPTOMS Grey-brown necrotic patches develop on the foliage and may also appear on shoots. The tree or shrub appears scorched. Leaves drop prematurely and stem infection may cause extensive dieback. Conifers are commonly affected.
CAUSE The fungus *Pestalotiopsis*; most frequently *P. guepini*, which is encouraged by high humidity.
CONTROL Remove affected leaves, prune out infected stems.

Petunia Foot Rot

SYMPTOMS The plant wilts and dies, often quite rapidly. The stem base may also become blackened and atrophied.
CAUSE Various species of the soil-borne fungus-like *Phytophthora*, which produces

resilient resting bodies that enable it to persist in the soil for several years.
CONTROL No control is available. Remove infected plants promptly, together with the soil in the immediate vicinity of the roots. Do not grow petunias on that site for several years; instead choose other bedding plants such as zonal pelargoniums, French and African marigolds, begonias, and heliotrope, since these are not susceptible.

Petunia Viruses

SYMPTOMS Yellow flecking, spotting, mosaicing, and streaking on the foliage. The leaves may also be stunted and generally chlorotic and may be deformed. The whole plant appears stunted and the flowers may show colour breaking (pale streaking) in the petals.
CAUSE Various viruses, some of which may be transmitted by aphids and other sap-feeding pests; others may be spread by handling or when propagating from cuttings.
CONTROL Remove and dispose of affected plants promptly. Always wash hands thoroughly after handling infected plants. Control pests which are potential vectors.

Phellinus ignarius

SYMPTOMS Pale brown to buff-coloured fungal bracket, with a rounded outline, develops on the branches or trunk of the tree. There may be no apparent loss of vigour of the tree but internally the wood is weakened.
CAUSE The polypore fungus *Phellinus ignarius*.
CONTROL No control is currently available. Infected trees may stand safely for many years, but they may be rendered dangerous, so it is better to employ a reputable

tree surgeon to examine the tree and advise as to the best course of action.

Phellinus pomaceus

SYMPTOMS Hoof-shaped or rounded fungal fruiting bodies develop on the bark of dead or living stems or trunks of various deciduous trees (*see p.55*). They have a woody texture and measure about 5cm in diameter. The upper surface is pale grey-brown and may show concentric furrows. The margin of the developing fruiting body is velvety grey, later becoming cinnamon-coloured. The wood within the stem may decay, turning the centre of the stem or trunk crumbly and white with a dark brownish purple zone around the edge.
CAUSE The fungus *Phellinus pomaceus* (syn. *Fomes pomaceus*) which produces spores from minute pores on its undersurface. The spores infect via wounds caused by pruning or other mechanical injury.
CONTROL Remove fruiting bodies to minimize spread of spores. However, the decay will continue within the branch or trunk. Have the tree inspected to determine its overall health and safety.

Phlox Eelworm

SYMPTOMS In early to midsummer, phlox stems become abnormally swollen and are liable to split. Growth is stunted and leaves at the shoot tips are greatly reduced in width, sometimes being little wider than the midrib (*see p.49*). Badly infested plants rot and die.
CAUSE Microscopic eelworms or nematodes, *Ditylenchus dipsaci*, that feed within the stems and foliage. They overwinter in dormant buds at the base of the plant.

CONTROL Phlox eelworm does not attack the roots, so clean plants can be propagated by taking root cuttings. There are no effective pesticides available to gardeners, so affected plants should be removed before the pest spreads.

Phlox Powdery Mildew

SYMPTOMS White, powdery fungal growth develops on the upper and occasionally lower leaf surfaces. Infected leaves may be distorted and puckered and turn yellow and die off. The stems may show similar symptoms.
CAUSE The fungus *Sphaerotheca fuliginea*; this produces numerous spores, which are spread by water splash or on air currents; it is encouraged by dry soil conditions and humid air.
CONTROL Treat affected plants with a suitable fungicide (see p.94). Water more regularly, but avoid splashing water over the foliage.

Phomopsis Canker

SYMPTOMS Dieback of shoots, often accompanied by girdling of the stem by cankers. Girdled areas become constricted, leaving wider areas of stem above and below. On some species the cankers are accompanied by resin production. Many conifers may be attacked – in particular, Douglas fir, and occasionally various species of cedar, larch, and other firs.
CAUSE The fungus *Phacidiopycnis pseudotsugae*, which enters stems and branches through wounds.
CONTROL Prune out and burn affected areas where feasible. Prune carefully, avoiding leaving stumps, and carry out any extensive pruning during the active growing season.

Phormium Mealybug

SYMPTOMS A white, waxy substance accumulates at the ensheathing leaf bases on New Zealand flax plants. This is secreted by greyish-white, soft-bodied insects that measure up to 4mm in length. Particularly heavy infestations cause a loss of vigour and plants may eventually be killed.
CAUSE A sap-sucking insect, *Balanococcus diminutus*. Most mealybugs are greenhouse pests, but this species is not harmed by winter frosts. It breeds continuously throughout the warmer months of the year.
CONTROL Applying a systemic insecticide (see p.94) may reduce but not eliminate infestations. When purchasing a phormium, inspect the leaf bases carefully to ensure that it is free of this pest.

Phosphorus Deficiency

SYMPTOMS Growth is weak and reduced and leaves are small and may fall early. Leaves develop a blue-green or purplish discoloration, or scorching around the edges. Plants may show delayed or poor flowering and fruiting.
CAUSE A deficiency of phosphorus, which is most likely to occur on acid soils or after periods of heavy rain or watering.
CONTROL Apply fertilizers containing a high proportion of phosphorus, or apply super-phosphate at the rate and in the manner specified by the manufacturer.

Phyllosticta Leaf Spot and Dieback

SYMPTOMS Plants affected include holly, ivy, yew, and rhododendron, as well as many other woody plants. Numerous symptoms include brown blotches or spots on foliage

(see p.44), or browning of needles. The infected leaves often remain attached to the plant for a long time; however, infection is sometimes followed by leaf or needle drop and twig dieback. Close inspection of infected material reveals numerous tiny black dots, which are the fruiting bodies of the pathogen.
CAUSE Various species of the fungus *Phyllosticta*. This fungus is usually associated with dead, fallen leaves beneath trees, but will occasionally attack healthy living leaves and twigs, causing significant damage.
CONTROL Remove infected leaves and twigs growing on the plant, as well as fallen leaves on the ground beneath it. This will limit sources of infection and the spread of spores, which likely occurs through rain splash. Plants that are not well cared for seem to be more prone to attack, so regular but not excessive watering, feeding, and mulching is advisable.

Phytophthora kernoviae

See RAMORUM DIEBACK p.174.

Phytophthora Root Rot

SYMPTOMS Yellowing, sparse foliage and branch dieback of woody plants, often followed by plant death. Conifer foliage often becomes dull and greyish before turning brown and dying. However, these symptoms could be caused by many other factors. Below ground symptoms will be seen much more rapidly and these include a small root system with many of the finer roots having rotted off and with larger roots decaying; internally the larger roots show brown or blackish discoloration and may be soft and easily broken. In a severe infection, where the

Phytophthora has invaded the base of the stem or collar of the plant, there is a blackish or brown discoloration, often in shape of an inverted "V", clearly visible beneath the bark. There may also be discoloration or oozing on the outer surface of the bark. As most symptoms could be caused by other factors, laboratory examination is needed for conclusive accurate diagnosis.
CAUSE Species of *Phytophthora*, a fungus-like micro-organism that is able to move short distances in the water films between soil particles. Once the plant has been infected the soil can be contaminated for several years, since *Phytophthora* releases resting spores into the soil as the roots decay. *Phytophthora* is spread when spores, or the water that contains them, are carried in run-off water or drainage, or when contaminated plants, soil, or compost are moved.
CONTROL Improve soil drainage to reduce the risk of the disease developing. If the disease is identified, remove and dispose of the infected plant and the soil in the vicinity of its roots. The soil should then be replaced with fresh, uncontaminated soil and plants known to be very susceptible should be avoided. These include *Aucuba*, box, *Ceanothus*, *Chamaecyparis lawsoniana*, holly, horse chestnut (*Aesculus hippocastanum*), lavender, *Malus*, maple, *Prunus*, rhododendron, *Ribes*, *Rubus idaeus*, *Sorbus*, *Viburnum*, and yew.

Phytotoxicity

Plants can be damaged by insecticides or fungicides; this is known as phytotoxicity. Follow the manufacturer's instructions carefully to ensure that the correct dilution is being applied. Plants known to be harmed by a chemical will be listed by the manufacturer, but this information is not known for all potential plant/chemical interactions. The risk of spray damage occurring can be minimized by not treating plants that are already under stress as a result of exposure to bright sunlight, extremes of temperature, or dryness at the roots. Seedlings will be more susceptible than mature plants, and petals are more likely to be scorched than foliage. If in any doubt, and there are many plants to be sprayed, it is sensible to treat a few and watch for adverse reactions over the next few days. Damaged plants may suffer a significant check in growth, but usually make a recovery.

Pieris Lacebug

SYMPTOMS The main host plants are *Pieris* species. but *Rhododendron* can also be attacked. The insect lives on the underside of the foliage and causes a coarse pale mottling on the upper surface. In heavy infestations the foliage is extensively bleached and the underside of the leaves is covered in sticky tar-like spots of excrement. The adult bugs measure 4mm in length and have wings with many lace-like veins. The wings are held flat over the insect's body and black markings on the wings form a distinctive "H" shape pattern when the bug is at right angles to the observer. The nymphs are wingless dark brown insects with spine-covered bodies.
CAUSE A sap-sucking insect (*Stephanitis takeyai*) that originates from Japan but became established in Europe during the 1990s. This insect can be confused with the closely related rhododendron lacebug (*see p.176*), which only attacks rhododendrons.

CONTROL Check regularly, and spray with a systemic or contact insecticide (*see p.94*) when the pest is seen. It is active in most months of the year except late winter and early spring.

Pigeons

The wood pigeon, *Columba palumbus*, can cause damage at any time of year, but they are particularly troublesome in the winter when snow or frost encourages them to feed in gardens. Brassicas, such as cabbages, Brussels sprouts, and cauliflowers, can have their leaves stripped to the midribs (see *p.46*). The foliage of peas, lilacs, and some other plants can also be damaged.
CONTROL Shooting is usually not a feasible proposition in gardens, so other means are needed to protect brassicas. Netting is the only certain way of keeping birds away. Various scaring devices, such as scarecrows, humming tapes, glitter strips, and models of cats or birds of prey, are of limited benefit. Birds quickly become used to scaring devices, particularly when hungry and other food supplies are scarce.

Pine Adelgid

SYMPTOMS The new shoots of pines, especially Scots pine, develop a fluffy, white, waxy coating in mid- to late spring (*see p.58*). Black aphid-like insects can be found underneath the wax. The insects and wax persist through the summer but gradually become less obvious.
CAUSE Sap-sucking insects, *Pineus pini*, which overwinter on the tree as immature nymphs and have several generations during the summer.
CONTROL Spraying is usually unnecessary, and although the waxy secretions may be

unsightly, the adelgids have little obvious effect on the tree's vigour once it has become established. Young trees could be sprayed with a contact insecticide (*see p.94*) on a dry, mild day in late winter against the overwintering nymphs before they produce their waxy covering. Treatment later in the spring is less effective, since the wax acts as a protective barrier that is difficult to penetrate with pesticides.

Pine, Needle Blights and Casts

See NEEDLE BLIGHTS AND CASTS OF PINES *p.159*.

Pineapple Gall Adelgid

SYMPTOMS In early summer, spruce trees (*Picea* species), especially the Norway spruce or traditional Christmas tree, develop swellings at the shoot tips that resemble green pineapples (*see p.57*). These galls contain many small aphid-like insects. In late summer the galls begin to dry up and turn brown. Slits open up in the gall as it dries, allowing the winged adult insects to emerge. Old galls can persist on trees for many years and spoil their appearance, particularly if they are to be used as Christmas trees.
CAUSE Sap-sucking insects, *Sacchiphantes abietis*, which initiate the galls by secreting chemicals as they feed on the buds in the spring. These chemicals prevent normal shoot development and, instead, the bases of the leaves become swollen and fused together, while leaving cavities within which the adelgid nymphs continue their feeding and growth.
CONTROL Pineapple gall adelgids have to be tolerated on trees that are too tall to be sprayed. Apart from the visual

effect of the old galls, little real damage is done. On small trees, galls can be cut off or, as a preventive measure, trees can be sprayed on a mild, dry day in late winter with a contact insecticide (*see p.94*). This treatment is aimed at overwintering nymphs before they mature and lay eggs on the buds.

Plane Anthracnose

SYMPTOMS A brown, discoloured patch develops at the base of the leaf bud in winter. This may not be easy to see, but in spring affected buds die without producing a leaf. Adjacent areas of stem turn brown and die back and a small cankered area may appear on the stem. Buds that do open into leaves may show more obvious symptoms as the newly-produced growth is killed rapidly. Very dark brown, necrotic areas develop on leaves, primarily adjacent to the main veins. Throughout the summer, leaves that started off perfectly healthy may develop symptoms due to the leaf blight stage of the disease and this is the stage most frequently noticed.
CAUSE The fungus *Apiognomonia veneta* syn. *Gnomonia platani*, which is encouraged by cool weather conditions in the spring and which may overwinter on fallen infected leaves.
CONTROL None available. If the infection is mild and the tree well-sized, it may be possible to prune out all infected areas. Rake up and dispose of fallen leaves.

Plum Blossom Wilt

SYMPTOMS Blossom withers and dies shortly after appearing. Trusses of dead blossom remain hanging on the tree. Infection may then spread to adjacent leaves, causing them to wilt, turn brown, die, and remain hanging on the branch. Tiny, pinprick-sized, raised, buff-coloured fungal pustules develop on infected areas. Localized dieback may occur.

CAUSE The fungus *Monilinia laxa*, which may overwinter as cankers on infected stems, or pustules on infected flowers and foliage. The spores are wind-borne and spread is most rapid during damp weather.

CONTROL Prune out infected flower trusses, preferably before the infection spreads to leaves or into the spur. Grow relatively resistant plums such as 'Jefferson', 'President', and 'Marjorie's Seedling'.

Plum Leaf Gall Mite

SYMPTOMS Whitish-green, pouch-like swellings or galls develop on the leaves of plums, damsons and gages during the early summer, mainly on the upper leaf surface, especially around leaf edges (*see* **p.29**).

CAUSE Microscopic gall mites, *Eriophyes similis*, which suck sap from the underside of the leaves during the spring. They also secrete chemicals into the leaf, causing the growth of the galls, which are hollow and enclose the mites.

CONTROL Nothing currently available to gardeners is effective against gall mites. Fortunately, the mites have little harmful effect, even when galls are numerous, so their presence can be tolerated.

Plum Leaf-curling Aphid

SYMPTOMS The foliage on plums, damsons and gages becomes tightly curled and crinkled during the spring (*see* **p.41**). Green insects, up to 2mm long, and their whitish cast skins can be found on the underside of the leaves. Winged aphids develop during late spring and early summer and fly away to spend the rest of the summer on various herbaceous plants. After that time, the infestation on plums dies out and trees start to produce normal foliage again.

CAUSE A sap-sucking aphid, *Brachycaudus helichrysi*, which migrates back to plums in late summer to lay over-wintering eggs near the buds.

CONTROL Small trees can be sprayed with a systemic insecticide (*see p.94*) in early spring. A plant oil winter tree wash applied in early winter to control overwintering eggs can reduce the need to spray in spring.

Plum Moth

SYMPTOMS A pinkish-white caterpillar, up to 12mm long, with a brown head feeds inside a ripening plum fruit. Most feeding occurs around the stone and the damaged area becomes filled with the caterpillar's brown excrement pellets (*see p.14*). Damaged fruits tend to ripen prematurely and often have a depressed area on the surface where the fully-fed larva has eaten its way out before seeking an overwintering site under a loose flake of bark.

CAUSE Caterpillars of a moth, *Grapholita funebrana*, which usually has one generation a year, but may have two in hot summers.

CONTROL Pheromone traps (*see p.88–89*) are available for detecting the presence of this pest by capturing male plum moths. On isolated plum trees, pheromone traps may disrupt the females' mating success sufficiently to reduce the number of maggoty fruits. Traps can also help predict the correct time to spray. Spray with a contact insecticide (*see p.94*) in early summer, with a second application three weeks later,

if this pest has been a problem in previous years.

Plum Pox (Sharka)

SYMPTOMS The fruit develops discoloured, usually dark brown streaks or rings, which are easiest to see on red-fleshed plums. On darker-coloured or yellow fruits there is clear pitting on the surface. Affected fruits may fall early and have an unpleasant and very acidic taste. Pale green blotches or rings develop on the foliage. Although very damaging, this disease is much less common in Britain than the rest of Europe. It may be confused with physical injury caused by very heavy rainfall or hail.

CAUSE The plum pox virus, which is transmitted by aphids and may be introduced on plum rootstocks.

CONTROL None available. In Britain, suspected outbreaks in private gardens no longer need to be reported to the Animal and Plant Health Agency.

Plum Rust

SYMPTOMS The upper leaf surface of plums or anemones is covered with bright orange-yellow spots from midsummer onwards. In severe cases the whole leaf may appear yellow. Numerous rust-brown, fungal pustules full of brown spores develop in corresponding areas beneath the leaf (*see* **p.36**). Towards late summer and early autumn spores darken and appear black. Affected leaves fall prematurely and if severe infections occur for several years in succession the tree may be weakened.

CAUSE The fungus *Tranzschelia pruni-spinosae*, which is encouraged by damp weather. Despite the link between the disease on plums and on anemones, it may appear in the same garden on one without any sign of it on the other.

CONTROL Rarely necessary on plums; rake up and dispose of fallen infected leaves. Feed the tree to maintain vigour. Remove and destroy any infected anemones nearby. Avoid growing the plum 'Victoria', which seems particularly susceptible.

Plum Sawfly

SYMPTOMS Unlike plum moth, this pest has caterpillar-like larvae that attack plum fruitlets, rather than mature fruits. The larvae are creamy white with brown heads and are up to 10mm long. Damaged fruitlets have a round hole bored in them, from which black excrement pellets may protrude. These fruits fall from the tree in late spring and early summer.

CAUSE Larvae of a sawfly, *Hoplocampa flava*, which lays eggs on the blossoms during mid-spring.

CONTROL Plum sawfly is of local occurrence and so preventive measures are only worthwhile if this pest has been troublesome in previous years. If control is necessary, spray seven to ten days after petal fall with a contact insecticide (*see p.94*). On small trees, remove damaged fruitlets before the larvae leave to go into the soil where they will overwinter. Some plums, such as 'Victoria' and 'Czar', are more susceptible than others.

Pocket Plum

SYMPTOMS As infected fruits of plums and damsons develop, they become deformed, twisted, and often banana-shaped. They contain no stone, are hollow like pockets, and pale green and smooth. A white spore layer then develops over the surface and the plums start to shrivel.

CAUSE The fungus *Taphrina pruni*, which appears to be

capable of overwintering in twigs. The spores are carried on air currents or by water splash. **CONTROL** None available. Remove affected plums promptly, preferably before they turn white. Although causing a loss of crop, this disease rarely affects all the fruit in one year, and may not reappear for several years.

Pollen Beetles

SYMPTOMS Black or bronze-green beetles, measuring 1–2mm in length, cluster in flowers (see **p.61**) in the spring and again in midsummer. Yellow flowers, such as daffodils, are affected in the spring but many others, including sweet peas, runner beans, marrows, roses, dahlia, and shasta daisy, attract these beetles during the summer months. They can be a nuisance, especially when flowers are cut for display in the house, but in general little damage is caused. When abundant, however, the beetles will sometimes cause damage by eating their way into unopened rose buds, and they may also discolour cauliflowers by feeding on the developing flowerhead. **CAUSE** There are many pollen beetles, *Meligethes* species, most of which are entirely harmless, as they feed on wild rather than cultivated flowers. The adult beetles feed on pollen, but do not eat enough to interfere with pollination; they may even assist with pollination if they move from one flower to another and carry pollen on their bodies. The species that are of concern to gardeners are generally those that develop as larvae in the flower buds of brassica seed crops. These include the widely grown agricultural crop, oil seed rape. In midsummer, adult pollen beetles that have bred in oil seed rape fields disperse into the countryside and gardens, where they seek out

flowers. They will not breed in garden flower buds, but they can be a temporary nuisance during the summer, and again in the spring when the over-wintered adults re-emerge. **CONTROL** Pollen beetles on flowers in the garden have to be tolerated. Insecticides should not be used on them, as they will also kill bees, butterflies, and other pollinators. When cut flowers are taken indoors, pollen beetles have an annoying habit of leaving the flowers and crawling about the room. However, placing cut flowers in a shed or garage for a few hours during the day before they are brought into the house, can have the effect of persuading the beetles to leave the flowers without being a nuisance.

Poor Pollination of Raspberries and Strawberries

SYMPTOMS Fruits develop but are distorted and may bear dry areas. On raspberry, individual drupelets or small groups of them may be dry and brown. Adjacent drupelets may be perfectly normal. The overall growth of the plant is normal and it may bear other, perfect fruits. **CAUSE** Poor pollination, which results in uneven swelling and development of the fruit. A lack of bees, or weather conditions that prevent their flying and pollinating activities. **CONTROL** Grow susceptible plants in an area which is provided with shelter. Grow plenty of other plants and flowers to attract bees and other pollinating insects.

Poplar Bacterial Canker

SYMPTOMS In spring a dense, off-white slime is exuded from cracks in young branches. The damaged areas may be

girdled and so die back. On older branches, the slime may also be apparent, and roughened areas of bark develop, up to 3cm in diameter. Larger, more "open" cankered areas may develop and stand out clearly from the branch as they become surrounded by raised callus. **CAUSE** The bacterium *Xanthomonas populi*, which is believed to enter the tree via wounds, such as those around bud scales, leaf scars, frost crack, and damage caused by insects feeding. The precise means of spread is not known, but rain splash and insects are probably the main ones. **CONTROL** Prune out infected areas, where feasible, to limit the spread of the bacteria.

Poplar Yellow Blister

SYMPTOMS In late spring or early summer raised, rounded blisters develop on upper, and occasionally lower, leaf surfaces. The concave, lower surface of each blister turns bright yellow. In damp weather young shoots may also be attacked. Although leaves are disfigured, the overall vigour of the tree is not affected. **CAUSE** The fungus *Taphrina populina*, which overwinters as spores on the bud scales. **CONTROL** No control measure is available, and none is justified.

Potassium Deficiency

SYMPTOMS Most frequently poor flowering and/or poor fruit set. Flowers may be undersized. On tomatoes this deficiency is at least partially responsible for the disorder known as blotchy ripening. Potassium is also important in the ripening of wood and in the structure of the stem, and too little may increase the risk of frost or similar damage. Leaf tips

may appear scorched around the edges and show brownish-purple spotting beneath (see **p.39**). **CAUSE** A deficiency of potassium in the soil, which is most common on light, sandy soils and those with a low clay content. **CONTROL** Apply sulphate of potash in spring and autumn.

Potato Black Leg

SYMPTOMS Foliage is chlorotic (yellow) and stunted; the leaves are small and slightly incurled. These symptoms are easily confused with other problems, but with black leg the stem base is blackened and rotted at ground level. The vascular strands within the stem are similarly discoloured, showing as distinct black spots when the stem is cut across. The parent tuber is completely rotted. The plant may be killed before a crop is produced. **CAUSE** The bacterium *Pectobacterium atrosepticum*, which is encouraged by wet soil conditions. It is often introduced on symptomless but mildly affected seed tubers. When these are planted they cause affected plants to appear scattered through the crop of other perfectly healthy plants. The bacteria may enter tubers via wounds either while in the soil, or at lifting. **CONTROL** Take great care at lifting time to keep damage to a minimum. If possible, do not lift crops during wet weather as infection is more likely then. Store only those tubers that appear quite healthy. Avoid growing 'Majestic', 'Arran Pilot', 'Maris Bard', 'Desirée', and 'Estima', as these cultivars seem to be particularly susceptible.

Potato Blight

SYMPTOMS Necrotic brown patches develop on the leaves,

largely on the leaf tips and around the edges (see *p.43*). As the spots enlarge, the leaves wither and die. In wet weather or under humid conditions slightly fluffy, white fungus-like growth may be visible around the edges of the spots. This is most common on lower leaf surfaces. The haulms of infected plants develop brownish-black patches and may collapse. Infected tubers develop slightly sunken, dark patches on the skin with a reddish-brown discoloration beneath. The discoloration may then spread into the flesh (see *p.22*). Secondary organisms often invade, turning the dry, brown areas into unpleasant-smelling, slimy wet rot. The whole tuber is then rotted.

CAUSE The fungus-like *Phytophthora infestans*, also responsible for blight on tomatoes. Spores are produced from the leaf lesions and carried by rain splash or on air currents. For infection to occur, the fungus needs two consecutive 24-hour periods, each with a minimum temperature of 10°C (50°F) and each with a minimum of at least 11 hours where the relative humidity is 89% or more. Spores from the infected top growth are washed down through the soil and may infect tubers.

CONTROL If a "Blight Infection Period" is mentioned on the radio for your area, it is advisable to check plants daily. Reduce the chances of the infection developing on the tubers by earthing up deeply. Remove haulms as soon as they show signs of infection. The tubers should then be lifted 1–2 weeks later, so allowing time for the tuber skin to toughen up. A new strain of blight has developed and observations suggest some of the cultivars that had shown good resistance in the past may no longer be quite so useful.

These include 'Cara', 'Estima', 'Kondor', 'Lady Balfour', 'Maris Peer', 'Pentland Crown', 'Record', 'Rembrandt', 'Romano', 'Sante', and 'Valour'. Cultivars 'Sarpo Axona' and 'Sarpo Mira' probably still offer the best resistance. Ensure that all remains of the potato crop, including tubers, are removed and binned or burned. Operate a crop rotation system (*see p.104*), ideally a four-year rotation. Do not compost any plant remains.

Potato Common Scab

SYMPTOMS Raised, roughened scabby patches develop on the skin of the tuber (see *p.22*). The skin ruptures, leaving the scabs with ragged edges. Damage may be superficial or result in cracking. The flesh is usually undamaged, although just beneath the scabby patches it may become discoloured. Potatoes are often infected, but beetroot, turnip, swede, and radish may also be attacked.

CAUSE *Streptomyces* species, which are particularly common on sandy, light soils with a low organic matter content. They are present naturally in most soils and potatoes planted on ground that was previously grassland are particularly prone to attack. Soil that has been limed is also likely to have more *Streptomyces* in it, whereas acid soil has less.

CONTROL Improve the organic matter content of the soil. Water regularly, as there appears to be a link between this disease and dry soil conditions. Do not lime soil prior to planting potatoes. Use acidic materials such as sulphate of ammonia to reduce scabbing. No fungicide treatment is available. Grow resistant cultivars such as 'Arran Comet', 'Arran Pilot', 'Golden Wonder', 'King Edward', 'Lady Balfour', 'Maris Peer', and 'Rembrandt'.

Potato Cyst Eelworms

SYMPTOMS Potato plants die prematurely and hence produce a poor crop of small potatoes. Affected plants die during mid- to late summer, with leaves progressively yellowing and drying up from the bottom of the stems upwards. When cyst eelworms first occur in a garden, small patches within the potato rows show symptoms. These patches gradually enlarge each time potatoes are grown in that part of the garden until eventually it is impossible to grow a worthwhile crop of potatoes. Tomato plants are also attacked. If infested plants are carefully dug up, spherical eelworm cysts, up to 1mm in diameter and white, yellow, or brown in colour can be seen on the roots (see *p.21*). These are the bodies of mature female eelworms and each one can contain up to 600 eggs.

CAUSE Two species of cyst eelworm attack potato and tomato. The golden cyst eelworm, *Globodera rostochiensis*, has female cysts that are white at first, but turn yellow then brown when mature. The white cyst eelworm, *G. pallida*, changes from white to brown without an intermediate stage. Both species develop inside the roots, where their feeding disrupts the uptake of water and nutrients. When mature, the females become swollen and burst through the root wall.

CONTROL There are no pesticides available to home gardeners for controlling eelworms. Potato cyst eelworms are particularly difficult because their eggs remain viable inside the cysts for many years. The eggs are stimulated to hatch by chemicals that are secreted into the soil by the roots of suitable host plants. Adopting a system of crop rotation can delay the buildup of damaging

infestations, but once a serious eelworm problem has arisen, the usual three or four-year rotation is not long enough to eliminate the pest. Some potato cultivars have some resistance to the golden cyst eelworm. These include 'Accent', 'Nadine', 'Pentland Javelin', 'Rocket', 'Swift' (earlies); and 'Alhambra', 'Cara', 'Nicola', 'Maris Piper', 'Picasso', and 'Stemster' (main crop). In addition there are the maincrop cultivars 'Maxine' and 'Sante' that are resistant to golden cyst eelworm and have some tolerance of the white cyst eelworm. The latter will, however, breed successfully in the roots, so infestations will increase.

Potato Dry Rot

SYMPTOMS The skin of the potato tuber becomes wrinkled at one end, followed by rapid shrinking, the development of concentric rings of wrinkles, and discoloration (see *p.23*). The affected tissues become black, grey, or brown. Pink, white, or blue-green pustules of fungal spores develop on the shrivelled tissues. The symptoms are likely to be found on stored tubers only, though it may also cause poor growth after planting.

CAUSE Species of the fungus *Fusarium*, which infect tubers while they are still in the ground or as they are being lifted. The fungus enters through the lenticels (breathing pores), eyes, or wounds caused by abrasions, pests, or scab infection.

CONTROL Handle potatoes very carefully at lifting and when storing. Minimize other problems such as wireworm, slugs, and scab infections. Ensure that the potatoes are properly mature at harvest time and that they are stored in a frost-free but cool and dry place. Do not grow 'Arran Pilot', which is notably susceptible, but consider 'King Edward' or 'Pentland Crown' instead, which are notably resistant.

Potato Early Blight

SYMPTOMS Despite its name, this disease usually occurs late in the season in Britain and Europe, though it is an early season infection in North America. Dark brownish-black spots, often angular in outline and showing concentric ringing, appear on the foliage of both potato and tomato. Sunken spots may develop on the stems and a dark rot and associated furry fungal growth may occur at the sepal end of the fruits. Inadequately fed plants may also be more susceptible. Although causing some damage, this disease rarely affects cropping.
CAUSE The fungus *Alternaria solani*, which can carry over from year to year on infected plant debris.
CONTROL Rarely necessary, but the treatments applied to control potato or tomato late blight (*Phytophthora infestans*) should give incidental control.

Potato Gangrene

SYMPTOMS Small areas of damage, slightly sunken with clearly defined edges, develop on the tubers after lifting (*see p.23*). The skin may become slightly wrinkled. The affected tubers rot, turning wet and pale pink, and then darken. There is no visible fungal growth. If planted, a gangrene-infected potato produces a poor, gappy crop or no crop at all.
CAUSE The fungus *Boeremia foveata*.
CONTROL Avoid damaging tubers on lifting and inspect all tubers carefully before planting and throughout storage. Dispose of any tubers showing symptoms.

Potato Hollow Heart

SYMPTOMS The tubers may look perfectly normal on the outside, as does the plant itself. When cut open, the tuber has an internal cavity – this may be simple and surrounded by corky, somewhat darkened tissue; it may be star-shaped, or it may consist of two or more interconnecting cavities. Large potatoes are particularly susceptible; small ones are rarely affected.
CAUSE This disorder occurs when potatoes have been growing rapidly (perhaps due to overwatering, or excessive fertilizer use) and are then subjected to a sudden check in growth, usually caused by lack of water. Conversely, it may also occur if the plants are subjected to an over-dry soil, followed by a sudden increase in the soil moisture due to heavy rain or watering. In some instances it appears to be induced by extremely high temperatures.
CONTROL Grow plants in uniform stands and ensure regular but not excessive watering, particularly during dry weather.

Potato Internal Rust Spot

SYMPTOMS The tubers and the whole plant look perfectly normal, but when cut open, the flesh of the tuber is marked with many irregular, small rust-brown spots.
CAUSE The precise cause is unknown, but unsuitable soil conditions are probably largely responsible. It may occur on sandy and clay soils but is usually most severe on light soils or loam soils with a low organic matter content. Low potassium and lime levels may also be responsible.

CONTROL Improve the condition of the soil prior to planting – incorporate plenty of humus and water regularly as the crop grows. The soil could also be limed but this must be done well in advance of planting. Grow 'King Edward' or 'Arran Consul', which appear to show a degree of resistance.
• *see also* POTATO COMMON SCAB *p.171*.

Potato Powdery Scab

SYMPTOMS Small, nearly circular scabby patches develop on the tuber. Each has a raised margin. The scabs later burst open and produce masses of brown spores that are released into the soil (*see p.22*). Occasionally a canker form of this disease occurs when affected tubers are seriously deformed. They may then be incorrectly assumed to have potato wart disease.
CAUSE The fungus *Spongospora subterranea*, which is particularly common in wet seasons and on heavy soils, especially those which have grown many crops of potatoes.
CONTROL Dispose of infected tubers; do not incorporate them into the compost heap. Do not grow potatoes on that site for at least three years. Improve soil aeration before planting potatoes. Avoid growing 'Pentland Crown', as this is particularly susceptible.
• *see also* POTATO WART DISEASE *below*.

Potato Silver Scurf

SYMPTOMS Inconspicuous silvery markings, usually lines, develop on the skin of the tuber (*see p.22*). The flesh of the tuber is not affected and this disease is of little practical consequence unless the tubers are to be used for showing.

The condition usually only becomes apparent when the tubers are stored, particularly if the storage conditions are too humid. The markings may turn black if the fungus produces spores.
CAUSE The fungus *Helminthosporium solani*.
CONTROL None necessary.

Potato Spraing

SYMPTOMS The flesh of tubers is marked by tan-brown or red-brown rings or arcs of discoloration (*see p.22*). There may be development of some corky tissue associated with the discoloration, too. Tubers may, sometimes, be distorted and the stems and foliage may be mottled yellow.
CAUSE Tobacco rattle virus and occasionally potato mop top virus, which are spread by free-living nematodes (eelworms) in the soil. Infection may be started if infected tubers are planted.
CONTROL No chemical control available. Remove affected plants, as tobacco rattle virus has a wide host range and may affect sweet pepper, hyacinth, gladiolus, tulip, China aster, tobacco plants, and many common weeds. Practise good weed control. Grow potatoes on a fresh site. 'Pentland Dell' appears to be particularly susceptible to this problem, so other cultivars should be grown in preference.

Potato Viruses

SYMPTOMS These vary with the precise combination of viruses involved, the potato variety and the growing conditions. Leaf symptoms include yellow flecking, streaking or mosaicing, dark spotting and leaf distortion, upward rolling, and stiffening of the leaflets. If home-saved

seed potatoes are used for many years in succession, there may be a gradual but noticeable decline in the plant vigour and cropping and this is often due to the build-up of viral diseases.
CAUSE Many different viruses, including potato virus X, potato virus Y, and potato leaf roll virus.
CONTROL None is available. Always grow potatoes on a fresh site and do not save tubers for seed. Buy tubers certified as virus-free.

Potato Wart Disease

SYMPTOMS The tubers are covered in dense, warty outgrowths, at first off-white, and later black. Affected tubers may rot. Tubers may only start to show symptoms in store. Occasionally small, pale green growths rather like those of crown gall may develop on the stem at soil level. This disease is rare, but very damaging. The symptoms may be confused with the distorted form of powdery scab, or even with fusarium rot of potato.
CAUSE *Synchytrium endobioticum*, which is usually found in wet years or on heavy soils, and infects tubers through the eyes. Spores are produced by the wart-like growths. These spores are very resilient and can remain viable in the soil for about thirty years, even in the absence of a potato crop. They are easily spread from one area to another in particles of soil adhering to tools and footwear.
CONTROL Infected tubers should be reported to the local Department for Environment, Food and Rural Affairs. Potatoes must not be grown on infested land.
• *see also* POTATO POWDERY SCAB *p.172*.

Powdery Mildews

SYMPTOMS White, powdery fungal growth develops on the leaf surface (*see p.37*). Upper leaf surfaces are usually affected first and most severely but the symptoms may spread to the lower leaf. The mildew may also develop on all other above ground parts of the plant; the precise location of its growth depends on the host plant and the species of mildew involved. Affected leaves or other plant parts may yellow and become distorted. Distortion is particularly common if infection occurs on young foliage. Affected fruits may crack and split because they are unable to expand normally (*see p.15*). Sometimes buff or even pale brown fungal growth may be seen; this is most common on rhododendron (where the mildew is often restricted to the lower leaf surface), gooseberry, and laurel. The mildew may kill off small areas of leaf tissue, which then drop away, causing a shothole effect. Growth may be poor and in extreme cases dieback or even death may occur following premature leaf fall.
CAUSE Various fungi, in particular many species of *Oidium*, *Microsphaera*, *Podosphaera*, *Uncinula*, *Erysiphe*, and *Phyllactinea*. Each usually only infects a single genus or closely related host plants. These fungi are often encouraged by plants growing in dry soils and by humid or damp air around the top growth.
CONTROL Remove infected leaves promptly. Consider spraying infected ornamental plants with a suitable fungicide (*see p.94*) or plant oil (*see p.89*). Keep plants adequately watered but avoid overhead watering. Where available, grow resistant varieties.
• *see also* APPLE POWDERY MILDEW *p.109*; BEGONIA POWDERY MILDEW *p.115*;
GERANIUM DOWNY MILDEW *p.139*; HYDRANGEA POWDERY MILDEW *p.147*; OAK POWDERY MILDEW *p.160*; PANSY POWDERY MILDEW *p.162*; PEA POWDERY MILDEW *p.163*; PHLOX POWDERY MILDEW *p.167*; QUINCE POWDERY MILDEW *p.174*; RHODODENDRON POWDERY MILDEW *p.177*; ROSE POWDERY MILDEW *p.179*.

Privet Thrips

SYMPTOMS During the summer months, privet and lilac leaves develop a dull green appearance, becoming increasingly silvery brown on their upper surface (*see p.36*). Elongate, narrow-bodied insects up to 2mm long occur on the leaves. The adults are black with three white bands across wings that are folded back over the body. The nymphs are similar in shape but are creamy yellow and wingless.
CAUSE Sap-sucking insects, *Dendrothrips ornatus*, which are more troublesome in hot, dry summers.
CONTROL Light infestations have little effect on the plant, but if damage starts becoming noticeable, especially before late summer, spray with a systemic or contact insecticide (*see p.94*).

Proliferation

SYMPTOMS Only the flower or bud is affected. Buds develop within the centre of developing buds (*see p.64*). These may never mature further or may produce fully grown flowers, which grow out from the centre of the original flower. Occasionally a "triple-decker" effect is seen, with three flowers on top of each other. Stems and foliage may also appear between flowers. On fruiting plants, each flower may develop into a fruit.
CAUSE Most commonly, physical injury such as that
caused by a late frost or insect attack, which damages developing buds. Occasionally virus infection is responsible. If symptoms occur all over the plant and for several successive years, then virus would seem a likely cause and the affected plant should be removed.
CONTROL Rarely necessary, though affected shoots could be pruned out.

Pyracantha Leaf Miner

SYMPTOMS An oval, silvery-brown patch develops on the upper leaf surface of *Pyracantha*, or firethorn, where the moth's caterpillar has eaten the internal tissues. As the caterpillar matures, it spins silk threads that cause the leaf to fold in half upwards so that the mine is no longer visible. Leaves at the bottom of the plant are often more heavily attacked than those elsewhere.
CAUSE Caterpillars of a tiny moth, *Phyllonorycter leucographella*, which can have three generations a year, with leaf damage being most obvious over the winter period. It was first recorded in Britain in 1989, but is now widespread in England and Wales. It originates from southern Europe, but occurs in gardens in many northern European countries.
CONTROL Pyracanthas can tolerate heavy infestations, so control measures are unnecessary.

Pyracantha Scab

SYMPTOMS Greyish-black or dark khaki-coloured, scabby patches develop on the leaves and berries (*see p.15*). Infected leaves become chlorotic and may fall prematurely. Infected berries may remain small, start to crack, or split and fall. Loquat (*Eriobotyra*) and medlar (*Mespilus*) may also be attacked.

CAUSE The fungus *Venturia inaequalis* f. sp. *pyracanthae*, which is encouraged by damp weather. It does not appear to overwinter on fallen leaves (unlike apple and pear scab), but overwinters on infected leaves which remain on the plant, and as tiny pustules on the shoots.

CONTROL Prune out severely infected areas, and twigs bearing the pustules. Some pyracantha cultivars show a degree of resistance to this infection (and also to fireblight) and these include 'Golden Charmer', 'Orange Charmer', 'Shawnee', and 'Teton'.

Pythium Root Rot

SYMPTOMS The whole plant wilts and dies back. Seedlings are particularly prone, but older plants may also succumb if otherwise lacking in vigour. Roots are killed and may disintegrate, giving the plant the appearance of having only a few roots.

CAUSE The fungus-like organism *Pythium*, which is soil- or water-borne.

CONTROL Remove infected plants promptly, together with the soil or compost in the immediate vicinity of the roots. Do not water seedlings with water from a water butt or use unsterilized compost or soil for raising seedlings. Use mains water for watering seedlings.

Q

Quince Leaf Blight

SYMPTOMS Numerous irregular, small spots appear on the foliage. These are dark red at first but soon turn black and may join together (*see p.13*). The leaves turn yellow and then brown and fall prematurely. Occasionally,

similar spotting develops on the fruits and these may show some distortion. Shoot tips may also be infected and then develop dieback. On rare occasions other members of the *Rosaceae* family may be affected, such as hawthorn, medlar, pyracantha, and pear.

CAUSE The fungus *Diplocarpon mespili*, which may overwinter on infected shoots and leaves.

CONTROL Rake up and dispose of fallen leaves. Prune out infected stems.

Quince Powdery Mildew

SYMPTOMS Powdery, white deposits on the upper leaf surface and occasionally on the lower surface.

CAUSE The fungus *Podosphaera leucotricha*, which can also affect apple.

CONTROL Keep quinces well watered, as dry soil seems to encourage mildew development. Apply a mulch to the moist soil to preserve soil moisture. Prune out severely infected stems. Prune established trees to improve air circulation within the crown.

• *see also* APPLE POWDERY MILDEW *p.109.*

R

Rabbits

Rabbits (*Oryctolagus cuniculus*) can feed on a very wide range of plants. Herbaceous plants may be grazed down to ground level, while the foliage and soft shoots of woody plants are eaten up to a height of about 50cm. Rabbits also gnaw the bark from the base of trunks, especially on young apple trees, and the tree dies if bark is lost from most or all of the trunk's circumference. Bark feeding occurs at any time of year, but trees are particularly at

risk when snow or frost makes other food items unavailable. Rabbits are inquisitive animals that often feed on newly planted plants, even if they have previously ignored similar established plants in the same garden. They were greatly reduced by a viral disease, myxomatosis, during the 1950s and 60s, but they now have greater resistance and are often numerous in rural areas.

CONTROL Direct control measures, such as shooting and trapping, are usually not suitable in domestic gardens. A better solution is to keep rabbits away from plants by netting. Fences around the perimeter of a garden need to be at least 120–140cm tall with a further 30cm sunk below soil level, and angled outwards to discourage rabbits from burrowing underneath. The wire netting needs a maximum mesh size of 25mm, otherwise young rabbits may squeeze through. Where individual plants or small flower beds are being protected, it is usually sufficient to enclose them with netting 90cm high without the need to bury part of the netting. Gates to gardens or enclosed flowerbeds also need to be rabbit-proof and kept closed when not in use. Tree bark can be protected by placing wire netting or spiral tree guards around the base of the trunks. Animal repellent products cannot be relied on to give long-term protection. Although most plants can be eaten, there are some that are usually left alone; these feature in various published lists of "rabbit-proof plants". Looking at other nearby gardens will also provide suggestions for plants that local rabbits ignore.

Ramorum Dieback

SYMPTOMS These vary according to host plant and

pathogen species; both are aerial pathogens and do not cause root decay. On shrubs such as camellias, *Kalmia*, rhododendrons, and *Pieris*: leaves develop spreading brown lesions, usually starting at the tip, petiole, or leaf edges. A "V"-shaped, discoloured patch may be seen as the lesion spreads down the midrib. Stems develop lesions or cankers with brown discoloration beneath the bark. Stems wilt and die back. On viburnums the infection usually starts at the stem base. On trees such as holm oak (*Quercus ilex*) and ash, only the leaves show symptoms, with usually browning and death around the margins. On beech trees, bleeding cankers develop on the bark, which ooze a brownish-black liquid. If the bark is removed there is a mottled brownish decay of the inner bark. On larch trees, needles turn a purple-black colour and are shed prematurely. Buds fail to open properly and shoots wilt and die back. Bleeding cankers may also develop.

CAUSE The fungus-like organisms *Phytopthora ramorum* and *P. kernoviae*. *Phytophthora ramorum* was first found in the UK in 2002 and has since become widespread. Most reports are still from nurseries on hosts such as beech, camellia, rhododendron, and Viburnum, but there have also been many outbreaks in gardens. More recently it has also become significant on species of larch (*Larix*). It is also commonly found attacking holm oak, *Leucothoe*, and *Pieris*. *Phytophthora kernoviae* was first reported in the UK in 2003. It commonly attacks beech, holm oak, magnolia, and rhododendrons, and causes similar symptoms to *Phytophthora ramorum*, but is more aggressive on rhododendrons. Spread is encouraged by wet or humid conditions, and the pathogens may also be spread in contaminated soil or compost and in water. They produce resilient and long-lived resting structures, which are found in contaminated soil in pots, growing beds, on vehicle tyres, and on animal and human feet.

CONTROL Both pathogens are notifiable: suspected outbreaks must be reported to the Animal and Plant Health Agency.

Raspberry Beetle

SYMPTOMS All cane fruits, including raspberry, blackberry, loganberry, and tayberry, are susceptible. The ripe fruits have dried-up patches at the stalk end of the berries (*see p.15*). When fully grown, the larvae are up to 8mm long and are creamy white with pale brown markings on their upper surface. The grubs feed initially at the base of the berries but later feed in the inner core or plug. They are often seen crawling around the bowl after the fruits have been picked.

CAUSE The larval stage of a small greyish-brown beetle, *Byturus tomentosus*. Eggs are laid on the flowers during early and midsummer.

CONTROL Spray with a contact insecticide (*see p.94*). Timing of control measures is important in order to control the newly-hatched grubs before extensive damage has occurred. On raspberries, spray when the first pink fruits are seen and again two weeks later; on blackberries and hybrid berries, such as loganberries, spray at 80 per cent petal fall. Avoid spraying when plants are in full flower.

Raspberry Cane Blight

SYMPTOMS The shoots or the whole cane start to die back during the summer. This is due to an infection at the base of the canes, which causes them to become brown and the bark to rupture. They also become very brittle and may snap off readily at or just above ground level. Minute, black, pinprick-sized, fungal fruiting bodies develop on dead areas and exude spores. Other spore forms develop later and these forcibly eject their spores into the air.

CAUSE The fungus *Paraconiothyrium fuckelii* (syn. *Leptosphaeria coniothyrium*), which is carried both by rain or watering splash and on air currents. Infection most frequently takes place through wounds, such as those caused by cane midge attack, late spring frosts, or following pruning. The fungus can persist in the soil on infected plant debris.

CONTROL Prune and train the plant carefully to avoid infection. Protect from frost and control any pests that may cause wounds through which the fungus can gain entry. If infection occurs, prune out affected canes and cut away any discoloured wood from the crown.

Raspberry Cane Spot

SYMPTOMS Purple spots with silvery-white central areas develop on the stems of raspberries and hybrid cane fruits. They may also appear on the leaves and flower stalks (*see p.50*). On loganberries, the fruits may be infected. The bark may split as the spots enlarge and whole canes may be killed.

CAUSE The fungus *Elsinoe veneta*, which usually appears first on the canes in early summer, and may then spread to the foliage and the fruits.

CONTROL Prune out infected canes. Avoid growing the raspberry cultivars 'Norfolk Giant' and 'Lloyd George' as these are particularly prone to this infection.

Raspberry Leaf and Bud Mite

SYMPTOMS Pale yellow, rounded blotches develop on the upper surface of raspberry leaves (*see p.33*) from late spring onwards. On the underside of the leaves, corresponding with the yellow blotches, are slightly darker areas, compared with the usual silvery-green colour. By midsummer the foliage may be extensively discoloured and leaves at the shoot tips may be distorted. These symptoms can be confused with those of some virus diseases. Canes affected by the mite normally grow to their usual height and produce an adequate crop, whereas virus-infected plants lack vigour and crop badly.

CAUSE Microscopic, sausage-shaped mites, *Phyllocoptes gracilis*, which suck sap from the underside of leaves. In the autumn they hide in or near buds but cause no damage in the winter.

CONTROL None of the pesticides available for garden use is effective against this mite. Fortunately, its impact on the plant's health and cropping ability is not as great as the symptoms might suggest. Raspberry cultivars vary in their susceptibility; 'Malling Jewel' is often attacked, while 'Leo', 'Glen Prosen', 'Glen Lyon', and 'Terrie Louise' have some resistance. Plants growing in warm, sheltered places are also likely to have heavier infestations.

Raspberry Rust

SYMPTOMS Bright orange, raised pustules full of fungal spores develop on leaves of raspberries and blackberries in spring or early summer. They develop first on the upper leaf surface. Towards the end of summer or in early autumn, dark brown-black spores develop on the lower surface. Premature leaf fall occurs (*see p.57*).

CAUSE The fungus *Phragmidium rubi-idaei*, which can overwinter on infected fallen leaves and on the canes.

CONTROL Rake up and dispose of infected leaves. The raspberry cultivar 'Glen Prosen' shows a degree of resistance. Avoid growing 'Glen Clova', 'Glen Moy', 'Malling Delight', 'Malling Jewel', and 'Malling Joy'.

Raspberry Spur Blight

SYMPTOMS Dark purplish patches develop around the buds of new canes. These increase in size and spread up and down the canes, causing extensive discoloration. In autumn and winter they change colour, becoming greyish-silver and are covered in numerous minute raised, black, pinprick-sized, fungal fruiting bodies. In spring, infected canes bear few viable buds and produce weak shoots, which wither and die, or persist but crop very poorly. Loganberries may also be affected.

CAUSE The fungus *Didymella applanata*, which is usually most troublesome in wet weather, but occurs in most seasons regardless of the weather. Plants that have been fed excessive quantities of nitrogen or which are overcrowded are more prone to infection and encourage rapid disease spread.

CONTROL Prune out any badly infected canes. Grow cultivars showing a good degree of resistance, such as 'Glen Lyon', 'Glen Moy', 'Glen Rosa', 'Julia', 'Leo', and 'Malling Admiral'.

Raspberry Viruses

SYMPTOMS Yellow patterning of the foliage (*see p.33*), most commonly mosaic patterns often combined with slight distortion, stunting, and reduction in vigour and cropping. There may also be downward curling of the leaves. The foliage symptoms may be confused with those caused by raspberry leaf and bud mite, but the other symptoms are not present.

CAUSE Various viruses, either alone or in combination with each other. The most common are raspberry mosaic, curly dwarf, and raspberry yellow dwarf.

CONTROL Remove and dispose of infected plants. Control aphids and other sap-feeding pests as these may spread the virus. Do not grow replacement raspberries on the same site, as some viruses are transmitted by soil-living nematodes.

Rats

Rats cause damage to stored fruits and vegetables, and also to growing root crops and sweetcorn. Rats spoil foodstuffs by contaminating it with their excrement and urine. Rats often carry in their urine the bacterium that causes Weil's disease; this can cause serious illness in humans. The brown rat, *Rattus norvegicus*, is widespread in Britain. Pest control contractors will undertake rat control in homes and gardens. Otherwise they can be controlled with rat traps or poisoned baits (see *p.89*). Poisoned baits and traps need to be placed where rats are active, but they must always be covered up so that children and pets do not have access to them.

Red Berry Mite

SYMPTOMS Blackberry fruits ripen unevenly, with some parts of the berries remaining red and hard (see *p.14*). This problem is more frequently seen in hot summers. The first fruits to develop usually ripen normally,

but later fruits show progressively more severe symptoms.

CAUSE Microscopic mites, *Acalitus essigi*, that feed by sucking sap from the flowers, foliage, and fruits. Only the fruits suffer damage, however.

CONTROL None of the pesticides available to home gardeners is effective against this mite. Some of the overwintering mites can be eliminated by pruning old fruiting canes to the ground and burning them.

Red Spider Mites

See BEECH RED SPIDER MITE *p.114*; BOX RED SPIDER MITE *p.118*; CONIFER RED SPIDER MITE *p.128*; FRUIT TREE RED SPIDER MITE *p.138*; GLASSHOUSE RED SPIDER MITE *p.141*.

Red Thread

See TURF RED THREAD *p.191*.

Replant Problems

See ROSE SICKNESS *p.179*.

Reversion

See CURRANT REVERSION *p.130*.

Rhododendron Bud Blast

SYMPTOMS Flower buds fail to open, and turn brown and dry (see *p.59*). They may remain on the plant for several years. The surface of the bud is covered in numerous tiny, black, bristle-like fungal outgrowths. These are the spore-bearing parts of the fungus, and it is the presence of these that differentiates between this and dry bud with which it is often confused. Quite often infected buds and healthy ones are seen growing side by side. Leaf, stem, and general growth are unaffected.

CAUSE The fungus *Seifertia azalea* is spread by rhododendron leafhoppers (see below) as they lay their eggs in the bud scales.

CONTROL The leafhoppers first appear in mid- to late summer, so pick off and dispose of infected buds before that time of year. Control of the leafhopper is possible but unlikely to be significantly effective in controlling bud blast spread, because the pest simply re-invades from nearby gardens.

• *see also* BUD DROP *p.121*; BUD, DRY *p.121*.

Rhododendron Lacebug

SYMPTOMS During summer the upper leaf surface of rhododendrons, especially those growing in sunny positions, develop a coarse yellowish mottling (see *p.35*). The underside of the leaves is covered in rusty-brown excrement spots deposited by the lacebugs. These are present between late spring and early autumn. The wingless nymphs are yellowish-brown, while the adults are 4mm long, brownish-black and with wings that are folded flat over their backs. The transparent wings have many veins, giving them a lace-like appearance – hence the name lacebug.

CAUSE Sap-sucking insects, *Stephanitis rhododendri*, which have one generation a year. In late summer the adult lacebugs lay overwintering eggs in the midribs of the leaves. Rhododendrons can also be attacked by pieris lacebug (see *p.168*).

CONTROL Avoid planting rhododendrons in warm, sunny spots. If the pest does occur, spray the foliage with a systemic or contact insecticide (see *p.94*). Damaged leaves do not regain their green colour and can persist on the shrub for more than a year; new growth

develops normally once the pest has been dealt with.

Rhododendron Leaf Spot

SYMPTOMS Spots with clearly defined edges develop on the leaves. The older leaves are usually worst affected and, if severely damaged, some may fall prematurely. The spots are brownish-purple, often with a distinct black or very dark purple ring around the edge (see *p.26*). The spots often have concentric ringing with numerous pinprick-sized, raised, black fungal fruiting bodies on the dead areas.

CAUSE The fungus *Gloeosporium rhododendri*, which, although causing unsightly damage to leaves, is usually only likely to reach serious levels on a plant lacking vigour or in poor health.

CONTROL Remove severely affected leaves. Take steps to improve the general vigour of the plant by regular feeding, watering, and mulching.

Rhododendron Leafhopper

SYMPTOMS The creamy yellow nymphs are present on the underside of rhododendron leaves from late spring to midsummer, but cause no noticeable damage. The adult insects are more readily seen, as they often sun themselves on the upper leaf surface between midsummer and mid-autumn. They are 8–9mm long and turquoise-green with two orange stripes on each forewing. When disturbed, they jump off the leaf and fly a short distance before returning to the plant.

CAUSE A sap-sucking insect, *Graphocephala fennahi*, which causes no direct damage to rhododendrons. The females, however, insert overwintering eggs into the flower buds and this is thought to make entry wounds for the fungal disease rhododendron bud blast (see p.176).

CONTROL Spraying against adult leafhoppers in late summer and early autumn may reduce the incidence of bud blast. A systemic or contact insecticide (see p.94) can control leafhoppers, but more may fly in from nearby gardens, especially during hot weather, which increases leafhopper activity.

Rhododendron Petal Blight

SYMPTOMS Small spots develop on the petals – these are white on coloured flowers and pale brown on white blooms (see **p.60**). The spots increase in size and appear water-soaked. The entire flower rapidly collapses and becomes slimy and shapeless. Affected flowers do not fall but remain on the plant. Dry weather may then cause the dead flowers to dry out and they may remain on the plant in this state until the following spring.

CAUSE The fungus *Ovulinia azaleae*, which is encouraged by mild, humid growing conditions. The spores develop on the petals as they deteriorate and are spread by insects and on air currents.

CONTROL Remove and dispose of infected flowers promptly to minimize spread of the fungus.

Rhododendron Powdery Mildew

SYMPTOMS Yellow blotches develop on the upper leaf surface. Beneath each is a corresponding patch of felty, buff-coloured fungal growth (see **p.28**). Premature leaf fall occurs. Occasionally the fungal growth occurs on the upper surface and is then more off-white and powdery than buff and felty.

CAUSE The precise identity of the fungus or fungi is as yet undetermined, but the most important is probably a species of *Erysiphe*. Dry soil and wet conditions around the foliage in late summer and autumn encourage both its development and spread.

CONTROL Pick off severely affected leaves. Consider spraying with a suitable fungicide (see p.94). Keep plants well watered and mulched but avoid overhead watering. Avoid growing *R. cinnabarinum* and its hybrids, and hybrids with species from the subsections *Campylocarpa* or *Fortunea* as their parentage.

Rhododendron Rust

SYMPTOMS The upper leaf surface develops bright orange-yellow spots with corresponding yellow and brown fungal pustules appearing beneath. The disease is not very common and may be confused with the damage caused by rhododendron lacebug (see opposite).

CAUSE The fungus *Chrysomyxa ledi* var. *rhododendri*, which persists within the affected leaves as long as they remain on the plant. It is encouraged by damp conditions.

CONTROL Remove infected leaves promptly, pruning out entire shoots if necessary. Some success may be achieved by spraying with a suitable fungicide (see p.94).

Rhubarb Gumming

SYMPTOMS The sticks develop normally but some are spoiled by drops of resin-like exudation that appear along their length. Sticks may crack around the areas of gumming, allowing the entry of secondary organisms which then cause rotting.

CAUSE The precise cause is not known but poor growing conditions, in particular an erratic supply of moisture, are believed to be involved. Inadequate feeding may also play a part.

CONTROL Improve growing conditions and pay particular attention to watering and feeding the plant.

Robin's Pin Cushion

SYMPTOMS Wild hedgerow roses, sucker growth on grafted roses and some species roses develop roughly spherical galls, up to 60mm across, on stems (see **p.56**) and occasionally foliage in mid- to late summer. The gall consists of a hard central core in which many white grubs live and feed. The exterior of the gall is covered in a dense mass of moss-like growths that are reddish or yellowish-green in colour. In the autumn the mossy covering dries up, leaving the inner core in which the grubs overwinter before pupating in the spring.

CAUSE A small gall wasp, *Diplolepis rosae*, which lays eggs in buds in summer. The galls that subsequently develop are also sometimes known as bedeguar galls.

CONTROL Apart from creating the galls, this insect has little adverse effect on the plant and so control measures are unnecessary. If the galls are considered unsightly they can be pruned out.

Root Aphids

SYMPTOMS Infested plants tend to have reduced vigour and are more likely to wilt during sunny weather. The aphids, up to 2–3mm long and usually a dirty cream colour but sometimes bluish-green, live among the roots or at the stem bases of host plants (see **p.21**). Root aphids often secrete a white, powdery or fluffy wax from their bodies and this gives infested roots and nearby soil particles a white coating.

CAUSE Several different species of root aphids occur in gardens. These include *Smynthurodes betae* on runner and French beans, *Pemphigus bursarius* on lettuce (see LETTUCE ROOT APHIDS p.146), *Thecabius auriculae* on auriculas, *Trama troglodytes* on Jerusalem artichoke, *Maculolachnus submacula* on roses (see ROSE ROOT APHIDS p.179), *Aphis sambuci* on pinks, *Dysaphis crataegi* on carrots and parsnips.

CONTROL Root aphids are difficult to control because they are hidden by the soil. With edible plants, crop rotation helps avoid infestation by aphids that have overwintered in the soil on the remains of the previous year's crops. Keeping plants watered helps them tolerate root aphid infestations. Container-grown plants, such as auriculas, can be treated with a systemic insecticide potting compost drench (see p.94).

Root Knot Eelworms

SYMPTOMS Many plants can be attacked by root knot eelworms. These microscopic animals live within the roots and cause knobbly swellings (not to be confused with the nitrogen-fixing nodules normally found on the roots of legumes). They disrupt the uptake of water and nutrients, so that plants lack vigour and have poor leaf colour. In northern Europe these pests are mainly encountered in greenhouses, but can attack garden plants on light sandy soils.

CAUSE Several *Meloidogyne* species may infest a wide range of plants.

CONTROL There are no effective pesticides available

to gardeners. Infested plants should be destroyed, together with the soil around the roots.

Root Mealybugs

SYMPTOMS Root mealybugs are mainly a problem on pot plants, particularly cacti, succulents, pelargonium, ferns, African violet, and fuchsia, causing them to lack vigour. The mealybugs are up to 2mm long and they cover themselves and the roots with a white, waxy powder (*see p.21*). They have flatter, more elongate bodies than root aphids (*see p.177*), which have a globular body shape.
CAUSE Several *Rhizoecus* species occur in greenhouses and on the roots of house plants. They tend to be more troublesome on plants that are normally grown in a dryish potting compost.
CONTROL Root mealybugs are difficult to eliminate. Container-grown plants can be treated with a systemic insecticide potting compost drench (*see p.94*). With many plants, however, it is better to discard the old infested plants.
• *see also* GOLDEN ROOT MEALYBUG *p.141*.

Rose Aphids

SYMPTOMS The younger leaves, shoot tips, and flower buds are often covered in dense colonies of green, yellowish-green, or pink insects up to 2–3mm long (*see p.61*). Heavy infestations cause stunted growth and poor quality blooms, while the foliage becomes sticky with a sugary honeydew excreted by the aphids. Sooty mould may develop on the honeydew, which also traps the white cast skins shed by the aphids as they grow.
CAUSE Several species of aphid occur on roses; the

most troublesome is usually *Macrosiphum rosae*.
CONTROL Look for aphids from mid-spring onwards, and spray with a systemic or contact insecticide (*see p.94*) before heavy infestations have a chance to develop. Organic sprays (*see p.89*) are also effective but may need more frequent use.
• *see also* ROSE ROOT APHID *p.179*.

Rose Balling

SYMPTOMS Flower buds fail to open or start to do so but never open fully (*see p.64*), generally in damp weather. The outer petals become pale brown, dry, and papery. The inner petals are perfectly normal but hidden beneath the outer casing of dead ones. If damp weather persists, the bud rots off, usually becoming covered in grey mould growth. The rose's general growth is not affected; healthy buds and flowers may appear among those showing balling.
CAUSE Most commonly rain followed by bright or hot sun on the petals. The petals become slightly scorched and form an outer casing which prevents undamaged petals from appearing.
CONTROL Little can be done to prevent this, but do not water roses on hot, bright days. Water instead during the evening and directly at the base of the plant. Prune out blooms showing the symptoms before they have a chance to develop grey mould, as this secondary infection may then result in dieback.

Rose Black Spot

SYMPTOMS Diffuse purple-black spots or blotches develop on the leaves (*see p.26*), shortly followed by yellowing, and then premature leaf drop. The spots do not have sharply defined edges and can be seen

to be made up of tightly packed strands of fungal growth growing through the surface of the leaf. The spots enlarge and may sometimes join together. If black spot occurs several years in succession, particularly if infection occurs early in the summer, the rose may be severely weakened. Much smaller purple-black spots (usually only a millimetre or two across) may also develop on the stems, particularly on species roses.
CAUSE The fungus *Diplocarpon rosae*, which overwinters on stem lesions, on the bud scales, and on fallen leaves.
CONTROL Dispose of infected leaves promptly. During spring, prune out stems showing the lesions associated with black spot. Spray with a suitable fungicide (*see p.94*) either immediately after spring pruning, or just before the leaves start to break, whichever is appropriate to the pruning regime. Repeat applications of fungicide according to the instructions. It is usually necessary to make a number of applications in order to achieve reasonable control. Winter treatments of soil tend not to be very effective, because of the way in which the leaves are curled and packed on the ground. Affected leaves should, however, be raked up and disposed of as soon as possible. Check current catalogues and choose roses that are described as showing good disease resistance.

Rose Canker and Dieback

SYMPTOMS Stems discolour and die back or fail to produce growth in the spring. They may have a purple or blackish patch of discoloration, which becomes uniform if the stem is allowed to die off totally (*see p.53*). In extreme cases, the rose may be

killed. During damp weather fuzzy grey fungal growth may develop from areas damaged by *Botrytis* (grey mould) and minute raised, fungal bodies may develop on the areas killed by the other fungi.
CAUSE Various fungi, in particular *Leptosphaeria coniothyrium* and *Botrytis cinerea*. These gain entry to the stems, usually at the base, via wounds or other points of injury. These diseases are all encouraged by poor growing conditions such as over-wet soil, too deep planting, and excessively high mulching (when the mulch is placed too close to the stems and at some height up them).
CONTROL Prune out affected shoots completely, cutting right back into healthy tissue. Clear mulch or excess soil away from the base of the plant. Provide good growing conditions – water, feeding, and mulching – to ensure good, sturdy replacement growth.

Rose Grey Mould

See GREY MOULD *p.142*; ROSE CANKER (*above*).

Rose Leaf-rolling Sawfly

SYMPTOMS During late spring and early summer, rose leaflets become tightly rolled in a downwards direction (*see p.41*). The rolled leaflets often hang down compared with unaffected parts of the leaf. In the following weeks the leaflets are eaten by pale green caterpillars, up to 10mm long, which are hidden within them.
CAUSE The leaf curling is brought about by chemicals injected into the leaves by female sawflies, *Blennocampa phyllocolpa*, as they insert their eggs. There is one generation a year, with egg laying occurring in late spring to early summer.

CONTROL It can be difficult to prevent the damage occurring, especially when hot weather coincides with the sawfly's flight period. This increases their activity and the number of eggs laid. Light infestations can be dealt with by picking off rolled leaflets. Pesticides available to home gardeners are unlikely to control this insect.
• see also LARGE ROSE SAWFLIES p.150.

Rose Leafhopper

SYMPTOMS Rose leaves develop a coarse, whitish mottling on their upper surface (see **p.35**) during the summer. Roses growing against a wall or in other sheltered positions are particularly susceptible and may be heavily infested; by late summer much of the green colour will have gone from their leaves. Pale yellow insects, up to 3mm long, live on the underside of the leaves and readily jump off when disturbed. They have narrow bodies that are broadest at the head end and taper to the rear. The nymphs are creamy white and crawl about on the lower leaf surface.
CAUSE Sap-sucking insects, *Edwardsiana rosae*, which have two or three generations during the summer.
CONTROL The insecticides used against rose aphids (see p.178) also control leafhoppers.

Rose Powdery Mildew

SYMPTOMS Powdery, white fungal growth develops on the upper leaf surface (see **p.33**). Leaves may be attacked when very young and extensive distortion then results. On young foliage the fungal growth frequently develops on both leaf surfaces. The fungal growth is superficial

and can be rubbed off but the leaf tissue beneath is discoloured. Powdery mildew growth may also develop on the stems, flower buds, and thorns. When flower buds are severely infected, the buds may fail to open fully. Affected leaves fall early.
CAUSE The fungus *Podosphaeria pannosa*, which overwinters as stem infections and in dormant buds. The spores are air-borne. Powdery mildew is encouraged by dry conditions around the roots and by moist air around the foliage. Roses growing up against walls are often particularly badly damaged because of the drying effect of the bricks and the rain-shadow effect.
CONTROL Prune out badly infected stems and consider spraying with a suitable fungicide (see p.94). Keep roses adequately watered, particularly during hot, dry weather, but avoid wetting the foliage. Apply a good bulky mulch to encourage soil moisture retention. Check current catalogues and choose roses described as showing good disease resistance.

Rose Root Aphid

SYMPTOMS This dark brown aphid, which measures 3-4mm in length, feeds by sucking sap from rose roots. It would probably go unnoticed if it did not climb up the stems to deposit eggs there in the autumn. Recently-laid eggs are brown, but they soon become shiny black (see **p.51**). The eggs are about 1mm long, shaped like rugby balls and may be thickly clustered on the lower part of the stems. They will hatch in early spring and soon afterwards the young aphids go down into the soil.
CAUSE A sap-sucking aphid, *Maculolachnus submacula*.

The root-feeding activities of this aphid seem to have little impact on roses and infestations can vary considerably in numbers from year to year.
CONTROL It is not usually necessary to control this pest. Some of the eggs can be removed when roses are being pruned. Where this is not possible, the aphids can be sprayed like other rose aphids (see p.178) when they have hatched.

Rose Rust

SYMPTOMS Bright orange spots develop on the upper leaf surface, with corresponding bright orange spore masses beneath (see **p.29**). Severely infected leaves may fall prematurely. The leaf infections develop in early summer and are the most common and most obvious. Later in the summer, or in very early autumn, dark brown spore masses develop beneath the leaf as the winter spores are produced to replace the summer ones. Defoliation at this stage is almost inevitable. Occasionally, spring infections may develop, but these are usually only seen on species roses as a rupturing of the stem and the production of large quantities of bright orange spores, which burst out from within the stem. The stem above may then wither as the crack allows secondary organisms to invade.
CAUSE The fungus *Phragmidium*, in particular *P. tuberculatum* and, less frequently, *P. mucronatum*. Moist air is essential for the development and spread of the infection, as the spores are carried on air currents. The fungus may persist within the stems of plants that have shown spring infections. The spores can also overwinter on the soil surface and on fallen debris, stems, fences, and stakes.

CONTROL Prune out stems showing spring infections. This should be done both promptly and thoroughly, so that no fungus is left within the adjacent tissue. Improve air circulation by pruning and by avoiding planting too closely. Spray with a suitable fungicide (see p.94).

Rose Sickness/ Replant Problems

SYMPTOMS Newly planted roses fail to thrive and may show signs of dieback. There is no obvious cultural problem – such as poor planting, very dry or very wet soil – or any signs of pest attack. The roots may appear poorly developed, compact, and dark in colour and there may also be rotting of the finer roots. Roses are most often affected by this, but other plants, in particular apple, cherry, peach, pear, quince, and certain plums may be affected too. When planted on a site previously occupied by the same species, plants can fail to thrive. However, if they are removed promptly and replanted on a fresh site they may recover.
CAUSE The precise cause is uncertain and may well vary from plant to plant and also with location. The main causes are soil-living nematodes (eelworms), which may transm it viruses. Soil-borne fungi such as *Thielaviopsis basicola* and *Pythium* may also play a part, as may nutritional factors, such as nutrient depletion.
CONTROL If possible, do not allow the problem to arise, as there is no reliable control other than clearing soil from the roots, then replanting on a fresh site. If necessary, change the soil in the bed or area before planting. Soil must be changed to a depth of at least 45cm, preferably deeper. If replacing individual roses, change a

cube of soil measuring at least 45 x 45 x 45cm. Feeding plants with a high nitrogen fertilizer and incorporating materials high in nitrogen into the planting hole may also help minimize damage. Treating the soil with a mycorrhizal fungus may help to reduce the incidence of replant problems. Roses grafted onto *Rosa laxa* rootstocks seem to be less susceptible.

Rose Slugworm

SYMPTOMS Yellowish-green, caterpillar-like larvae, up to 14mm long with pale brown heads, feed on the foliage during summer. They graze away the lower leaf surface, causing the remaining damaged tissues to dry up and turn whitish-brown (see **p.32**).
CAUSE Larvae of a sawfly, *Endelomyia aethiops*, which has two generations during the summer. Larvae of the second generation in mid- to late summer are often more numerous and therefore capable of more extensive damage.
CONTROL Slugworm larvae are relatively easy to control with most contact insecticides, (see *p.94*).
• see also LARGE ROSE SAWFLIES *p.150*.

Rose Viruses

SYMPTOMS Symptoms show up most clearly on leaves, usually as vein clearing, yellow flecking, or mottling (see **p.35**). The markings tend to be quite mild, and not as obvious as on most other virus-infected plants, but the plant may show some distortion and stunting. Symptoms may be confused with those caused by weedkiller damage, the latter being far more common than virus symptoms.
CAUSE Various viruses, including strawberry latent ringspot and rose mosaic.

CONTROL None is currently available. Some rose viruses are spread by soil-living nematodes, but other methods may be involved as well. Infected plants showing severe symptoms should be removed and destroyed.

Rosemary Beetle

SYMPTOMS Foliage at the shoot tips and flower spikes of rosemary (*Rosmarinus officinalis*), lavenders (*Lavandula* spp.), and thyme (*Thymus* spp.) are eaten during late summer to early summer. The larvae are greyish white, soft bodied grubs which are up to 8mm long. The adult beetles are 7-8mm long and are metallic dark green with five purple stripes on each wing case. Both adults and larvae cause damage to their host plants.
CAUSE A leaf beetle, *Chrysolina americana*. Despite its name, this insect originates from southern Europe but is now widespread in northern Europe. It became established in Britain during the 1990s. It is present on its host plants as adult beetles throughout the year. It begins feeding and laying eggs in late summer to autumn, with both larvae and adults being present during the autumn–spring period.
CONTROL Damage only occurs where a heavy infestation has developed. Hand picking can reduce infestations. If necessary spray with a contact or systemic insecticide (see *p.94*), avoiding the flowering period. On herbs being used for culinary purposes, use an insecticide approved for use on edible herbs. Young larvae are more susceptible to insecticides than the adult beetles.

Rosy Apple Aphid

SYMPTOMS Pinkish-grey insects, up to 2mm long, cluster

on the young foliage of apples in the spring. Their feeding causes leaves at the shoot tips to become curled and yellowish. The aphids also feed on the young fruitlets and this can result in the fruits staying small, with a pinched appearance around the eye end (see *p.13*). The distribution of damaged fruits can be very variable, with some branches producing normal fruits, while others have stunted ones at harvest time.
CAUSE A sap-sucking insect, *Dysaphis plantaginea*, which overwinters on the tree as eggs. These hatch at bud burst and the aphids are active on the tree from then until early or midsummer, when they migrate to their summer host plants, which are wild flowers known as plantains.
CONTROL Spray against the newly-hatched aphids as the foliage emerges from the buds, using a systemic or contact insecticide (see *p.94*). Poor control is likely once the foliage has become curled. A plant oil winter tree wash applied in early winter to control overwintering eggs can reduce the need to spray in spring.

Rusts

SYMPTOMS The symptoms may develop on the foliage and stems, depending both on the host plant and the rust involved. Spores, either as spore masses or pustules, are usually bright orange or dark brown, and their colour may vary with the time of year. Rusts may produce several distinct spore stages, commonly known as spring, summer, and winter spores. Those produced early in the year are generally orange or bright yellow and those later in the year usually brown. Occasionally the rust produces gelatinous masses containing the spores, or "cluster cups". Infected areas usually discolour

and may wither and die off. Certain rusts have alternate hosts: although unrelated, both plants are needed for the fungus to complete its life cycle. On others, the life cycle needs only one host.
CAUSE Various fungi, the most common being species of *Puccinia*, *Uromyces*, *Phragmidium*, *Melampsora*, and *Gymnosporangium*. The spores need a moist environment to germinate and infect, and so rust infections are generally at their most severe in damp conditions.
CONTROL Remove infected leaves. Consider spraying ornamental plants with a suitable fungicide (check the labels for the suitability of each chemical for the host plant that you have in mind; see *p.95*). Improve air circulation around the plant. Where available, hgrow resistant cultivars.
• see also BEAN RUST *p.113*; BLUEBELL RUST *p.117*; FUCHSIA RUST *p.138*; HOLLYHOCK RUST *p.144*; IRIS RUST *p.148*; LEEK RUST *p.152*; PERIWINKLE RUST *p.166*; PLUM RUST *p.169*; RHODODENDRON RUST *p.177*; ROSE RUST *p.179*.

S

Salt Damage

SYMPTOMS Leaf browning and curling, often associated with the leaf veins. Leaves may be undersized and fall prematurely. Dieback may also occur. Most tree species could be affected, but symptoms are particularly common on sycamore, maple, spruce, alder, lime, and plane.
CAUSE Sodium chloride, usually as a result of a coastal situation, excessive use of salt on roads in winter, or to the presence of a salt dump for road salting being placed too close to trees. Occasionally, stormy

weather may produce salt-laden winds, which cause salt damage a considerable distance inland, as well as to trees growing by the sea.
CONTROL Avoid contamination and where feasible, alert the local council to potential problems.

Sawflies

Sawflies have caterpillar-like larvae, but they belong to the insect order Hymenoptera. They are more closely related to ants, bees, and wasps than to butterflies or moths, which belong to the Lepidoptera order. Sawfly larvae can be distinguished from moth and butterfly larvae by counting the clasping prolegs on the abdomen. Both types of larvae have three pairs of jointed legs on the thorax, but sawfly larvae have at least seven pairs of prolegs, while Lepidoptera caterpillars have five or less. Sawfly larvae are mainly leaf-feeders and many of the pest species are capable of causing extensive defoliation. Other feeding habits include:
boring into developing fruits – see APPLE SAWFLY *p.109;* PLUM SAWFLY *p.169;*
causing galls or leaf distortion – *see* ROSE LEAF-ROLLING SAWFLY *p.178;* WILLOW BEAN GALL SAWFLIES *p.197;*
leaf-mining – *see* GEUM SAWFLIES *p. 140.*
• *see also* AQUILEGIA SAWFLY *p.110;* ARUNCUS SAWFLY *p.111;* BERBERIS SAWFLY *p.115;* GERANIUM SAWFLY *p.139;* GOOSEBERRY SAWFLY *p.142;* IRIS SAWFLY *p.148;* LARGE ROSE SAWFLIES *p.150;* PEAR AND CHERRY SLUGWORM *p.164;* ROSE SLUGWORM *p.180;* SOLOMON'S SEAL SAWFLY *p.185.*

Scale Insects

Scale insects are sap-sucking pests that attack the foliage and stems of many plants in greenhouses and gardens. The insects are hidden underneath waxy shells or scales that they secrete over their bodies. Newly-hatched scales crawl around looking for suitable places to feed, but they soon settle down and remain immobile for most of their lives. The colour of the protective scale is usually brown or greyish-white and it may be flat or domed. The size of the scale varies according to the species and is usually in the range of 1–6mm. Some scale insects excrete a sugary liquid called honeydew, which makes the foliage sticky and permits the growth of sooty moulds. Most scales lay their eggs under the protection of their own bodies, but the cushion scales deposit eggs among white, waxy fibres. Newly-hatched nymphs are the stage in the life cycle that is most vulnerable to systemic or contact insecticides (see *p.94*). With most outdoor species of scales, nymphs are present in midsummer, but on indoor plants breeding may be continuous throughout the year.
• *see also* BEECH BARK SCALE *p.114;* BROWN SCALE *p.120;* CUSHION SCALE *p.130;* EUONYMUS SCALE *p.135;* FLUTED SCALE *p.137;* HEMISPHERICAL SCALE *p.144;* HORSE CHESTNUT SCALE *p.146;* HYDRANGEA SCALE *p.147;* JUNIPER SCALE *p.149;* MUSSEL SCALE *p.158;* OLEANDER SCALE *p.160;* SOFT SCALE *p.184;* VIBURNUM SCALE *p.192;* WISTERIA SCALE *p.198;* WOOLLY VINE SCALE *p.199.*

Sciarid Flies

See FUNGUS GNATS *p.138.*

Sclerotinia

SYMPTOMS Affected tissues become discoloured, brown, and wet as they rot. Large quantities of fluffy, cotton wool-like fungal growth develop and are scattered with large, black sclerotia (fungal resting bodies), which may be several millimetres long. The fungal growth and sclerotia may also develop within the stems. This disease may affect most above ground plant parts, but fruit and stem base infections are most common (see *p.52*). In addition, it may develop on corms and tubers kept in store.
CAUSE The fungus *Sclerotinia sclerotiorum*, which is encouraged by cool, damp conditions. The sclerotia fall into the soil where they remain dormant until spring. They then produce spore-producing, cup-shaped fungal growths (called apothecia) which are responsible for the following year's infections.
CONTROL Remove and burn affected plant parts promptly. Dispose of all debris at the end of the season. Do not grow susceptible plants on that site for at least four years following an outbreak of the disease.

Scorch

SYMPTOMS Pale, scorched-looking patches develop, which, are usually pale brown or bleached. Damaged areas usually dry out, becoming crispy and brown. Leaves or petals are most commonly affected but occasionally other above-ground plant parts may be damaged, too. Stems or trunks of trees and shrubs may show scorching only on the exposed side. Other growth remains normal.
CAUSE Hot or bright sun is the most common cause. The problem is exacerbated by droplets of moisture on the leaf surface – the droplets help to further magnify the sun's rays and increase the damage. Soft- or hairy-leaved plants are often most severely affected. Similar damage may occur with chemical contamination; either an unsuitable pesticide is used, or as a result of injury by a contact weedkiller. Less frequently, damage is due to the presence of moisture of any sort on the foliage, particularly if the leaves remain wet overnight as temperatures drop. On bark, the injury is usually the result of the scorching effect of bright sun on thin or young bark.
CONTROL Take care when watering, and water in early evening when leaves are not exposed to direct sunlight, yet have time to dry off before nightfall. Ensure that chemicals are selected properly and used with great care. In greenhouses and conservatories, provide adequate shading during the summer months.
• *see also* HORSE CHESTNUT MARGINAL LEAF SCORCH *p.146;* IRIS SCORCH *p.148;* NARCISSUS LEAF SCORCH *p.158.*

Seiridium Canker

SYMPTOMS Foliage of cypresses (*Cupressus*), junipers (*Juniperus*), Leyland cypresses (× *Cupressocyparis leylandii*), and thuyas (*Thuja*) loses colour, becoming slightly yellowed and losing its lustre. It then turns brown and dies (see *p.51*). This deterioration is due to the development of a canker on the affected stem. The bark becomes slightly roughened and resin is produced from around the cankers. Close inspection reveals black, raised, pinprick-sized fungal bodies around the cankered area. These may be rounded or may have burst, leaving minute craters. The infection is often first noticed as many small areas of dying foliage, but it may attack larger branches and cause extensive dieback. The tree may be rendered very unsightly and can even be killed.

CAUSE The fungus *Seiridium cardinale* (syn. *Coryneum cardinale*), which enters branches through twig crotches or fine cracks in the bark, and attacks most readily when the tree is growing slowly. The spores are produced from within the fungal bodies on the cankers and are carried on the wind. The fungus kills by girdling the affected stems as the canker enlarges.

CONTROL Prune out infected areas promptly, as this may prevent further damage occurring. There is no chemical control available. Check other susceptible trees in the area, such as *Cupressus macrocarpa*, *C. sempervirens* var. *horizontalis*, *C. glabra*, and *Thuja plicata*, and remove any infected areas.

Sempervivum Leaf Miner

SYMPTOMS The outer leaves on the rosettes of house leeks (*Sempervivum* species) become discoloured and collapse due to the inner tissues being eaten by the larvae of a hoverfly. This pest has two generations in early and late summer, when it makes the plants look as though they are rotting off.

CAUSE The larvae of a leaf-mining hoverfly, *Cheilosia caerulescens*, which has spread across Europe in recent years and is now established in Britain. The adult flies are 8–10mm long and have black bodies with short white hairs. They lack the yellow and black stripes on the abdomen found on many of the hoverflies that have aphid predator larvae. When fully grown, the legless creamy white maggots are 8–10mm long.

Each larva will feed on several leaves before it is ready to pupate in the soil.

CONTROL On small plantings, it is feasible to search for mined leaves and remove the grubs before they cause too much damage. A potting compost drench systemic pesticide (*see p.94*) used for vine weevil control, could be applied to container-grown house leeks in early and late summer.

Shanking of Grapes

SYMPTOMS Individual berries within the bunch fail to colour up normally – those of black cultivars turn red, those of white cultivars remain translucent. They then become wrinkled and, if left on the bunch, may start to resemble raisins. They have a watery or unpleasantly sour taste.

CAUSE A disorder that may result from many different cultural problems. Under- or over-watering are the most common causes; stagnant soil conditions and underfeeding may also be factors.

CONTROL Cut out the affected berries and spray the foliage regularly with a foliar feed to boost the vine's vigour. Ensure adequate but not excessive watering and feeding. Check that drainage is not impeded.

Sharka

See PLUM POX *p.169*.

Shieldbugs

When viewed from above, these insects have a shield-like shape and a broad, flattened beetle-like appearance, with sucking mouthparts. Some prey on other insects, but the majority feed on plant sap, although their impact on cultivated plants is usually negligible (but *see* SOUTHERN

GREEN SHIELDBUG *p.185*). Several species occur in gardens; the more common species include green shieldbug (*Palomena prasina*), hawthorn shieldbug (*Acanthosoma haemorrhoidale*), birch shieldbug (*Elasmostethus interstinctus*), and dock shieldbug (*Coreus marginatus*). They are up to 12mm long and are green, brown, or a combination of yellowish-green and reddish-brown. They are most often seen between late summer and early autumn, when they tend to sun themselves on plant leaves before seeking sheltered places in which to overwinter. Eggs are laid between mid-spring and early summer. The nymphs have a more rounded body shape, but increasingly come to resemble the adults as they mature.

CONTROL Not required.

Shothole

SYMPTOMS Holes develop on the leaves (*see p.45*), usually with an inconspicuous brown ring around the edge of each hole. Brown spots of dead tissue (*see p.31*) are present on the leaf.

CAUSE Various fungal and bacterial infections, including bacterial canker and occasionally powdery mildews, may be responsible. The pathogens kill off areas of leaf, producing leaf spots. The damaged areas then fall away, leaving holes.

CONTROL Determine the primary cause of the leaf damage and take the appropriate action.

Shothole Borers

SYMPTOMS Branches and sometimes the trunk of fruiting and ornamental forms of plums, cherry, almond, and less frequently, apple and pear, have many round holes, 1–3mm in diameter, in the bark.

Underneath the bark is a maze of tunnels (*see p.33*) where white, legless grubs have been feeding. Some species also bore into the heart wood. Sawdust may be seen coming from the shotholes in the bark, where the dark brown, cylindrical adult beetles, 2–4mm long, have emerged in early summer.

CAUSE Several beetle species are collectively known as shothole borers on fruit trees. These include *Scolytus rugulosus*, *S. mali*, and *Xyleborus dispar*. All are likely to attack branches or trees that are already in poor condition, and they are not a problem on vigorous trees.

CONTROL Pesticide control is not available since much of the life cycle is protected under the bark. Prune out and burn infested branches. Improve the tree's health by feeding in the spring and keeping it watered in dry summers. Unsuitable growing conditions, such as poorly drained soil or a weak root system arising from bad planting, will make trees prone to dieback and shothole borers.

Silver Fir Adelgid

SYMPTOMS The smaller stems on silver firs such as *Abies procera*, *A. grandis*, and *A. balsamea* develop lumpy growths where small black insects covered with white, waxy fibres are feeding.

CAUSE A sap-sucking insect, *Dreyfusia piceae*, which is active between late spring and autumn.

CONTROL Little can be done on large trees, but those small enough to be sprayed can be treated with a contact or systemic insecticide (*see p.94*) in the spring.

Silver Leaf

SYMPTOMS Single or small numbers of branches develop foliage, which has a silver sheen. On some hosts, such as rhododendron, the silvering is almost imperceptible, or non-existent. If affected stems more than 2.5cm in diameter are cut across, they have a central brownish stain. Other limbs appear perfectly normal, but may show symptoms at a later date. Affected limbs die off or fail to leaf up the following spring. The symptoms may be confused with those caused by other problems, such as adverse weather conditions, drought, malnutrition, or insect attack. Fruiting bodies, often grouped together and adhering fairly closely to the bark, may develop on dead wood. Their exposed surface is a dark purple-grey colour.

CAUSE The fungus *Chondrostereum purpureum*, a fresh-wound parasite that invades freshly created wounds on deciduous trees. Pruning cuts or pest attack are common points of entry. The spores are produced from the fruiting bodies found on infected limbs of living trees, or on fallen timber. The infection is spread on air currents or on pruning tools. The fungus produces a toxin that causes the upper leaf surface to separate from the main leaf blade. The air that accumulates between the two layers alters the light-reflecting qualities of the leaf, causing it to appear silvered. The leaves do not carry the infection.

CONTROL There is no control available, but occasionally mildly affected trees may recover. Generally, the infection spreads throughout the tree unless infected branches are removed to a point at least 15cm past the point at which the staining in the wood ends. Infection is least likely to occur during the summer months, so all pruning of susceptible trees should be carried out at this time. Although no longer recommended in most cases, trees that are particularly prone

to silver leaf infection should have all pruning wounds painted with a wound treatment. This should be carried out at the same time as the pruning. Locate any possible sources of infection and dispose of them. Plums and cherries are the trees showing greatest susceptibility to this disease, 'Victoria' plum being the most susceptible cultivar. The rootstock may also influence the likelihood of infection occurring: some, such as 'Brompton', are very prone, whereas others, such as 'Pixie', show a good degree of resistance.

Sirococcus Blight

SYMPTOMS The tips of shoots die back, most commonly on cedar trees, and premature defoliation follows. Several shoot tips may be affected on one tree. The first symptoms are a slightly pink discoloration of the needles in the spring (*see p.44*), with needles browning later on. Cankers may develop on the branches (*see p.49*), leading initially to a greenish- or reddish-purple discoloration. The cankers then cause the branch to appear to shrink inwards, and may also exude resin. From summer onwards, tiny, dot-like fungal fruiting bodies may be visible on these cankers and on fallen needles. The sap-conducting phloem within infected areas may be stained brown. When older trees are attacked they often live for many years, but if younger trees succumb and the fungus girdles the branches or the trunk, they may be killed more rapidly.

CAUSE The fungus *Sirococcus tsugae*, first recorded in Britain in 2013. It is most likely to be seen on *Cedrus atlantica* and *C. deodora*, but has occasionally been reported on *Tsuga* spp. (hemlock) as well.

CONTROL There is no control available for this disease, but

infected trees often live and remain perfectly safe for many years. It would, however, be prudent to have an infected tree inspected by an experienced tree surgeon or arborist.

Slime Flux and Wetwood

SYMPTOMS Watery substances, sometimes slightly viscous and orange, pink, or white in colour, ooze from the stems of affected plants. The slime flux usually first appears a few inches above soil level and may have an unpleasant odour. The bark beneath may be killed and the upper parts of the stem wilt and die. Not all stems on the plant are affected at the same time.

CAUSE Injury to the stem early in the season when the sap pressure is high causes sap to leak out. The high sugar content of the sap means that it is rapidly colonized by various yeasts, fungi, and bacteria which are responsible for the thickening and discoloration.

CONTROL Cut out affected stems, pruning back to perfectly sound wood, even if this means going below soil level. Attempt to prevent stem injury.

Slime Mould

SYMPTOMS Grey, off-white, yellow, or orange growth develops, most commonly on stems, or on other plant parts. It is usually in the form of numerous tiny spheres which may appear to dissolve when wet, as the spores are released. The plant's growth is not affected, however, and if the slime mould is washed or scraped off, the plant beneath is generally unharmed. Slime moulds are particularly common on grass (*see p.66*).

CAUSE Various species of slime mould – non-parasitic organisms, believed to be most closely related to fungi.

CONTROL No chemical control is available, but none is justified, as the slime moulds are unlikely to cause much damage. Wash off with a strong jet of water.

Slugs

SYMPTOMS Irregular holes are eaten in foliage (*see p.45*), flowers (*see p.61*), and stems (*see p.59*), but examination of plants by torchlight on a mild evening may be necessary to find slugs feeding. Slugs secrete a slimy mucilage from their bodies, which may leave a distinctive silvery deposit where they have been feeding. Some slugs live mainly underground and damage potato tubers (*see p.23*) and tulip bulbs.

CAUSE Several species of slugs occur in gardens. These include the grey field slug (*Deroceras reticulatum*), the large black slug (*Arion ater*), garden slug (*A. hortensis*), and keeled slugs (*Tandonia* and *Milax* species). They can damage a wide range of plants and are present throughout the year, continuing to feed in winter if temperatures are above 5°C (40°F).

CONTROL Slugs cannot be eliminated from gardens, so control measures should be concentrated on protecting vulnerable plants, such as seedlings and soft new growth on herbaceous plants. Slug pellets (*see p.96*) should be scattered thinly amongst the plants. Metaldehyde-based pellets cause slugs to produce an excessive amount of slime and die of dehydration, but they may recover under wet conditions. Slug pellets can kill cats and dogs if a quantity is eaten, so take care in their use; pellets containing ferric phosphate are less toxic. Pellets are best applied in mild, damp weather in the spring and autumn. Poor results will be achieved during cold or dry weather, when slugs are relatively inactive. Non-chemical controls include hand-removal after dark on mild evenings, and use of beer traps. Jam jars or similar containers half filled with beer are sunk in the ground almost to ground level. Slugs, attracted by the odour, fall in and drown. Barriers that slugs and snails are reluctant to cross are available. These generally consist of moisture-absorbent minerals for surrounding plants, and copper tapes for putting around plant pots. Biological control in the form of a nematode, *Phasmarhabditis hermaphrodita*, can be used. It needs to be applied to soils that are moist, well-drained, and above 5°C (40°F); the best times for use are spring and autumn. The microscopic nematodes penetrate the bodies of slugs that have gone into the soil to seek shelter during the day, and release bacteria that cause a fatal disease in infected slugs. Under suitable conditions it gives a significant reduction in slug numbers for at least six weeks. This is a useful method of tackling underground slugs that damage bulbs and potato tubers. These slugs come to the soil surface infrequently and are poorly controlled by slug pellets. Damage to potato tubers can be reduced by lifting the crop as soon as the tubers have matured. There is some variation in the susceptibility of potato cultivars to slugs. The cultivars 'Maris Piper', 'Cara', 'Rocket', 'Kondor', and 'Maris Bard' are all prone to be heavily attacked, while 'Pentland Ivory', 'Pentland Dell', 'Wilja', 'Stemster', 'Charlotte', and 'Estima' are less susceptible.

Small Bulb Fly

See NARCISSUS BULB FLIES *p.158*.

Small Ermine Moths

SYMPTOMS The stems and foliage of various plants, including hawthorn, blackthorn, bird cherry, willow, apple, *Euonymus*, and *Sedum*, become covered with dense, whitish, silk webbing in summer (*see p.58*). Whitish-yellow caterpillars, up to 20mm long, with a row of black dots down each side of their bodies feed on leaves. Heavy infestations cause severe defoliation, with the plant festooned with webbing. The adult moths are silvery-white with many black spots on their wings; they emerge from pupae in the webbing in late summer.
CAUSE Caterpillars of several species of small ermine moth, *Yponomeuta* species. Most have one generation a year but *Y. sedella*, which attacks *Sedum telephium* and its hybrids, has two.
CONTROL Light infestations can be dealt with by pruning out the webbed shoots. Otherwise spray forcefully with a contact insecticide (*see p.94*) to penetrate the webbing.

Smuts

SYMPTOMS Affected areas of the plant are discoloured and may be distorted (*see p.51*). In a few cases, such as dahlia smut (*see p.130*), slight leaf spotting is all that is seen. In other plants, swellings develop and erupt to reveal large masses of soot-black spores (*see p.62*). Stems, leaves, flowers, and seeds may be affected. Some smut infections (for example, anther smuts) are systemic within the plant, but most are restricted only to discrete areas of the plant.
CAUSE Various species of the fungi *Entyloma*, *Ustilago*, and *Urocystis*, which are spread by rain or water splash from nearby infected plant matter, or from the soil surface.

CONTROL Remove affected areas from plants as soon as seen, preferably before the spores are released. Avoid splashing water around plants and clear up all debris at the end of the season. Grow new plants on a fresh site.
• *see also* DAHLIA SMUT *p.130*; DIANTHUS ANTHER SMUT *p.131*; WINTER ACONITE SMUT *p.198*.

Snails

The most common snail pest is the garden snail, *Cornu aspersum*. Banded snails, *Cepaea* species, are less troublesome. Many plants are eaten by snails between spring and autumn. Irregular holes are rasped in the foliage (*see p.45*), and the stems and flowers (*see p.61*), particularly of annuals and herbaceous plants. Snails are mostly active after dark or in wet weather. They secrete a slimy mucilage from their bodies and this often dries to leave a silvery deposit. Snails are less numerous in areas with acid soils, which lack the calcium necessary to form a snail's shell.
CONTROL Snails can be controlled by the methods used against slugs (*see p.183*). Biological control is less effective as snails live mainly above soil level and are unlikely to be infected by the bacterium-carrying nematodes.

Snowdrop Grey Mould

SYMPTOMS Fuzzy, grey fungal growth, typical of a grey mould type of infection, develops around the foliage and/or flower stems of infected snowdrops. The plant starts to rot and as the infection spreads back, the bulb may be killed. Infection often first appears as the foliage comes above ground. Small black sclerotia (fungal resting bodies) develop

around the bulb and foliage. The sclerotia drop off and overwinter in the soil.
CAUSE The fungus *Botrytis galanthina*, which is encouraged by mild winters and wet spring weather. The spores are spread by rain splash and on air currents.
CONTROL Remove infected plants and the soil in their immediate vicinity. Do not grow snowdrops on that site for at least five years.

Soft Rot

See BACTERIAL SOFT ROT *p.112*.

Soft Scale

SYMPTOMS Yellowish-brown, flat, oval scales, measuring up to 3-4mm long, occur on the underside of leaves and are often clustered along the larger leaf veins. Bay, citrus, ivy, ferns, *Ficus* species, and *Schefflera* are frequently attacked, but this pest can infest a wide range of indoor and some outdoor plants. They excrete a sugary substance called honeydew that makes the upper leaf surface sticky and allows the growth of sooty mould.
CAUSE A sap-sucking insect, *Coccus hesperidum*, which breeds continuously throughout the year on house plants and in heated greenhouses.
CONTROL Spray the underside of the leaves of non-edible plants with a systemic or contact insecticide (*see p.94*). Culinary herbs, like bay, fruiting citrus, and also ornamental plants, can be sprayed with an organic insecticide (*see p.89*). Heavily infested plants may need three applications of organic insecticides at two-week intervals in order to control nymphs as they hatch from eggs that are concealed beneath the adult scales' bodies. Alternatively, in greenhouses in the summer, a parasitic wasp,

Metaphycus helvolus, can be introduced for biological control.

Solomon's Seal Sawfly

SYMPTOMS Greyish-white caterpillar-like larvae with black heads and up to 20mm long eat the foliage of Solomon's seal (*Polygonatum*) in early to midsummer, often reducing plants to bare stalks (*see p.47*). The adult insects have black bodies and two pairs of smoky-grey wings. They emerge when the plant is coming into flower, when females lay rows of eggs in the stems, causing purplish scars up to 25mm in length.
CAUSE Larvae of a sawfly, *Phymatocera aterrima*.
CONTROL In spite of severe leaf loss, affected plants normally make adequate growth and flower the following year, so control may not be necessary. The larvae can be removed by hand or sprayed with a contact insecticide (*see p.94*) when damage begins in early summer.

Sooty Mould

Sooty mould is a black or sometimes greyish-green non-parasitic fungus that grows on foliage and other surfaces that have an accumulation of honeydew on them. Various sap-sucking insects, including aphids, whiteflies, mealybugs, and some scale insects and suckers, excrete honeydew. This falls onto the upper surface of leaves below where the pests are feeding. Sooty mould can therefore be found on plants that are themselves free of pests, but which are growing beneath trees that are infested with aphids. Heavy coatings of sooty mould spoil a plant's appearance and reduce the amount of light and air reaching the foliage. There is no direct control for sooty mould, but dealing with

the pest that is causing the problem will allow new growth to remain clean.

Southern Green Shieldbug

SYMPTOMS Pale green shield-shaped insects about 10–12mm long or their nymphs are found clustered on the seed heads of ornamental plants, runner beans, tomatoes, and raspberries, mainly in late summer to early autumn. The nymphs have a more rounded body shape than the adult and they are black or green with many round spots, which may be white, pink, or yellow, on the upper surface. The spotty nature of the nymphs can result in them being misidentified as ladybirds. Their feeding can affect the development of bean pods and fruits on tomato and raspberry.
CAUSE A sap-sucking shieldbug, *Nezara viridula*, that is native to southern Europe but has been established in south east England since 2003. There is a native green shieldbug, *Palomena prasina*, that is more widespread in Britain and which does not cause damage to garden plants. The nymphs of the latter are green with black markings but no rounded spots. Adult native green shieldbugs are pale green with a distinctive blackish brown diamond-shaped area at the rear end of the body. Southern green shieldbugs are uniformly pale green when adult.
CONTROL Britain is probably at the northern edge of this bug's range and the climate may have to get warmer still before this becomes a significant pest. At present, the bugs do not become numerous until early autumn, by which time its host plants are coming to the end of their growing and cropping season. If control is required spray with a contact insecticide (*see p.94*).

Sparrows

The house sparrow, *Passer domesticus*, will peck the flowers of crocus (*see p.65*), primrose, and polyanthus to shreds. Sparrows also eat the flower buds of many other plants, including runner bean and wisteria, although loss of flower buds on these plants is more frequently due to dry soil conditions than bird damage. Freshly sown grass seed is also liable to be eaten, and sparrows can cause an uneven soil surface by dust-bathing in seed beds. Seedling lettuce and beetroot plants can be defoliated.
CONTROL Netting is the only certain means of protecting flowers, seeds, and seedlings from damage caused by sparrows. An alternative is to place sticks in the soil and criss-cross the area with black cotton. Sparrows have difficulty in seeing the cotton and are deterred from feeding. Encourage rapid germination and growth through the vulnerable seedling stage by sowing in warm soil and keeping the young plants watered in dry weather.

Splitting

SYMPTOMS Fruits and occasionally stems may show cracking. This usually runs longitudinally and initially the rest of the plant appears normal. The splitting allows other potential pathogens to enter, and so dieback or rotting may occur later. The cracks may not necessarily become infected and may dry or heal over.
CAUSE An erratic supply of water, or great fluctuations in temperature are the most common causes. Erratic availability of nutrients may be involved.
CONTROL Concentrate on improving growing conditions for plants by regular watering and feeding, and use a mulch

to help prevent soil drying out completely. Keep a watch on damaged areas, and in the case of affected fruits, remove them so they are not infected by secondary organisms.

Spotted-wing Fruit Fly

SYMPTOMS Ripening fruits of raspberry, blackberry, blueberry, currants, strawberry, cherry, plum, peach, and nectarine have sunken patches and holes in the outer skin (*see p.15*). White maggots up to 4mm long are present inside the fruits.
CAUSE A small fly, *Drosophila suzukii*, which lays eggs in a wide range of ripening fruits on wild and cultivated plants. The adults are about 2mm long and are pale brown flies with a darker abdomen. The males, but not the females, have a black spot on their wing tips. Other *Drosophila* species in Britain develop in over-ripe fruit, but only *D. suzukii* attacks fruits that are still ripening. This pest has several generations between April and October, and overwinters as adult flies.
CONTROL Trap adult flies in bottle traps baited with apple cider vinegar. Remove and destroy any unpicked or damaged fruits to prevent the fly breeding in them. Fruits can be sprayed at the onset of ripening with a contact insecticide (*see p.94*), but check the product label to see if it is suitable for the type of fruit being sprayed. Protect low-growing fruits by covering them with ultra-fine netting (mesh size of 0.98mm or under) – put this in place after flowering but before ripening starts.

Springtails

SYMPTOMS Active white insects, up to 2mm long, live in the compost of pot plants (*see p.20*), especially those

growing in peat-based types. They are most frequently seen when plants are being watered, as this washes them out of the bottom of the pot or flushes them up onto the soil surface.
CAUSE There are many species of springtails, but those associated with pot plants are generally *Orthonychiurus* species. These are wingless arthropods with a prominent pair of antennae on their heads.
CONTROL Springtails feed on decaying plant material and fungi. No control measures are required, as they will not harm established plants.

Squirrels

The grey squirrel, *Sciurus carolinensis*, can be delightful to watch as it scampers around a garden, but squirrels can be very destructive. They dig up and eat crocus corms and tulip bulbs, eat flower buds and shoot tips, take nuts (*see p.18*), strawberry, and tree fruits, and also strip the bark from trees such as sycamore, beech, and ash. If bark is lost from all or most of the circumference of the trunk, the growth above that point will dry up and die. Squirrels can also cause annoyance by taking food put out for birds.
CONTROL Shooting, traps, and poisoned baits are used in forests against squirrels; however, these methods are impractical and undesirable in gardens. They are also unlikely to be effective unless used over a wider area than an individual garden. Squirrels are very mobile animals and others will soon move in to occupy the territory of those which have been removed. Netting can give protection to fruits and flowering shrubs such as magnolia and camellia during periods when squirrels are showing an interest in them.

Wire netting is better used for permanent structures such as fruit cages, as squirrels can easily bite through plastic. Netting can also be placed over areas where bulbs and corms have been planted, to deter squirrels from digging them up. Squirrel-proof bird feeders are available from most garden centres.

Stachys Case-bearing Caterpillar

SYMPTOMS In early summer, the foliage of *Stachys* species develops a number of brown blotches where the internal tissues have been eaten out. The small caterpillars responsible for this damage spin silk cases around their bodies. Hairs and other fragments of the host plant's leaves are incorporated into these cases. The caterpillars are almost entirely hidden but, when feeding, the larvae will partly emerge in order to bore into the leaf. The feeding area is limited to how far the caterpillar can reach inside the leaf without leaving its protective case.
CAUSE Caterpillars of a tiny moth, *Coleophora lineolea*, which feed on various types of *Stachys* and related plants. It is most frequently seen on lambs' ears, *Stachys byzantina* (formerly *S. lanata*), where the caterpillars incorporate leaf hairs into the case to give them a silvery-white, furry covering measuring up to 4mm long and 2–3mm wide.
CONTROL The caterpillars can be removed by hand or controlled by spraying with a contact insecticide (*see p.94*) if detected in late spring.

Stem and Bulb Eelworm

This microscopic nematode, *Ditylenchus dipsaci*, occurs as several biological races which have different ranges of host plants. The types that are most likely to be encountered in gardens are narcissus eelworm (*see p.158*), onion eelworm (*see p.161*), and phlox eelworm (*see p.166*). These pests cause distinctive forms of distortion on the host plants indicated, and often cause their death. No control is available.

Storage Rots

SYMPTOMS Fruits, roots, and tubers kept in store may begin to rot. In some cases the infection may already have taken place while the plant was growing, but the storage conditions are such that rotting is encouraged. In other instances, previously perfectly healthy plants or plant parts may actually become infected while they are in store, most often from neighbouring diseased material. The rotting may also be accompanied by the development of fungal growth on the surface of the fruit, roots, or tubers or spreading within the plant.
CAUSE A very wide range of fungi and bacteria, some of which are described as secondary, enter only through injured areas; others, which are primary pathogens, cause damage to previously sound and uninjured tissue.
CONTROL Store only perfect fruits, roots, and tubers. Inspect stores frequently and remove anything showing signs of deterioration, however slight. Provide storage conditions as close as possible to those advised.

Strawberry Black Eye

SYMPTOMS Strawberry flowers develop a distinct dark brown or black centre. The petals and growth of the rest of the plant usually appear quite healthy and normal.
CAUSE Frost damage to the young flower which injures the centre. Flowers damaged in this way do not later form fruit.
CONTROL Use cloches, horticultural fleece, fleece-covered pull-out tunnels, or similar means to protect crops, particularly early cultivars, from frost. Pinch out affected flowers.

Strawberry Green Petal

SYMPTOMS The flowers are reduced in size and bear green petals. Plants produce few, if any, fruits and those which are formed are very small and often distorted. The plant itself is generally somewhat stunted and distorted and may show yellow markings on the leaves. Once flowering is over, the older leaves may turn red.
CAUSE A phytoplasma, known as strawberry green petal, which is transmitted by leafhoppers.
CONTROL Remove and burn infected plants promptly and do not propagate from them.
• see also LEAFHOPPERS *p.151*.

Strawberry Grey Mould

SYMPTOMS Fruits develop normally, but as they start to ripen and colour, pale brown necrotic patches develop on their skin. The fruit then deteriorates rapidly, becoming soft and brown. Shortly afterwards a fuzzy, grey fungal growth develops over the surface (*see p.15*). Numerous spores are given off if the fruit is disturbed.
CAUSE The fungus *Botrytis cinerea*, which is very common on both living and dead plant material. The spores enter the open flowers but remain dormant until the fruits start to ripen. The spores are then

readily spread by either rain or water splash and on air currents. Strawberries, which are not infected at the flowering stage, but have formed healthy fruits, which have been injured in some way (by slugs, for example), may subsequently become infected by spores from these rotting fruits. The fungus may carry over from year to year either on plant debris or as sclerotia in the soil.

CONTROL Control is very difficult, because the fungus is so widespread. Remove all dead and injured plant parts and fruits before they become infected. Remove infected fruits promptly. Do not leave plant debris lying around the garden. Avoid wetting the strawberry plant foliage and flowers when watering.

Strawberry Leaf Spot

SYMPTOMS Reddish-purple spots develop on the leaves (*see p.24*), each with a grey centre. Off-white fungal growth and, occasionally, numerous tiny, raised, black fungal fruiting bodies may develop on the spots. The spots may join together and cover much of the leaf surface (*see p.31*). They may also then spread to the flower and leaf stems. Although it spoils the appearance of the foliage, this infection usually has little effect on the vigour of the plant – it may occur in spring but more often appears fairly late in the season.

CAUSE The fungus *Mycosphaerella fragariae*, which forms sclerotia (black fungal resting bodies) which may overwinter in the soil and then cause the earlier infections in the spring.

CONTROL Remove badly infected leaves. Clear up all infected debris at the end of the season.

Strawberry Mite

SYMPTOMS Strawberry plants make poor growth in the latter half of the summer, with new leaves appearing puckered and failing to expand to their full size. Similar symptoms can be caused by some virus diseases. Strawberry mites, which are whitish-brown and less than 1mm long, live in the crown of the plant and feed on the developing leaves. A x10 hand lens is necessary to see them.

CAUSE A tarsonemid mite, *Phytonemus pallidus*, which feeds by sucking sap. It tends to be more troublesome in hot, dry summers.

CONTROL None of the pesticides available to home gardeners will control strawberry mite. Destroy affected plants and start a new bed with certified stock strawberry plants that will be free of problems such as strawberry mite and virus diseases.

Strawberry Red Core

SYMPTOMS Groups of stunted plants appear in late spring, bearing reddish-brown, strangely stiff leaves. The roots are discoloured and have a red inner core.

CAUSE The fungus-like organism *Phytophthora fragariae*, which is encouraged by heavy, wet soils and produces resistant spore stages from the deteriorating root material, releasing these into the soil where they may persist for more than ten years. The fungus is easily introduced on contaminated plants or on soil adhering to plants, footwear, tools, etc.

CONTROL None available. Remove infected plants immediately, together with the soil in the vicinity of their roots, and burn them. Do not

grow strawberries on the same site again and avoid moving soil from that area to other parts of the garden.

Strawberry Seed Beetles

SYMPTOMS Seeds are removed from the surface of ripening strawberry fruits, causing small, dried-up brown patches. The culprits are fast-moving black beetles, about 15mm long, with slender reddish-brown or black legs. They are mainly active after dark. In addition to eating the seeds, they sometimes eat into the fruits, but it is difficult to distinguish this form of damage from that caused by slugs.

CAUSE Carabid or ground beetles, *Pterostichus* species and *Harpalus rufipes*. At other times of the year these beetles feed on weed seeds.

CONTROL Keep the strawberry bed free of weeds to discourage a build-up of seed beetles. Jam jars sunk into the soil will act as pit-fall traps, but they will also capture beneficial ground beetles that are predatory in their feeding habits. Because ripe strawberries have such thin skins, insecticide controls are undesirable while ripe fruit is present, and nor should plants be sprayed when in flower.

Strawberry Viruses

SYMPTOMS Stunting and distortion of the whole plant, usually combined with stunting, distortion, and yellowing of the leaves. The discoloration is seen as distinct patterns – mosaic, ring spot, streaking, or yellowing around the leaf edge. Infected plants fail to thrive and may not flower or fruit properly.

CAUSE Various viruses, most commonly strawberry yellow edge or little leaf virus, crinkle

(spread by aphids), arabis mosaic, strawberry ring spot, and tomato black ring (spread by eelworms).

CONTROL None currently available. Remove and burn infected plants as soon as the symptoms are noticed. Control aphids and leafhoppers with appropriate insecticides. Do not grow strawberries on that site again. Do not propagate from infected plants, as their runners will also be infected, even if without symptoms. Always buy certified stocks of strawberries, guaranteed to be virus-free when purchased. Certain cultivars are known to be particularly susceptible to particular viral infections and these include 'Royal Sovereign' and 'Redgauntlet' (yellow edge).
• *see also* APHIDS *p.108;* LEAFHOPPERS *p.151.*

Strawberry Yellows

SYMPTOMS Leaves turn yellow. The entire plant may be affected or it may appear blotched yellow with part of the leaves remaining green. Symptoms usually appear most severe in midsummer.

CAUSE The precise cause is not known, but it is thought to be largely the result of unsuitable soil conditions. A soil which is excessively wet or which is lacking adequate nutrients will generally encourage these symptoms.

CONTROL Determine which adverse soil conditions are involved and attempt to improve problems such as poor drainage. Apply a fertilizer containing trace elements.

Suckers

These are small sap-sucking insects that cause damage on various garden plants during the spring and/or summer months. They are also known as psyllids and have very distinctive

immature nymphs that appear to be squashed flat when viewed from the side. The adult insects have a more rounded body shape and resemble winged aphids. They are 2–3mm long when adult. An immature nymph has wing pads on its upper surface, which become larger each time the nymph sheds its skin, until the adult stage is reached with fully formed wings. Some suckers cause gall formation on their host plants while others live exposed on the leaf surface.
• *see also* ALDER SUCKER *p.106;* APPLE SUCKER *p.110;* ASH SUCKER *p.113;* BAY SUCKER *p.113;* BOX SUCKER *p.118;* ELAEAGNUS SUCKER *p.134;* EUCALYPTUS SUCKER *p.135;* PEAR SUCKER *p.165.*

Sweet Chestnut Gall Wasp

See ORIENTAL CHESTNUT GALL WASP *p.162.*

Sweetcorn Smut

SYMPTOMS Individual kernels on the cob become hugely swollen and deformed (*see p.17*). Each is pale grey and ruptures to release large quantities of powdery black spores. These spores may become mixed with rain to produce a black liquid, which runs down the plant. Occasionally, other parts of the plant may be affected. The infection is not systemic and healthy and infected cobs may develop on the same plant.
CAUSE The fungus *Ustilago maydis*, which is quite common when summer temperatures are high. The spores are carried on air currents and by rain and water splash, and either infect other sweetcorn plants directly or persist in the soil.
CONTROL None available. Remove infected cobs before

the swellings have a chance to rupture. Remove and burn all infected plant debris at the end of the season. Do not grow sweetcorn on that site for at least five years.

Swift Moths

SYMPTOMS Long, slender, white caterpillars, up to 35mm long, with brown heads are found in the soil (*see p.19*). They feed on plant roots, especially of herbaceous perennials, and can also damage bulbs, corms, and rhizomes. Heavy infestations cause plants to lack vigour or may be killed. Some species take two years to complete their larval development, so the caterpillars may be unearthed at any time of year. The adult moths emerge and lay eggs from late spring to late summer.
CAUSE There are several species of swift moth, but the most troublesome in gardens are the ghost swift moth, *Hepialus humuli*, and the garden swift moth, *Korscheltellus lupulina*.
CONTROL Good weed control can help avoid infestations, as the moths often deposit eggs where there is a dense cover of vegetation. Any caterpillars found during digging or division of perennials should be crushed and thrown to the robins. Pesticide control is difficult, as the older and more damaging larvae are more tolerant. Pathogenic nematodes are available for controlling soil-dwelling caterpillars.

Sycamore Sooty Bark Disease

SYMPTOMS Leaves wilt and die back on infected branches, but remain attached to the tree. If the bark is pared off, green or dark yellow staining may be visible (*see p.52*). Although these are the first symptoms, the problem may only become apparent at a later stage, when

black spore masses or blister-like outgrowths develop on the bark of affected stems and when the tree does not come into leaf in the spring. These outgrowths enlarge and elongate and may be associated with the extensive death and breaking up of the bark. Death of the tree usually occurs within a couple of years.
CAUSE The fungus *Cryptostroma corticale*, which appears to be encouraged by warm temperatures and, in particular, by hot summers.
CONTROL Remove infected trees promptly to limit the spread of the disease.

T

Tar Spot of Acer

SYMPTOMS Large, slightly raised, shiny black blotches develop on the leaves (*see p.25*), each of which can measure up to 15mm in diameter.
CAUSE The fungus *Rhytisma*, in particular *R. acerinum*, which can overwinter on fallen infected leaves. In spring, it produces spores in a gelatinous coating which adhere to young leaves as they are produced. Leaves may fall prematurely, but this is not always the case and the vigour of the tree is rarely significantly affected.
CONTROL None available, but it is not necessary. Rake up and dispose of affected leaves as they fall.

Tarsonemid Mites

SYMPTOMS Mainly greenhouse and house plants are attacked by minute, whitish-brown mites that infest the developing leaves at shoot tips and the flower buds. Their feeding activities cause distorted flowers with white or brown flecks on the petals. New foliage fails to expand to its full size

and becomes progressively more stunted and distorted as the infestation develops (*see p.42*). In severe attacks, growth comes to a complete halt and the stems have a brownish scarring. Tarsonemid mites can be a particular problem on cyclamen, begonias, pelargoniums, busy Lizzies, and many other pot plants.
CAUSE Microscopic mites that feed by sucking sap. These include broad mite, *Polyphagotarsonemus latus*, and cyclamen mite, *Phytonemus pallidus*. The latter also has races that attack some garden plants – see MICHAELMAS DAISY MITE *p.156* and STRAWBERRY MITE *p.187.*
CONTROL None of the pesticides available to gardeners control tarsonemid mites. Infested plants should be destroyed. This pest usually comes into a greenhouse when an infested plant is acquired. Take care to avoid this by purchasing plants from reputable suppliers.

Thrips

Thrips are narrow-bodied, elongate insects up to 2mm long. They feed by sucking sap and generally cause a fine silvery-white mottled discoloration of the upper leaf surface. Thrips often feed from the upper leaf surface, unlike most sap-feeding pests that remain on the underside of leaves. Adult thrips are usually black with two pairs of wings that are heavily fringed with hairs. The wings are folded back along the body when not in use and so are not readily visible. Some thrips have black and white areas on their wings, which give the insects a banded appearance. The immature nymphs are wingless and creamy yellow in colour. Some species, like western flower thrips, can spread certain plant virus diseases on their mouthparts. To control specific thrips commonly found in gardens and greenhouses, *see* BANDED PALM THRIPS *p.113;* GLADIOLUS THRIPS *p.140;* GLASSHOUSE THRIPS *p.141;* ONION THRIPS *p.161;* PEA THRIPS *p.163;* PRIVET THRIPS *p.173;* WESTERN FLOWER THRIPS *p.196.* Others may be controlled by spraying with a contact or systemic insecticide *(see p.94)* when signs of activity are seen.

Thuja Blight

SYMPTOMS Small, black fungal fruiting bodies, measuring about 1 mm across, are embedded in the foliage of *Thuja plicata.* These later split open as the spores are released, leaving small, brown sunken pits, or holes, in the leaves. Foliage, particularly towards the base of the plant, may turn brown *(see p.43)* and die back if the infection is extensive.
CAUSE The fungus *Didymascella thujina.* The spores are air-borne and most infections take place in late summer or early autumn, especially in humid or damp weather. They may overwinter in mucilage attached to the damaged areas.
CONTROL Prune out infected areas promptly, before the spores are released. Remove severely infected stems.

Thunderworms

Thunderworms derive their common name from the fact that they are often seen after heavy rain, particularly between late spring and midsummer. Under damp conditions the nematodes, *Mermis* species, emerge from the soil and climb onto plants where they can be seen writhing about. They are less than 1mm in diameter, but can be up to 60mm long. They are creamy white, although blackish body contents can often be seen through the semi-transparent skin. Eggs are deposited on foliage and if the eggs and leaf are subsequently eaten by a suitable animal, such as a caterpillar, grasshopper, or earwig, the eggs will hatch and develop as parasites in the insect's gut. Eventually the fully-formed nematode emerges and goes into the soil, killing the host. Thunderworms can only develop inside insects and so pose no threat to humans or pets.

Toadstools

See TURF TOADSTOOLS *p.192.*

Tomato Blight

SYMPTOMS Brown patches develop on the leaves, causing them to dry and curl. The stems may also show blackening. Infected fruits develop a brown discoloration and may appear to shrink inwards. They may rot, either on the plant or a few days after being picked. Outdoor tomatoes are affected more frequently than those grown under glass. Potatoes may also be attacked *(see* POTATO BLIGHT *p.170).* Tomatoes usually succumb slightly later than potatoes.
CAUSE The fungus *Phytophthora infestans,* which is encouraged by moist or wet weather, particularly if it occurs in late summer. The spores are carried on air currents or by water splash, and need a minute film of moisture on the foliage if they are to cause damage.
CONTROL In a wet yet warm season check plants regularly and bin or burn any infected material; if seen early and removed promptly, and weather conditions become drier, it may be possible to save an infected plant. Grow tomatoes in a greenhouse or pop-up clear polythene shelter to reduce the risk of blight infection. Grow cultivars showing a good degree of resistance, such as 'Losetto' and 'Crimson Crush'.

Tomato Blotchy Ripening

SYMPTOMS Patches of hard, yellow or green flesh remain unripened *(see p.16).* These areas may be anywhere on the fruit and, unlike tomato greenback *(see p.189),* this disorder does not follow a regular pattern. The patches are not clearly demarcated. The damage is only visible on mature fruits, and those on the lower trusses are generally the most frequently affected.
CAUSE This disorder is believed to be caused by some form of malnutrition, and a deficiency of potassium is probably the single most important factor. Dry soil or compost or a poorly functioning root system may also play a part. High temperatures in the greenhouse also increase the damage.
CONTROL Keep the plants well fed, in particular ensuring that a high potash feed is used. Keep the greenhouse well ventilated and the plants well watered.

Tomato Cyst Eelworms

See POTATO CYST EELWORMS *p.171.*

Tomato Fruit Splitting

SYMPTOMS Tomatoes develop normally but start to split, usually just before picking *(see p.16).* The splits may become infected by secondary organisms such as grey mould *(see p.142),* and may then rot. Occasionally the split surface temporarily dries up and the fruit does not rot.
CAUSE Erratic growth and fruit swelling. This is usually caused by an irregular supply of water or wildly fluctuating temperatures.
CONTROL Ensure that plants are adequately watered at all times and are not subjected to temperature extremes. In greenhouses, provide shading and ventilation. Remove affected fruits before secondary organisms invade.

Tomato Ghost Spot

SYMPTOMS On unripe fruits, pale green or yellow rings of discoloration appear. As the fruits develop and colour, the rings gradually turn yellow or pale orange. The fruit's development is not affected and it is perfectly edible.
CAUSE The fungus *Botrytis cinerea.* Spores are readily spread by rain or water splash and on air currents. The fungus is widespread on both living and dead plant matter. The rings of discoloration result from a hypersensitive reaction to the fungal spores.
CONTROL As affected fruits rarely deteriorate, there is no need to remove them, but ensure you eliminate sources of infection in the area.

Tomato Greenback

SYMPTOMS As greenhouse tomatoes begin to ripen, a partial or complete ring of clearly demarcated leathery tissue remains around the stalk end. It does not ripen properly, remaining either green or yellow *(see p.16).* The rest of the plant develops normally.
CAUSE A deficiency that is under genetic control, and is brought about by several different factors either alone or, more commonly, in combination. Low levels of potassium and phosphorus can be involved but hot, bright sunlight, which causes heat injury, is probably the most important single factor.

CONTROL Provide adequate shading for the plants, particularly once the fruits start to develop. Do not remove too much of the plant's foliage, as this acts as a good shading material. Ventilate the greenhouse to prevent excessively high temperature levels. Keep the plants well fed, ensuring adequate supplies of potash and phosphate in particular. Choose tomatoes that show resistance to this disorder. The following are suitable, but in some cases it appears that the resistance may be breaking down, and even these may show some symptoms in very hot weather: 'Craigella', 'Moneymaker', 'Shirley', and 'Tigrella'.

Tomato Leaf Mould

SYMPTOMS Yellow blotches develop on the upper leaf surface. Beneath each spot is a grey or faintly purple velvety fungal growth. The lower leaves are usually attacked first, but all leaves may eventually be affected. The leaves turn brown, wither, and dry but remain hanging on the plant. A secondary infection of grey mould may mask the symptoms of leaf mould. While greenhouse tomatoes may be frequently affected, those that are grown out of doors rarely develop this disease.
CAUSE The fungus *Passalora fulva*, which spreads rapidly in warm, humid conditions. The spores are carried on air currents and may also be spread by insects, and on clothing. The fungus overwinters on infected plant debris and on greenhouse structures.
CONTROL Remove and burn all infected plants at the end of the season and thoroughly clean out the greenhouse. Ensure that the ventilation is adequate and attempt to minimize excessively humid conditions by spacing plants out well and by carefully removing

some of the foliage. Grow tomatoes that show resistance to this disease, such as 'Shirley'.

Tomato Moth

SYMPTOMS The foliage and fruits of tomatoes are eaten by caterpillars. Newly-hatched larvae feed together by grazing away the lower leaf surface, causing the remaining upper surface to become papery-white. In greenhouses, such damage can often be seen first on leaves that are growing close to the glass. As they grow larger, the caterpillars disperse and eat holes right through the foliage, and also bore into green and ripening fruits. When fully grown, the caterpillars are up to 35mm long and either brown or pale green in colour, with a yellow line – marked by many tiny white dots – along each side of the body.
CAUSE Caterpillars of a moth, *Lacanobia oleracea*, which is also sometimes known as the bright-line brown-eye moth.
CONTROL Look for leaves bearing newly-hatched caterpillars and remove them. Otherwise spray with a contact insecticide *(see p.94)* when caterpillars are seen.

Tomato Pith Necrosis

SYMPTOMS The upper leaves wilt and turn yellow as the fruits start to ripen. Dry, dark brown to black patches develop on the stems. If cut longitudinally, the internal tissues of the stem can be seen to be discoloured, usually black in the pith cavity or in the pith itself.
CAUSE The bacterium *Pseudomonas corrugata*, which is probably spread by rain and water splash and by handling, and possibly on any tools used when removing side shoots.
CONTROL Remove and burn infected plants promptly

to reduce the risk of the infection spreading.

Tomato Viruses

SYMPTOMS Several different viruses may affect tomatoes and the symptoms they produce vary, not only with the virus involved, but also depending on the tomato cultivar and the prevalent growing conditions. Common leaf symptoms include yellow mottling or mosaicing, distortion, and the production of unusually small leaves. With tomato mosaic, in addition to these symptoms, the youngest leaves may curl downwards. With fern leaf, the leaf blades are extensively narrowed and malformed and may be mistakenly assumed to be damaged by weedkiller. Other symptoms of virus infection may include yellow streaking, stunting, and the bronzing of the young fruits, yellow streaking of the fruits, or failure of the plant to set fruit.
CAUSE The viruses most commonly involved, either alone or in combination, are tomato spotted wilt, streak, tomato mosaic, potato virus X, and cucumber mosaic. All these viruses are readily spread by handling and in some cases by aphids or other sap-sucking pests. Many of these viruses have very wide host ranges, so they may be introduced from unrelated plants in the garden, or may be spread to these from infected tomatoes.
CONTROL Remove and dispose of infected plants promptly. Avoid handling plants unnecessarily and always wash hands thoroughly after doing so. Control aphids and other sap-sucking pests. Remove and burn all crop debris at the end of the season. Grow tomato cultivars described as showing resistance. At present, only the

cultivar 'Shirley' is listed as showing resistance to tomato mosaic virus.

Tortrix Moths

See CARNATION TORTRIX MOTH, *p.123*.

Tulip Fire

SYMPTOMS As the foliage emerges above ground it may be withered, distorted, and covered in numerous buff-coloured flecks *(see p.36)*, or scorched. It soon becomes covered in fuzzy, grey fungal growth. Elliptical grey-brown marks also develop on the stalks. The plants may fail to mature, though some may flower, or flower buds may develop but fail to open. Those that open normally may develop numerous small, bleached spots on the petals, each one being caused by the plant's hypersensitive reaction to the spores. The flower stems may topple over and in wet weather the whole plant may become covered in the fungal growth and then rot rapidly. Bulbs may rot or may appear firm but bear black sclerotia (fungal resting bodies) clustered around the neck *(see p.21)*.
CAUSE The fungus *Botrytis tulipae*, which overwinters as pinhead-sized sclerotia in the soil or on infected bulbs. The spores produced on all the above-ground parts of the plant are readily spread on air currents and by water splash.
CONTROL Remove infected foliage promptly. Do not plant tulips on the same ground for more than two years in succession, as this may encourage build-up of the sclerotia in the soil. Remove and dispose of severely infected plants, together with the soil in the immediate vicinity of the roots. Check bulbs carefully

before planting and only use those which appear to be perfectly healthy and free from signs of sclerotia. Late planting may also decrease the chances of the disease developing. On sites where the disease has been identified, do not replant with tulips for at least three years.

Tulip Grey Bulb Rot

SYMPTOMS Bulbs fail to emerge above ground, leaving gaps in the bed. If shoots are produced, they are extensively distorted and soon wither and die off. There are no signs of fungal growth on the affected foliage. Infected bulbs turn grey and dry as they start to rot. The roots and the basal plate may be all that remain. Soil tends to adhere to infected bulbs, especially at the noses. A dense, white or grey felty fungal growth develops between the scales and on the outside of the bulb. Black sclerotia (fungal resting bodies) develop in among the fungal growth and may reach up to 7mm in size, remaining viable in the soil for anything up to five years.
CAUSE The fungus *Rhizoctonia tuliparum*, which may attack many bulbs, including alliums, amaryllis, brodiaeas, chionodoxas, colchicums, crocus, *Eranthis*, fritillaries, gladioli, hyacinths, iris, ixias, lilies, daffodils and narcissi, snowdrops, scillas, and tulips.
CONTROL Remove and burn affected plants, together with the soil around them. Do not grow susceptible bulbs in that soil for at least five years.

Turf Dollar Spot

SYMPTOMS Grass dies off in patches, becoming straw-yellow (*see p.66*). Patches usually measure between 7.5–10cm in diameter.

CAUSE The fungus *Sclerotinia homeocarpa*, which is most prevalent during warm, damp spells in early autumn, and on lawns which have become compacted.
CONTROL Improve drainage by spiking the lawn in autumn. Avoid using grass seed mixtures containing a high proportion of *Festuca* species (the fescues) as these are particularly prone to this infection.

Turf Fairy Rings

SYMPTOMS Rings develop on the turf, increasing in size as the fungus spreads. Around the periphery of the ring one or more zones of very lush, green grass develop. There is also an inner ring of similarly luxuriant grass. The area between these two rings is dead and brown, or the soil may be completely bare. Occasionally, incomplete rings or other formations develop. The fungus grows just beneath the soil surface and around the bases of the grasses, forming a dense, off-white fungal growth or mycelium. Pale brown, slender toadstools, each 4–10cm tall, may develop at the outer part of the middle zone in summer or autumn (*see p.67*).
CAUSE The fungus *Marasmius oreades*. The death of the middle zone of grass is caused by drought – the dense fungal mycelium is so water-repellent that water cannot penetrate to the grass roots in this area.
CONTROL Rake up and dispose of all toadstools, preferably before their caps open to release spores. Mow the infected area of grass separately and collect up and burn mowings from this area in order to limit the risk of the fungus spreading. Spiking the infected areas may help improve the appearance of the grass, as this will allow more moisture to penetrate. Control

by removal of all infected areas is possible, but an extreme undertaking. You would need to dig out the affected grass and topsoil to a depth of at least 30cm, and then, in addition, dig out the grass in the surrounding area, extending about 30cm beyond the edges of the ring and to the same depth. You would then need to dispose of all affected grass and soil, taking care not to contaminate surrounding areas of lawn or grass, and would finally need to re-fill the trench with topsoil and re-seed.

Turf Ophiobolus or Take-all Patch

SYMPTOMS Bleached patches develop and spread. Fungal growth may be visible in among dead, brown roots and stolons of the grasses.
CAUSE The fungus *Gaeumannomyces graminis*, which is very damaging on lawns composed largely of fine grasses. It is also very prevalent on chalky or limey soils or on lawns which have been limed.
CONTROL Remove affected areas and several inches of soil beneath the remains of the grasses. Use fresh topsoil to fill in, and then re-seed. Avoid liming the soil.

Turf Red Thread

SYMPTOMS Patches of either reddish or bleached grass develop, up to 7.5–8cm in diameter. In among the grasses and attached to them, dark pink, horn-like, gelatinous, branching strands of fungus develop (*see p.67*). These strands of fungus subsequently turn pale pink and slightly fluffy. The grass is rarely killed, but it is weakened and its appearance spoiled.
CAUSE The fungus *Laetisaria fuciformis*, which is most common on lawns composed

mainly of fine lawn grasses such as the fescues. Perennial rye grass and annual meadow grass are less susceptible and are rarely affected. Inadequate nitrogen levels and poor aeration encourage the disease. It is usually most prevalent in the year following a drought, because the dry weather will have made the usual fertilizing routine very difficult.
CONTROL Improve drainage by aeration and ensure the grass is adequately fertilized. Apply nitrogen as sulphate of ammonia and this may cure the problem. Consider using a suitable fungicide (*see p.94*).

Turf Snow Mould

SYMPTOMS Small patches of dying grass develop, gradually turn brown, then increase in size and may even join together. In damp weather the dying patches may become covered in a pale pink, fluffy fungal growth (*see p.67*). Snow mould is most commonly seen in the autumn months of the year and during mild spells in winter.
CAUSE The fungus *Microdochium nivale* and occasionally other fungi. Snow mould is often encouraged by poor aeration and by damp conditions. If the grass is walked on in winter after a fall of snow, the disease frequently develops on these areas.
CONTROL Carry out regular routine spiking and use any other methods for improving aeration. Brush off early morning dews and avoid walking on the grass during wet or cold weather. Consider using a suitable fungicide (*see p.94*).

Turf Thatch Fungal Mycelium

SYMPTOMS Bleached, yellowed, or occasionally reddish patches of grass develop on the lawn: the grasses are severely stressed or may be killed (see **p.66**). A dense mat of fungal growth grows around the bases of the grasses.

CAUSE Various fungi, often described as thatch fungi, which are not directly pathogenic but live on organic debris, including thatch (remains of old grasses, mowings, and other debris), which is found at the base of the grasses. These fungi form a strongly water-repellent layer, which causes the grass to die from drought stress. Lawns that have not been properly maintained, and newly-laid turf, are the most susceptible.

CONTROL Carry out all necessary lawn maintenance: scarification, spiking and weed and moss control. Do not allow mowings to be redistributed over the lawn. Buy turf only from a reliable, reputable supplier and check it carefully on delivery. Where necessary, the only solution may be the removal of badly infested areas followed by re-seeding or re-turfing.

Turf Toadstools

SYMPTOMS The grass grows normally, but toadstools develop in random arrangements on the lawn. Toadstools may also develop in straight lines, following the path of tree roots or other buried woody debris.

CAUSE Various fungi – including ink caps, field mushrooms, and puff balls – which do not attack the grasses but feed on organic debris in the soil.

CONTROL Brush off toadstools as soon as they appear, preferably before they release their spores. If the problem is recurrent, the only long-term solution may be to excavate beneath the grass and remove the tree root or other material on which the fungus is feeding.

Turnip Gall Weevil

SYMPTOMS The adult beetles lay eggs on the roots of turnip, swede and leafy brassicas such as cabbage, cauliflower, Brussels sprouts, and broccoli. The feeding activities of the white, legless grubs cause the plant to produce rounded swellings that enclose the grubs. This can be confused with roots swollen by infection with clubroot disease (see **p.127**). They can be distinguished by cutting the swellings in half. Turnip weevil galls are hollow and contain grubs, or have circular exit holes if the grubs have left to pupate in the soil, whereas clubroot galls are solid and less regularly rounded.

CAUSE Larvae of a beetle, *Ceutorhynchus assimilis*. Infestations on the roots of leafy brassica plants have little impact on their growth, but this pest is more troublesome when it is developing on edible roots such as turnip and swede.

CONTROL There is no effective pesticide control and so the presence of galls has to be tolerated.

Two-spotted Mite

See GLASSHOUSE RED SPIDER MITE *p.141*.

V

Vapourer Moth

SYMPTOMS Hairy caterpillars, up to 30mm long, occur during late spring to late summer on a wide range of trees and shrubs (see **p.47**). They are greyish-black with numerous orange-red spots on the body; there are four clumps of pale yellow hairs, resembling shaving brushes, on the upper body surface and long tufts of black hairs pointing forward from the head end, and backwards from the rear end. The caterpillars generally occur

in small numbers, but are sometimes abundant enough to cause significant damage.

CAUSE Caterpillars of a moth, *Orgyia antiqua*, which emerge in mid- to late summer. The female is wingless and after mating she lays her eggs on the silk cocoon from which she has emerged. The eggs are the overwintering stage.

CONTROL Hand picking will deal with light infestations. This pest is rarely so abundant that it is necessary to use a contact insecticide (see *p.94*).

Verticillium Wilts

SYMPTOMS Plants show wilting of the foliage (see **p.44**). Most or all of the leaves on an affected branch or stem show the symptoms, but the whole plant is unlikely to be affected simultaneously. The leaves may show interveinal discoloration (usually as yellowing or browning between the veins) and they then die. Stem death follows shortly, but it may be several years before a large shrub is killed completely. Smaller shrubs or herbaceous perennials and annuals may be killed within a single season. If the bark is removed from an affected stem, the vascular tissue (conducting tissue) is seen to be discoloured, usually by brown or purple-brown streaks, which run the length of the stem (see **p.50**), but are usually most apparent towards the base. Roots have a central core of discoloured conducting tissue.

CAUSE *Verticillium dahliae* and *V. albo-atrum*. These are both found in infested soil, on plant debris, and within the plant sap. *V. dahliae* forms fungal resting bodies, which may persist in the soil. The host ranges are wide, and as well as many cultivated plants, several common garden weeds may harbour the infection too.

CONTROL None available. Remove affected plants promptly, together with the soil surrounding their roots. As the infection can be spread on pruning tools, always thoroughly clean any tools after they have been used on an infected plant. Avoid replanting any of the following on the same site: maples, Indian bean tree (*Catalpa bignonioides*), Judas tree, smoke bush (*Cotinus*), golden rain tree (*Koelreuteria paniculata*), stag's horn sumach, lime, berberis, daphnes, quince, apples, pears, or *Prunus*. Some plants show good resistance, including beech, birch, *Cercidiphyllum*, *Gleditsia*, hawthorn, *Liquidambar*, mulberry, plane, and willow. Conifers are not susceptible to this pathogen.

Viburnum Beetle

SYMPTOMS The foliage of viburnums, particularly *Viburnum tinus*, *V. opulus*, and *V. lantana*, is reduced to a lacework of damaged leaves during the summer. Most of the damage is caused by the creamy-yellow, black-spotted larvae, up to 7mm long, during late spring to early summer. The greyish-brown adult beetles, 5–5mm long, emerge from pupae in the soil in late summer and cause additional, but less extensive, damage.

CAUSE Adults and larvae of a beetle, *Pyrrhalta viburni*. This pest overwinters on its host plant as eggs, which are placed in small pits in the bark of the stems.

CONTROL Look for signs of damage on the new foliage in late spring, and spray with a contact or systemic insecticide (see *p.94*) while the larvae are still small and not yet at their most damaging.

Viburnum Scale

SYMPTOMS The upper leaf surface on *Viburnum*

tinus and ivies (*Hedera* species) becomes sticky with honeydew excreted by this pest. Black sooty moulds often develop on the honeydew. The scales live on the underside of leaves, usually next to a leaf vein. They are pale yellow, flat, oval creatures up to 4mm long. When fully mature in late spring, the females are pale brown and they cover themselves with a fluffy, white waxy material in which they deposit their eggs.

CAUSE A sap-sucking insect, *Lichtensia viburni*. It is more troublesome on plants growing against a wall or in other sheltered places.

CONTROL Spray thoroughly with a systemic or contact insecticide (*see p.94*), especially on the underside of the leaves, in midsummer to control the more vulnerable newly-hatched scale nymphs.

Viburnum Whitefly

SYMPTOMS The only viburnum affected is laurustinus (*Viburnum tinus*), but this pest also attacks strawberry trees (*Arbutus*). In midsummer, white-winged insects, 1–2mm long, occur on the underside of the younger leaves. The immature nymphs are flat, oval, yellowish-green, scale-like objects found on the lower leaf surface in late summer. More easily seen is the overwintering pupal stage; these are black, oval objects 1mm long, which are encrusted with a white, waxy substance (*see p.27*). Heavy infestations may cause soiling of the foliage with honeydew and sooty moulds.

CAUSE Adults and nymphs of a sap-sucking insect, *Aleurotuba jelinekii*.

CONTROL Light infestations can be ignored, since the pest has little direct impact on its host plant's growth. If necessary, spray the underside of the leaves

with a contact or systemic insecticide (*see p.94*), when adult whitefly are present.

Vine Erinose Mite

SYMPTOMS During mid- to late spring, parts of grape vine leaves bulge upwards (*see p.28*), the underside of these deformed areas is covered in a dense mat of fine, whitish hairs. As the summer progresses, the hairs darken and become yellowish or brown. The upper surface of the blistered areas may also become discoloured.

CAUSE Microscopic gall mites, *Colomerus vitis*, which suck sap from the foliage and secrete chemicals into the leaf to induce the excessive growth of hairs. Apart from the leaf distortion, the mites seem to have little impact on the plant's vigour and ability to produce grapes.

CONTROL No effective pesticides are available. Removing affected leaves may check infestations, but as the mites are of little consequence, heavy defoliation would be more harmful than tolerating their presence.

Vine Weevil

SYMPTOMS Both the adult and larval stages cause damage. The larvae are plump, creamy white, legless grubs with brown heads. They are up to 10mm long and have slightly curved bodies. They live in the soil (*see p.19*) and feed on roots, which are either severed or, on woody plants, have the outer bark removed. This results in plants making slow growth, followed by wilting and death when insufficient roots remain to sustain the plant. The grubs also bore into the tubers of plants such as begonia and cyclamen. A wide range of garden and indoor plants is attacked and plants grown

in pots or other containers are particularly at risk. Larval damage is mostly seen between early autumn and mid-spring. The adult weevils are 9mm long, with dull black, pear-shaped bodies and antennae that are bent at right angles about half-way along their length. They are slow-moving beetles that emerge at night and eat irregular notches from the leaf margins of many plants (*see p.45*), but especially rhododendrons, *Euonymus japonicus*, and hydrangea. The adults can be found at any time from spring to autumn.

CAUSE Larvae and adults of a beetle, *Otiorhynchus sulcatus*. All the adults are female and capable of laying many hundreds of eggs over a period of several months. Other species, such as *O. armadillo*, *O. crataegi*, and *O. salicicola* also cause leaf damage on a wide range of plants.

CONTROL A systemic insecticide applied as a compost drench (*see p.94*) to container-grown ornamental plants in late summer will protect them from vine weevil grubs. Biological control with pathogenic nematodes, *Steinernema kraussei* or *Heterorhabditis megidis*, is an alternative means of protecting plants. Water nematodes into the potting compost or garden soil in late summer, before the grubs have grown large enough to cause serious damage. Nematodes are ineffective in heavy soils, or those which are dry or colder than 5°C (41°F). Adult weevils can be searched for by torchlight and removed from plants. The leaf damage caused by the adults is rarely sufficient to affect plant growth, but if it is causing concern, plants can be sprayed at dusk on mild evenings with a contact insecticide (*see p.94*). Biological control traps containing the

nematode *Steinernema carpocapsae* is also available.

Violet Gall Midge

SYMPTOMS Leaves become greatly thickened and fail to unfurl (*see p.41*). If broken open, orange-white maggots or white silk cocoons, 2mm long, will be found inside the galls. Galled leaves are present throughout the year, but are often found in the winter after normal foliage has died down.

CAUSE A small fly, *Dasineura affinis*, which has three or four generations during the summer when females deposit their eggs on the developing leaves.

CONTROL Heavy infestations may kill plants, but usually they survive. Affected leaves can be removed by hand, but it is difficult to prevent reinfestation. There are no insecticides recommended for this pest.

Violet Root Rot

SYMPTOMS Plants are stunted and chlorotic, but the most characteristic symptoms are seen on the affected roots, tubers, and rhizomes. Their surface is covered in a dense network of dark purple fungal strands and often have considerable quantities of soil adhering to them (*see p.23*). Closely packed areas of fungal threads form into sclerotia which may fall into the soil as the tissue beneath starts to rot and turns brown, but this may be largely the effect of secondary organisms.

CAUSE The fungus *Helicobasidion purpureum*, which is most prevalent in wet, acid soils and persists in the soil as sclerotia. It may also remain viable on plant tissue for a considerable time.

CONTROL Lift and burn infected plants to minimize

disease carry-over. Where possible, remove infected plants before they start to disintegrate and release sclerotia into the soil. Improve soil drainage. Do not grow susceptible plants on that soil.

Virus Vectors

Some virus diseases affecting plants can be spread by certain pests and these are described as virus vectors. It is mainly sap-sucking insects such as some aphids, leafhoppers, and thrips that are involved. When they insert their needle-like mouthparts into virus-infected plants, they will pick up some virus particles which can be transmitted when they move on to new host plants. The champion virus vector seems to be the peach potato aphid, *Myzus persicae*, which feeds on a wide range of plants and is capable of transmitting more than a hundred viruses. Some soil-dwelling eelworms, such as some *Xiphinema*, *Trichodorus*, and *Longidorus* species, can transmit virus diseases when they feed on roots. The spread of viruses can be limited by controlling insect pests where feasible, and by removing plants suffering damaging virus diseases. Raspberries, strawberries, and daffodils are susceptible to eelworm-vectored viruses. New plants should be put into fresh ground, since if they are planted where virus-infected plants were previously growing, eelworms may infect them with viruses when they start feeding on the roots.

Viruses

SYMPTOMS The most common and widely spread symptoms of viral infection are stunting of the whole plant, and distortion (*see pp.40, 42*). The leaves and occasionally other above-ground parts of the plant develop markings (*see p.38*). These are usually yellow, though on some plants such as orchids they may be dark brown or black. The markings are commonly seen as streaks, mosaic patterns, flecks, mottles, or ring spots. Flowers may show distortion or fail completely, or they may form but show "flower breaking" where pale streaks of colour develop on dark petals (*see p.60*), or coloured streaks develop on pale petals. Infected plants generally crop poorly or fail to crop at all and in many cases death is premature. Some viruses may infect certain plants without causing visible symptoms, and this is called latent infection. For example, cucumber mosaic virus (*see p.129*) can exist in a number of common weeds which, although looking perfectly normal, can then act as a source of viral infection for other plants.

CAUSE A wide range of viruses, most of which may be found alone or in combination with other viruses. There are various methods of transmission, the most common being by sap-sucking insects, in particular aphids. Leafhoppers, whitefly, thrips, scale insects, mites, and even beetles may also be involved. In most cases the infection is non-persistent, that is to say the insect can acquire the virus after feeding on an infected plant for only a few seconds, but can then transfer it to other plants only within a short period of time. Cucumber mosaic virus (*see p.129*) is an example of this. Some insect-borne viruses are persistent, that is, the insect cannot transmit the virus for several hours after feeding on an infected plant, but is then capable of transmitting it for a long time afterwards, and sometimes for the duration of its life. Tomato spotted wilt, which is transmitted by western flower thrips (*see p.191*), is an example of this. Some viruses are transmitted by soil-inhabiting nematodes and fungi. Viruses may also be transmitted by handling, by grafting, by tubers, bulbs, and corms, and during routine operations such as disbudding or removal of side shoots.

CONTROL None available. Remove and dispose of affected plants promptly. Wash hands and tools thoroughly after handling infected plants. Do not put replacement plants of a similar type on the same site, and control potential vectors. Good weed control is essential too. Do not propagate from infected plants. Where possible, choose virus-resistant cultivars and buy plants certified as virus-free.

Voles

Voles are rodents with shorter tails and more rounded heads than the mice and rats that they resemble. They feed on a wide range of plant material, including seeds, flower buds, bulbs, corms, and plant roots. They also gnaw bark from the roots, trunks, and branches of young trees and shrubs (*see p.52*). If bark is eaten away around the entire stem (girdling), this prevents the uptake of water and nutrients, leading to the death of branches or sometimes the whole plant. Several species occur in gardens, including the mouse-sized bank vole, *Myodes glareolus*, and the short-tailed field vole, *Microtus agrestis*. The water vole, *Arvicola amphibius*, is larger, and can be confused with a rat. Vole numbers can fluctuate widely, especially of the short-tailed field vole, and it is usually only in years when they are abundant that they cause serious damage to garden plants.

CONTROL Where field or bank voles are causing problems in a garden, some protection can be given by setting mouse traps in the places where voles are active. Pieces of dessert apple seem to be a particularly attractive bait for the traps. These need to be placed under covers made of logs or bricks to keep birds away and protect pets from harm. Water voles are scarce in Britain and are now a legally protected species.

Volutella Blight

SYMPTOMS Infected leaves of box plants turn brown and then fall. The fungus overwinters on these leaves and in the spring produces spores, which are then splashed on to the leaves and stems of the plant. The fungus enters via natural stem cracks and pruning or clipping wounds. Masses of pinkish spores are produced on the lower surface of infected leaves. Large patches of dead stems may develop, spoiling the plant's appearance.

CAUSE The fungus *Pseudonectria buxi* (syn. *Volutella*) which may attack *Buxus sempervirens* and other species of box, and is encouraged by wet weather. The symptoms it causes are less severe than those caused by the *Cylindrocladium* species (*see p.118*).

CONTROL Remove and dispose of all infected plant parts promptly, ideally during dry weather to minimize spread of the spores. There are no chemicals available to control this fungus.

W

Walnut Blister Mite

SYMPTOMS During early summer, leaves develop oval, upwardly bulging areas up to

20mm long and about 10mm wide. The underside of these blistered areas is covered with a dense mass of creamy white hairs that darken and become brown later in the summer. The upper surface of galled areas may become purplish-yellow.
CAUSE Microscopic gall mites, *Aceria erinea*, which suck sap from the underside of the foliage and cause the distorted growth. Most walnuts are affected and sometimes a large proportion of leaves will have galls by late summer.
CONTROL No effective pesticides are available to gardeners. Fortunately, the mite has no impact on the tree's vigour or cropping.

Walnut Blotch

SYMPTOMS Leaves develop brown, necrotic blotches, and premature leaf fall occurs. Similar lesions may develop on the leaf stalks and fruits, turning them from green to black.
CAUSE The fungus *Ophiognomonia leptostyla*, which is encouraged by damp summer conditions. It overwinters on fallen leaves.
CONTROL Rake up and dispose of affected leaves.

Walnut Soft Shell

SYMPTOMS The walnuts appear to develop normally, but the shell fails to form properly at one end. It may be thin or totally absent, so the nut within may become damaged or rotted by organisms that can gain entry through the faulty area.
CAUSE The precise cause is unclear, but soft shell is believed to be largely the result of poor feeding and an erratic supply of water at that stage when the nuts are developing.
CONTROL Keep trees fed, particularly if they have been cropping well. Mulch to help soil moisture retention.

Wasps

Social wasps, such as *Vespula vulgaris* and *V. germanica*, and tree nesting species, such as *Dolichovespula media* and *D. sylvestris*, are important predators of garden pests. Wasps can, however, be significant pests because of their interest in ripe fruits. On relatively soft-skinned fruits, such as grapes and plums, they are capable of initiating damage (*see p.14*), but on apple and pears they generally only enlarge on damage started by birds. Wasps eat the soft inner tissues and gradually hollow out large cavities (*see p.13*). Some species of wasp make their papery nests above ground level in the branches of trees and shrubs. This can make hedge-cutting hazardous.
CONTROL If wasp nests can be located, they can be controlled by placing a wasp nest insecticide dust or aerosol spray in the nest entrance at dusk when wasps have stopped flying (*see p.94*). Aerial nests are best dealt with by using an aerosol spray. There is no point in spraying fruit trees against wasps, since any killed will soon be replaced by others. Ripening fruits can be protected by enclosing some of the better fruit trusses in bags made from old nylon tights or muslin. *See p.88* for information on making a wasp trap.

Water Core

See APPLE WATER CORE *p.110*.

Water Lily Aphid

SYMPTOMS Olive-green or brownish insects, about 2mm long, cluster on the upper surface of water lily leaves and on the flowers during the summer. Other floating and emergent pond plants are also liable to be attacked. Heavy attacks result in poor growth and flowering, with water lily leaves appearing dirty due to the accumulation of aphids and their whitish cast skins.
CAUSE A sap-sucking aphid, *Rhopalosiphum nymphaeae*. It overwinters as eggs on wild and cultivated plums and cherries, feeding on the foliage of these plants in the spring before migrating to water lilies in early summer.
CONTROL Although many insecticides will control aphids, none of them can be used on pond plants, as they will kill fish, frogs, and other pond wildlife. Forceful spraying with plain water can reduce infestations by knocking aphids off the plants.

Water Lily Beetle

SYMPTOMS Damage starts in late spring and continues throughout the summer. Yellowish-brown beetles, 7mm long, eat irregularly shaped slots in the lily pads (*see p.45*). In midsummer the larvae, which have elongate bodies, up to 8mm in length, black on the upper surface but pale yellow underneath, continue this damage. Infested leaves are unsightly and can deteriorate rapidly as secondary rots develop on the damaged tissues. The adult beetles also feed on the flowers.
CAUSE Adults and larvae of a beetle, *Galerucella nymphaeae*. The entire life cycle of eggs, larvae, pupae, and adults takes place on the upper surface of water lily leaves, and there are usually two generations during the summer.
CONTROL Insecticides cannot be used, as they are harmful to fish and other animals in ponds. With small ponds, adult beetles, larvae, and pupae can be removed by hand. Forceful spraying with water will dislodge the larvae, some of which will fail to find their way back onto the leaves.

Water Lily Leaf Spot

SYMPTOMS Red-brown spots, sometimes with a distinct yellow band around the edge, develop on the leaves. The upper leaf surface is generally most affected, but spotting may also occur beneath the leaves. Concentric zoning and numerous tiny, black pinprick-sized fungal bodies may also be visible. As the spots mature, they darken, turning a brownish-black colour. The centre of the spot may disintegrate, leaving a hole.
CAUSE The fungus *Ramularia nymphaerum*, which is carried by water splash.
CONTROL Remove affected leaves promptly. No chemical control is available and none should be attempted, as it may pose a threat to pond life.

Waterlogging

SYMPTOMS The plant appears to be suffering from drought, with wilting foliage (*see p.50*) and, in extreme cases, flowers (*see p.79*). The leaves may be yellowed and may fall early. Inspection of the roots may reveal that they have disintegrated, often peeling readily, to leave just the dead vascular strands.
CAUSE Overwatering, excessive rain, or poor drainage may all be involved.
CONTROL Attempt to improve soil conditions and avoid over-application of water. On heavy soils ensure that a large area of soil is lightened by the incorporation of plenty of grit, bulky organic matter, and other materials before planting. The application of

a foliar feed to any remaining green leaves may help to counteract the effects of slight waterlogging; it stimulates root production to replace some of those roots that have been lost. Where feasible, plant on a slight mound to improve drainage.

Weedkiller Damage – Broadleaved

SYMPTOMS The leaf blades are distinctly narrowed (see **p.42**), often to such an extent that the leaf veins are so closely drawn together that they appear parallel. Petioles and stems may show spiral twisting, and the foliage may become cupped (see **p.42**). Fruits may not form or if they do, they may be distorted. Tomatoes produce plum-shaped fruits, which are hollow with few, if any, seeds and little flesh. The stems of brassicas develop roughened gall-like growths at or just above soil level and these may appear to be adventitious roots (see **p.55**). Affected plants may live, but they can fail to crop normally.
CAUSE The plant has taken up broadleaved weedkiller, probably applied to control weeds nearby, into its tissues. Contamination may occur in many ways, such as inadvertent spraying, spray drift, use of watering cans or spraying equipment that have been insufficiently washed out, and mulching with lawn mowings, which contain traces of weedkillers.
CONTROL Take great care when spraying and never attempt to do it on hot, windy, or gusty days. Protect nearby plants with polythene sheeting or bags. Where possible, use a watering can and dribble bar rather than a sprayer, as these produce larger droplets that are less likely to cause drift problems. Keep equipment used for applying

weedkillers clearly labelled and use it for this purpose only. Always wash it out thoroughly after use. Prune out severely damaged areas on trees or shrubs and feed and water to promote new replacement growth.

Weedkiller Damage – Contact

SYMPTOMS Brown scorched spots develop on the foliage (see **p.38**) and any other part of the plant on which the chemical lands. If the weedkiller is applied on bulbs before the leaves have died down properly, bleached yellow or white foliage is produced the following season, which soon withers and dies. It may take several years for the bulb to recover fully.
CAUSE Contamination by a contact weedkiller.
CONTROL As for broadleaved weedkiller damage.

Weedkiller Damage – Glyphosate

SYMPTOMS Stunted rosettes of short shoots and malformed, strap-like leaves develop in place of normal side shoots. The symptoms resemble those caused by a witches' broom infection (see **p.198**).
CAUSE The widely-used weedkiller glyphosate, which may cause problems on a range of plants, in particular those that form suckers.
CONTROL As for broadleaved weedkiller damage (left).

Weevils

Weevils are beetles that are characterized by having clubbed antennae that can be bent at right angles about halfway along their length. The front of the head projects forward to form a snout or rostrum that may be long, as in the nut weevil (see **p.159**), or short, as in vine weevil (see **p.193**).

• see also APPLE BLOSSOM WEEVIL **p.108**; BEECH LEAF-MINING WEEVIL **p.114**; FIGWORT WEEVILS **p.135**; LEAF WEEVILS **p.152**; PEA AND BEAN WEEVIL **p.163**; TURNIP GALL WEEVIL **p.192**; VINE WEEVIL **p.193**.

Western Flower Thrips

SYMPTOMS Mainly house plants and greenhouse plants are affected, but this insect can also attack outdoor plants during the summer. On some host plants, such as tomato and cucumber, the thrips feed mainly on the foliage, causing a silvery mottling of the upper leaf surface. On ornamentals such as African violet, streptocarpus, gloxinias, chrysanthemums, verbenas, pelargoniums, fuchsia, cyclamen, and busy Lizzies, it is mainly flowers that are attacked (see **p.60**). They cause a pale flecking on the petals, and heavy infestations cause rapid deterioration in the blooms or prevent the buds from opening. Adult western flower thrips are 2mm long and have narrow, elongate, brown bodies. The immature nymphs have a similar shape and are pale yellow. This pest can transmit tomato spotted wilt virus, which causes dieback and distorted growth on many plants.
CAUSE A sap-sucking insect, *Frankliniella occidentalis*, which originates from North America but became widely established in Europe after 1986. It breeds continuously in heated greenhouses and has many generations a year.
CONTROL Many insecticides used against other thrips have little impact on this species. Biological control is available with a predatory mite, *Amblyseius degenerans*, but this has to be introduced before heavy infestations arise. Alternatively, spray with a systemic insecticide (see **p.94**).

When buying house plants, check the flowers carefully and reject any with thrips or signs of feeding damage on the petals.

Whiptail of Brassicas

SYMPTOMS Leaves become mottled yellow and growth is reduced.
CAUSE An inadequate supply of molybdenum (a trace element) in the soil or compost. This is common on acidic soils and is most likely to affect seedlings, especially brassicas.
CONTROL Lime the soil. Apply a proprietary mixture of trace elements that includes molybdenum.

White Blister

SYMPTOMS Raised, shiny white pustules develop, primarily on the lower leaf surface of brassicas (see **p.27**). Occasionally these may develop on other above-ground plant parts. The pustules are arranged either singly or in concentric rings. Affected leaves may become puckered and distorted, and the upper leaf surface develops yellow, slightly sunken pits.
CAUSE The fungus-like *Albugo candida*, which is encouraged by humid or moist air. The spores are spread on air currents or by water splash, or occasionally by insects. A resilient resting spore stage may also be formed and this may persist in the soil.
CONTROL Remove infected leaves. Remove and dispose of severely infected plants. Where feasible, grow resistant cultivars.

Whiteflies

Whiteflies are sap-sucking insects, about 2mm long, which, when adult, have white wings. They live on

the underside of leaves and readily fly up when the plant is disturbed. They lay eggs and these hatch into flat, oval, scale-like nymphs. Like the adults, these nymphs excrete a sugary excrement called honeydew. This drops down onto the upper surfaces of leaves below, which become sticky and covered with black sooty moulds. The final nymphal stage, sometimes called a pupa, is thicker, and often has hairs or white, waxy secretions on the upper surface. Some garden plants, such as rhododendron, honeysuckle, and *Phillyrea*, are occasionally attacked by whiteflies specific to these plants. If necessary, control them as for azalea whitefly.
• *see also* AZALEA WHITEFLY *p.112;* CABBAGE WHITEFLY *p.122;* GLASSHOUSE WHITEFLY *p.141;* VIBURNUM WHITEFLY *p.193.*

Willow Anthracnose

SYMPTOMS Leaves curl, turn yellow and fall prematurely. Extensive defoliation may occur by midsummer. Pinhead-sized, dark brown spots may develop on infected leaves and these produce spores. Raised, rough lens-shaped cankers develop on affected stems and these may cause dieback if they enlarge sufficiently to girdle the stem. Although symptoms may appear alarming, and recurrent attacks will weaken the tree, this disease is unlikely to cause the death of an otherwise healthy tree.
CAUSE The fungus *Drepanopeziza sphaerioides,* which is encouraged by moist conditions. The spores from the leaves and the cankers are carried by water splash. The fungus overwinters in the stem cankers, on bud scales and on fallen leaves.
CONTROL Rake up and dispose of fallen infected leaves. Prune out severely cankered

stems, if possible, without spoiling the weeping habit of the tree. Consider growing the resistant Peking willow *Salix babylonica* var. *pekinensis* 'Pendula' or the hybrid *S. × sepulcralis.*

Willow Bark Aphid

SYMPTOMS Dense colonies of brownish-black aphids, which are up to 5mm long, form on the bark of willow trunks and branches, mainly in late summer (*see p.58*). The upper surface of the aphids is marked with small black dots, with two larger tubercles towards the rear of the abdomen. Large quantities of sugary honeydew are excreted by the aphids and this makes the stems and ground below sticky. Sooty moulds often grow on the honeydew, which also attracts wasps and flies.
CAUSE A sap-sucking insect, *Tuberolachnus salignus.*
CONTROL On willow trees small enough to be sprayed, this aphid can be controlled with a systemic or contact insecticide (*see p.94*).

Willow Bean Gall Sawflies

SYMPTOMS Hard swellings form in the leaf blades during the summer (*see p.29*). They are yellowish-green or red, and their rounded oval shape resembles that of baked beans. The galls are initiated when the female sawfly lays her eggs in the leaf. She secretes chemicals that induce the leaf to produce a gall around each egg. The egg later hatches into a caterpillar-like larva that gradually hollows out the gall.
CAUSE Several sawflies, *Pontania* species, cause this type of gall on willows, and there are often two generations occurring between late spring and early autumn.

CONTROL Apart from causing galls on the foliage, these sawflies have no harmful effect on willows. Affected leaves can be removed from small trees if they are unsightly, but spraying would not be justified.

Willow Black Canker

SYMPTOMS In late spring or early summer angular, red-brown spots develop on the leaves; they spread, causing the leaves to shrivel and turn black, but remain hanging on the stem. The fungus then spreads into the stem and may produce lens-shaped, flattened cankers. It may be confused with willow scab (*see right*) or willow anthracnose (*see left*).
CAUSE The fungus *Glomerella miyabeana,* encouraged by warm temperatures and high humidity.
CONTROL Prune out affected shoots.

Willow Leaf Beetles

SYMPTOMS Both the adult and larval stages damage the leaves of willow, poplar, and aspen. The adult beetles are about 4mm long and metallic blue-black or brassy green. The larvae have elongate bodies up to 6mm long, and are mainly black on the upper surface, but creamy yellow below. Both the larvae and adult beetles graze away part of the leaf surface, causing the remaining damaged tissues to dry up and turn brown or white (*see p.31*). The larvae are gregarious and often feed side by side in groups of up to 30 or more.
CAUSE Several beetles and their larvae, including *Phratora vitellinae, P. vulgatissima,* and *Plagiodera versicolora,* cause damage to willows and poplars from late spring to early autumn. There are usually two generations.

CONTROL Only young trees suffer significant damage, although the foliage on older trees is often disfigured. Small trees can be sprayed with a contact or systemic insecticide (*see p.94*), particularly if infested in early summer.

Willow Scab

SYMPTOMS Similar to willow anthracnose and willow black canker (*see above*). In early spring irregular spots develop on the young foliage, spreading and causing it to turn black and die. Olive-brown pustules develop on the lower leaf surface and on affected stems. Black lesions develop on the stems and may cause girdling (*see p.59*).
CAUSE The fungus *Venturia saliciperda,* which overwinters as pustules on the stems.
CONTROL Prune out the affected stems.

Willow Watermark Disease

SYMPTOMS In mid to late spring, young leaves turn red and wither. Ooze, which contains the bacteria, is produced from the stems. Leaves deteriorate further and affected branches die back. The tree loses much of its growth, but is rarely killed. The wood develops a watermark-like stain internally and infected cut surfaces show complete or partial red or black rings. The disease is not common, occurring only in certain parts of the country. *Salix alba* var. *caerulea* and other white willows may be affected, and the disease has been reported on several other species of *Salix.*
CAUSE The bacterium *Erwinia salicis,* which persists in the wood and produces toxins that cause the leaf discoloration and dieback. The bacteria enter the tree via wounds, and are

probably spread by birds, insects, and on air currents. **CONTROL** Remove and burn affected trees.

Wilts

SYMPTOMS Wilting of the foliage followed by dieback. The whole plant is rarely affected simultaneously and may take several seasons to die. Smaller or herbaceous or annual plants may be killed in one season, however. Staining may develop, associated with the vascular or conducting tissues in stems and roots.
CAUSE Various fungi, including *Verticillium*, *Fusarium* and *Ceratocystis* species. Much damage is caused by the blocking of vascular elements and the subsequent deterioration of stems and leaves. The fungi may be spread by pruning, root contact or grafting, or by contaminated sap coming into contact with that of a healthy plant.
CONTROL None available. Where feasible, grow resistant plants. Remove and dispose of affected plants with the soil in the vicinity of the roots.
• see also ASTER WILT *p.111*; CLEMATIS WILT *p.127*; DROUGHT *p.133*; DUTCH ELM DISEASE *p.133*; FUSARIUM WILT *p.138*; PEONY WILT *p.166*; VERTICILLIUM WILTS *p.192*; WATERLOGGING *p.195*.

Wind Damage

SYMPTOMS Leaves appear scorched. One side or face of the plant is generally more severely affected than the rest. The most damage occurs on the side facing the prevailing wind. Flowers may be similarly affected (*see p.63*). If winds are salt-laden, the effect is usually much more severe and the whole plant may suffer extreme dieback or even be killed. Trees on particularly exposed sites may develop a lop-sided growth habit, known as "krumholtz", where they grow away from the prevailing wind.
CAUSE Wind. Often made worse if wind tunnels are created or if a windbreak is removed.
CONTROL Where possible, plant or erect windbreaks to prevent wind damage. If stems are killed, prune these out to prevent secondary organisms invading.

Winter Aconite Smut

SYMPTOMS Raised, blister-like grey outgrowths develop on stems and leaves, erupting to release large quantities of black, powdery spores. Severely affected parts die back and general growth and vigour is reduced.
CAUSE The fungus *Urocystis eranthidis*, which appears to be systemic (that is, found throughout all parts of the plant and not just on the area showing the symptoms) within the plant.
CONTROL Remove and dispose of affected plants promptly.

Winter Moth

SYMPTOMS Yellowish-green caterpillars with paler lines along their bodies feed on the foliage (*see p.47*) of many fruit trees and other deciduous trees and shrubs between bud burst and late spring. The caterpillars, which measure up to 25mm in length, walk with a looping action, as they have only two pairs of clasping legs on their abdomens. They bind leaves together loosely with silk threads and are often hidden from view. The caterpillars also eat fruit blossom and make holes in apple fruitlets, causing them to develop as badly misshapen fruits.
CAUSE Caterpillars of a moth, *Operophtera brumata*. The adult moths emerge between late autumn and mid-winter. Only males have normal wings; the female have swollen bodies with tiny wings and are incapable of flight. When they emerge from pupae in the soil, the females have to crawl up the tree trunk in order to lay their eggs on the branches.
CONTROL Place sticky grease bands around tree trunks, especially those of fruit trees, in mid-autumn to prevent female moths reaching the branches. Control newly-hatched caterpillars by spraying with a contact insecticide (*see p.94*) as the leaves emerge from the buds.

Wireworms

SYMPTOMS Wireworms are slender, orange-brown beetle larvae that have three pairs of short legs near the head end (*see p.20*). They can be up to 25mm long and have a small protuberance on the underside of the rear end of the abdomen. Wireworms live in the soil and can kill seedlings by biting into the stems just below soil level. In late summer, vegetables such as potato tubers (*see p.22*) and carrots can be extensively tunnelled by the larvae.
CAUSE The larval stage of some click beetles (*see p.127*), including *Agriotes lineatus, A. obscurus, A. sputator*, and *Athous haemorrhoidalis*. They are found in the greatest numbers in newly-dug grassland; their numbers will gradually decline over the next three or four years if the ground is kept in cultivation.
CONTROL Where wireworms are troublesome, dig up potato tubers as soon as they have matured to limit damage. There are no soil insecticides for wireworm control. Pathogenic nematodes are available for controlling wireworms.

Wisteria Scale

SYMPTOMS This pest is found mainly on wisteria but can also attack some *Prunus, Acer*, and *Parthenocissus* species. Domed, shell-like objects up to 10mm in diameter attach themselves to the bark of the woody stems (*see p.59*). Heavily infested plants lack vigour and may die back. Wisteria can also be attacked by the smaller brown scale (*see p.120*).
CAUSE A sap-sucking insect, *Eulecanium excrescens*. The adult scales are covered by a hard, blackish-brown shell, which may have a white, waxy dusting. The immature scales are less domed and are pale brown with white, waxy crystals on their upper surface. There is one generation a year, with insects overwintering on the woody stems as immature scales. They become adult in the spring, when eggs are laid under the shell-like covering. The eggs hatch in late May–June.
CONTROL The best time to control scale insects is when the more vulnerable, newly hatched nymphs are present. Spray with a contact or systemic insecticide (*see p.94*) in early summer. It can be difficult to spray a large wisteria effectively.

Witches' Brooms

SYMPTOMS Extensively branched twigs develop, clustered together to form "witches' brooms" in the crown of the tree (*see p.59*). In winter they may at first be mistaken for birds' nests and can spoil the outline of the tree or hedge, but in summer they appear as dense, leafy areas and so are less obtrusive.
CAUSE Various fungi, in particular species of *Taphrina*, that are spread by water splash and on air currents. Occasionally other factors, in particular mite infestations, may be responsible.

CONTROL Prune out the growths, if desired. Spread seems to be very slow, but removal will help to limit the problem.

• see also BIRCH WITCHES' BROOMS *p.115*; HORNBEAM WITCHES' BROOMS *p.145*.

Woodlice

SYMPTOMS Grey or pinkish-brown animals (see ***p.19***), with a hard segmented covering over their bodies. During the day, woodlice hide in dark places, such as under pots, logs, and stones, or in the surface layers of compost heaps. They can damage seedlings, but not established plants.

CAUSE Several species of woodlice are common in gardens, including *Oniscus asellus*, *Porcellio scaber*, and *Armadillidium vulgare*. The last-mentioned rolls itself up into a ball when disturbed and is known as a pillbug.

CONTROL Woodlice are so abundant in gardens that there is little prospect of achieving any lasting reduction in their numbers. Fortunately, they feed mainly on dead plant material rather than growing plants and are useful in the composting process. Good hygiene in greenhouses will help to keep their numbers down.

Woolly Aphid

SYMPTOMS A fluffy, white, waxy substance develops on the bark of apple trees and ornamental *Malus*, pyracantha, and *Cotoneaster horizontalis* in summer (see ***p.58***). The wax is secreted from the bodies of pinkish-brown aphids. During mid- to late spring they are mainly seen around old pruning cuts and in splits in older bark. Later in the summer the aphids spread to young branches where their feeding causes soft, knobbly swellings to develop.

These galls often split during frosty weather creating wounds that can become infected with apple canker (see *p.108*).

CAUSE A sap-sucking insect, *Eriosoma lanigerum*. Most aphids overwinter as eggs. This species, however, overwinters as immature nymphs which are found under loose bark.

CONTROL Watch for signs of woolly aphid during spring. Light infestations can be eliminated by scrubbing places where aphid colonies are developing to crush the aphids. More extensive infestations may require the use of a systemic or contact insecticide (see *p.94*) but avoid spraying during the flowering period. Heavy infestations in late summer are difficult to control.

• see also BEECH WOOLLY APHID *p.114*.

Woolly Vine Scale

SYMPTOMS This pest attacks many woody plants, including grape vine, alder, currants, gooseberry, and pyracantha. When mature in late spring, females deposit eggs in a thick pad of white, waxy fibres (see ***p.58***). The scales are dark brown and up to 6mm long. If they are removed by hand, the egg masses can be drawn out in long waxy threads. Heavily infested plants show a lack of vigour and the foliage is soiled by sooty moulds.

CAUSE A sap-sucking scale insect, *Pulvinaria vitis*. Nymphs hatch in early summer and feed first on the underside of leaves. Before leaf fall, however, they move back to the bark where they complete their development and overwinter as adults.

CONTROL Ornamental plants can be sprayed with with a systemic or contact insecticide (see *p.94*) in

midsummer against the newly-hatched nymphs. On edible plants, pick off the egg masses and spray with a contact or organic insecticide (see *p.94*).

Worm Casts

See EARTHWORMS *p.133*.

Worms

See CUTWORMS *p.130*; EARTHWORMS *p.133*; EELWORMS *p.134*; ENCHYTRAEID WORMS *p.134*; FLATWORMS *p.136*; THUNDERWORMS *p.189*; WIREWORMS *p.198*.

Y

Yellow-tail Moth

SYMPTOMS The adult moth has white wings and body, except for the end of the abdomen, which is covered with yellow hairs. It lays eggs on many trees and shrubs in late summer. It overwinters as young caterpillars, which feed during spring and early summer. The hairy caterpillars, up to 30mm long, are black with an orange-red double stripe running down their upper body and have white markings along their sides. They eat both the foliage and flowers of many plants.

CAUSE Caterpillars of a moth, *Euproctis similis*.

CONTROL Infestations are generally light and caterpillars can either be picked off or tolerated on established plants. If heavier attacks occur it may be necessary to spray with a contact insecticide (see *p.94*).

Yucca Leaf Spot

SYMPTOMS Circular brown spots develop on affected leaves. On variegated yuccas, the spots are usually most

prevalent on the variegated parts of the leaf (see ***p.31***). The spots show concentric ringing and numerous, pinprick-sized, raised, black fruiting bodies develop on them. The spots enlarge and may coalesce.

CAUSE The fungi *Cercospora concentrica* and *Coniothyrium concentricum*, which produce similar symptoms. The spores are spread by water splash.

CONTROL Remove severely affected leaves.

Z

Zucchini Yellow Mosaic Virus

SYMPTOMS Bright yellow mosaic patterns appear on the leaves of marrows, courgettes, and squashes. Foliage is also frequently distorted and the whole plant stunted. If plants reach their flowering and fruiting stage, fruits are very knobbly and distorted and the skin is blotched dark green.

CAUSE A mosaic virus, spread by aphids.

CONTROL Remove infected plants immediately.

Individual
Plant
Problems

Common problems to which garden plants
are subject are listed here grouped by plant type,
then under the common name of the plant within
its type. Where common names are not known, the
index (*pp.218–223*) lists these under plants'
botanical names, for ease of access. For each
individual plant problem, cross-references are
given both to its detailed entry in the *A–Z of Pests,
Diseases, and Disorders*, and to any illustrations of
the problem's effects in the *Gallery of Symptoms*.

Fruits and Nuts

Many gardeners are happy to use chemicals on ornamentals, but when it comes to cropping plants, most prefer to avoid them wherever possible. Good cultivation, preventive measures taken at the correct time, and the use of resistant cultivars will all help. Some problems affect many fruits (*see right*). Others are more host-specific; plants so affected are listed below. More resistant species or cultivars may be suggested in the entry for the problem in question in the *A–Z of Pests, Diseases, and Disorders*.

Vegetables and Herbs

Some problems affect a wide range of vegetables (*right*); others are more host-specific (*below*). Even more so than with fruit, the choice of pest or disease-resistant cultivars can help ensure healthy crops and minimize the need for chemicals. Resistant cultivars may be suggested in the entry for the problem in question in the *A–Z of Pests, Diseases, and Disorders.*

COMMON PROBLEMS

The most common non-host-specific ailments affecting vegetable crops are:

Page numbers in **bold** refer to colour illustrations

Garden Trees

Once established, trees are usually sufficiently vigorous for many problems to be insignificant. However, if seriously attacked by a pest or a disease, an established tree's size can make it difficult – sometimes impossible – to treat. Many problems can be avoided by careful maintenance. With deciduous trees, the autumn leaf-fall also helps, removing many pathogens on the foliage.

A number of common problems may affect a wide range of unrelated trees (*see right*). Some, however, are more host-specific; trees that are particularly prone to certain problems are listed below. Resistant species or cultivars may be suggested in the entry for the problem in question in the *A–Z of Pests, Diseases, and Disorders.*

COMMON PROBLEMS

The most common non-host-specific tree ailments are:

Algae 106	Lightning Damage 78
Aphids 108	Magnesium Deficiency 154
"Bleeding" after Pruning 116	Phyllosticta Leaf Spot and
Bracket Fungi 119	Dieback 167
Caterpillars 124	Phytophthora 167
Coral Spot 128	Powdery Mildews 173
Drought 133	Scorch 181
Frost Damage 138	Sooty Mould 185
Ganoderma 139	Verticillium Wilts 192
Honey Fungus 145	Viruses 194
Iron Deficiency and Lime-	Waterlogging 195
induced Chlorosis 148	Weedkiller Damage 196
Leaf Spots (Fungal) 152	Wind Damage 198

Page numbers in **bold** refer to colour illustrations

A

Acacia
Fluted Scale 137

Alder (*Alnus*)
Alder Leaf Beetles **45**, 106
Alder Sucker **48**, 106
Gall Mites 139
Phytophthora 167

Almond (*Prunus dulcus*)
Aphids 108
Bacterial Canker **51**, 112
Blossom Wilt **56**, 117
Brown Rot **13**, **14**, 120
Bullfinches **65**, 121
Fasciation **62**, 135
Honey Fungus **50**, 145
Iron Deficiency and Lime-
  induced Chlorosis **39**, 148
Peach Leaf Curl **40**, 164
Phytophthora 167
Powdery Mildews **37**, 173
Shothole **31**, **45**, 182
Shothole Borers **49**, 182
Silver Leaf 182

Amelanchier *see* **Shrubs and Climbers 206–7**

Ash (*Fraxinus*)
Ash Catkin Gall Mite 102
Ash Dieback **24**, **52**
Ash Sucker 111
Bacterial Canker **51**, 112
Dieback 132
Ganoderma 139
Inonotus hispidus 147
Nectria Canker 159

B

Bay (*Laurus nobilis*)
Bay Sucker **40**, 113
Carnation Tortrix Moth **34**, 123
Frost Damage (Foliage and
  Stems) **43**, 138
Horse Chestnut Scale **57**, 146
Powdery Mildews **37**, 173
Soft Scale 184

Beech (*Fagus*)
Beech Bark Disease **54**, 114
Beech Bark Scale **51**, 114
Beech Red Spider Mite **36**, 114
Canker 123
Coral Spot **56**, 120
Felt Gall Mites **33**, 135
Fomes Root and Butt Rot **53**, 137
Gall Midges 139
Ganoderma 139
Honey Fungus **50**, 145
Leaf Mining Weevil **39**, 114
Meripilus giganteus **55**, 155
Phytophthora 167
Powdery Mildews **37**, 173
Squirrels **18**, 186
Woolly Beech Aphid **58**, 114

Birch (*Betula*)
Aphids 108
"Bleeding" After Pruning 116
Bracket Fungi **53**, **54**, 119
Felt Gall Mites **33**, 135
Fomes Root and Butt Rot **53**, 137
Honey Fungus **50**, 145
Leaf Spot (Fungal) **25**, 152
Leaf Weevils 152
Phytophthora 167
Powdery Mildews **37**, 173
Rusts 180
Sawflies 181
Witches' Brooms **59**, 198

C

Catalpa
Magnesium Deficiency **37**, 154
Phytophthora 167
Powdery Mildews **37**, 173
Verticillium Wilts **50**, 192

Cedar (*Cedrus*)
Bracket Fungi **53**, **54**, 119
Cedar Aphid **48**, 124
Honey Fungus **50**, 145
Sirococcus Blight **44**, **49**, 183

Cercidiphyllum japonicum
Frost Damage (Foliage and
  Stems) **43**, 138
Verticillium Wilts **50**, 192

Cherry, flowering *see* **Prunus, ornamental**

Cherry laurel *see* **Laurel, cherry**

Crab apple (*Malus*) *see also* **Apple** in **Fruits and Nuts 201–2**
Apple Canker **52**, 108
Apple Scab **28**, 109
Blossom Wilt **56**, 117
Brown Rot **14**, 112
Coral Spot **56**, 117
Fireblight **51**, 135
Ganoderma 139
Honey Fungus **50**, 145
Leaf Miners **32**, 109
Leaf Weevils 152
Papery Bark **56**, 162
Powdery Mildews **37**, 137
Silver Leaf 182
Winter Moth **47**, **198**
Woolly Aphid **58**, **199**

Cypress, Leyland cypress (*Cupressus, x Cuprocyparis leylandii*)
Crown Gall **54**, 129
Cypress Aphid 130
Fomes Root and Butt Rot **53**, 137
Honey Fungus **50**, 145
Juniper Scale **53**, 149
Kabatina Shoot Blight 149
Phomopsis Canker 167
Phytophthora 167
Seiridium Canker 181

D

Dogwood (*Cornus*)
Dogwood Anthracnose 132
Horse Chestnut Scale **57**, 145
Phytophthora 167
Powdery Mildews **37**, 173

Douglas fir *see* **Fir, Douglas**

E

Elder *see* **Shrubs and Climbers 206–7**

Elm (*Ulmus*)
Bark Beetles **49**, 113
Coral Spot **56**, 128
Dutch Elm Disease 133
Gall Mites 139
Ganoderma 139
Honey Fungus **50**, 145
Horse Chestnut Scale **57**, 146
Leaf Spot (Fungal) **25**, 152
Phytophthora 167
Witches' Brooms **59**, 198

Eucalyptus
Eucalyptus Gall Wasp 134
Eucalyptus Sucker 135
Oedema **33**, 160
Phytophthora 167
Silver Leaf 182

F

False acacia (*Robinia pseudoacacia*)
Laetiporus sulphureus **54**, 150
Phytophthora 167
Sawflies 181
Verticillium Wilts **50**, 192

Fir, Douglas (*Pseudotsuga menziesii*)
Douglas Fir Adelgid 132
Fomes Root and Butt Rot **53**, 137
Phytophthora 167

Fir, Silver (*Abies alba*)
Honey Fungus **50**, 145
Silver Fir Adelgid 182

H

Hawthorn (*Crataegus*)
Aphids 108
Apple Canker **52**, 108
Cotoneaster Webber Moth **53**, 128
Crown Gall **54**, 129
Fireblight **51**, 135
Fomes Root and Butt Rot **53**, 137
Hawthorn Gall Midge 143
Honey Fungus **50**, 145
Leaf Spot (Fungal) **25**, 152

Shrubs and Climbers

Correct pruning is one of the keys to maintaining healthy shrubs and climbers, thinning overcrowded and old growth prevents a stagnant environment. Prompt recognition of problems is also vital; the removal of much badly affected growth can spoil an attractively shaped shrub or carefully trained climber.

Some common problems may affect many unrelated plants (*see right*). Others are more host-specific; plants particularly prone to certain problems are listed below. Resistant species or cultivars may be suggested in the entry for the problem in question in the *A-Z of Pests, Diseases, and Disorders.*

COMMON PROBLEMS

The most common non-host-specific ailments affecting shrubs and climbing plants are:

Algae 106
Aphids 108
Coral Spot 128
Deer 131
Frost Damage 138
Honey Fungus 145
Leaf Scorch 151
Leaf Spot (Fungal) 152
Iron Deficiency and Lime-
  induced Chlorosis 148

Magnesium Deficiency 154
Phyllosticta Leaf Spot and
  Dieback 167
Phytophthora 167
Powdery Mildews 173
Rabbits 174
Sooty Mould 185
Verticillium Wilts 192
Viruses 194
Weedkiller Damage 196

Page numbers in **bold** refer to colour illustrations

A

Actinidia
Glasshouse Red Spider Mite **36**, **38**, 141
Honey Fungus **50**, 145

Amelanchier
Fireblight **51**, 135

Aucuba
Frost Damage (Foliage and Stems) **43**, 138
Phytophthora 167

B

Bamboo
Aphids 108

Berberis (Barberry)
Aphids 108
Honey Fungus **50**, 145
Leaf Spot (Fungal) **25**, 152
Powdery Mildews **15**, 173
Rusts 180
Sawfly 115

Bay tree (*Laurus*)
Bacterial Canker **51**, 112
Bay Sucker **40**, 113
Carnation Tortrix Moth **34**, 123
Honey Fungus **50**, 145
Horse Chestnut Scale **57**, 146
Leaf Spot (Fungal) **25**, 152
Powdery Mildews **15**, 173
Soft Scale 184

Box (*Buxus*)
Box Blight 118
Box Red Spider Mite 118
Box Sucker **40**, 118

Box Tree Moth 118
Leaf Spot (Fungal) **25**, 152
Mussel Scale **56**, 158
Phytophthora 167
Rusts 180

Broom (Cytisus)
Broom Gall Mite **55**, 120
Frost Damage (Foliage and Stems) **43**, 138
Honey Fungus **50**, 145
Silver Leaf 182
Waterlogging **50**, 195

Buddleja
Aphids 108
Capsid Bugs **45**, **65**, 123
Earwigs **46**, 126
Figwort Weevils 128
Glasshouse Red Spider Mite **36**, **38**, 141
Honey Fungus **50**, 145
Leaf and Bud Eelworm **37**, 151
Leaf Spot (Fungal) **25**, 152
Mullein Moth 157
Viruses **38**, **42**, 194

C

Camellia
Aphids 108
Camellia Gall **41**, 122
Camellia Leaf Blight 123
Camellia Petal Blight 123
Camellia Yellow Mottle **36**, 123
Carnation Tortrix Moth **34**, 123
Cushion Scale **27**, 130
Honey Fungus **50**, 145
Leaf Spot (Fungal) **25**, 152
Pestalotiopsis 166
Phytophthora 167
Squirrels **18**, 186
Vine Weevil **45**, 193

Carpenteria
Leaf Spot (Fungal) **25**, 152

Caryopteris
Capsid Bugs **45**, **65**, 123
Honey Fungus **50**, 145

Ceanothus
Drought **62**, 133
Frost Damage **43**, **62**, 137, 138
Honey Fungus **50**, 145
Iron Deficiency and Lime-
  induced Chlorosis **39**, 148
Mussel Scale **56**, 158

Clematis
Aphids 108
Capsid Bugs **45**, **65**, 123
Caterpillars 124
Clematis Green Petal 127
Clematis Wilt 127
Earwigs **65**, 133
Glasshouse Red Spider Mite **36**, **38**, 141
Leaf Spot (Fungal) **25**, 152
Powdery Mildews **15**, 173
Slime Flux and Wetwood 183
Viruses **38**, **42**, 194

Cotinus
Powdery Mildews **15**, 173
Verticillium Wilts **44**, 192

Cotoneaster
Aphids 108
Brown Scale **58**, 120
Cotoneaster Webber Moth **53**, 128
Fireblight **51**, 135
Honey Fungus **50**, 145
Leaf Spot (Fungal) **25**, 152
Mussel Scale **56**, 158
Silver Leaf 182
Viruses **38**, **42**, 194
Woolly Aphid **58**, 199

D

Daphne
Cucumber Mosaic Virus **35**, **37**, 129
Fasciation **62**, 135
Grey Mould **51**, 142
Honey Fungus **50**, 145
Leaf Spot (Fungal) **25**, 152
Verticillium Wilts **44**, 192
Viruses **38**, **42**, 194

Dogwood (*Cornus*)
Dogwood Anthracnose 132
Horse Chestnut Scale **57**, 146
Mussel Scale **56**, 158
Phytophthora 167
Powdery Mildews **15**, 173

E

Elaeagnus
Coral Spot **56**, 128
Elaeagnus Sucker 134
Leaf Spot (Fungal) **25**, 152

Elder (*Sambucus*)
Blackfly 116
Glasshouse Red Spider Mite **36**, **38**, 141
Honey Fungus **50**, 145
Leaf Spot (Fungal) **25**, 152
Viruses **38**, **42**, 194

Escallonia
Brown Scale **58**, 120
Escallonia Leaf Spot **24**, 134
Silver Leaf 182
Viruses **38**, **42**, 194

Euonymus
Blackfly 116
Crown Gall **54**, 129
Cushion Scale **27**, 130
Euonymus Scale* **35**, 135
Powdery Mildews **15**, 173

Small Ermine Moths **58**, 184
Vine Weevil* **45**, 193
Viruses **38**, **42**, 194
(*E. japonica* only)

F

Fatsia
Capsid Bugs **45**, **65**, 123

Firethorn (*Pyracantha*)
Brown Scale **58**, 120
Coral Spot **56**, 128
Fireblight **51**, 135
Pyracantha Leaf Miner 173
Pyracantha Scab 173
Woolly Aphid **58**, 199
Woolly Vine Scale **58**, 199

Flowering currant (*Ribes*)
American Gooseberry Mildew **15**, 107
Brown Scale **58**, 120
Coral Spot **56**, 128
Crown Gall **54**, 129
Currant and Gooseberry Leaf Spot 129
Currant Blister Aphid **41**, 129
Dieback 132
Grey Mould **51**, 142
Honey Fungus **50**, 145
Iron Deficiency and Lime-
  induced Chlorosis **39**, 148
Rusts 180
Woolly Vine Scale **58**, 199

Forsythia
Bullfinches **65**, 121
Capsid Bugs **45**, **65**, 123
Forsythia Gall **56**, 137
Honey Fungus **50**, 145
Leaf Spot (Fungal) **25**, 152
Powdery Mildews **15**, 173

Fuchsia, hardy
Capsid Bugs **45**, 65, 123
Elephant Hawk Moth **46**, 134
Flea Beetles **45**, 136
Fuchsia Gall Mite 138
Fuchsia Rust **32**, 138
Glasshouse Red Spider Mite **36**, 38, 141
Leaf Spot (Fungal) **25**, 152

H

Hawthorn (Crataegus) see Garden Trees 204–5

Heathers (Erica, Calluna)
Honey Fungus **50**, 145
Leaf Spot (Fungal) **25**, 152
Phytophthora 167
Powdery Mildews **15**, 173

Hebe
Aphids 108
Carnation Tortrix Moth **34**, 123
Downy Mildew 132
Leaf Spot (Fungal) **25**, 152

Honeysuckle (Lonicera)
Honeysuckle Aphid 145
Leaf Spot (Fungal) **25**, 152
Powdery Mildews **15**, 173
Silver Leaf 182
Thrips 188
Whiteflies 196

Hydrangea
Aphids 108
Capsid Bugs **45**, 65, 123
Foot and Root Rots **21**, 137
Frost Damage (Foliage and Stems) **43**, 138
Glasshouse Red Spider Mite **36**, 38, 141
Grey Mould **51**, 142
Honey Fungus **50**, 145
Hydrangea Scale **28**, 147
Iron Deficiency and Lime-induced Chlorosis **39**, 148
Leaf Spot (Fungal) **25**, 152
Magnesium Deficiency **37**, 154
Powdery Mildews **15**, 173
Vine Weevil **45**, 193
Viruses **38**, 42, 194

Hypericum
Rusts 180

I

Ivy (Hedera)
Blackfly 116
Bryobia Mites 121
Leaf Spot (Fungal) **25**, 152
Phyllosticta Leaf Spot and Dieback **44**, 167
Soft Scale 184
Viburnum Scale 192
Vine Weevil **45**, 193
Viruses **38**, 42, 194

J

Japanese quince (Chaenomeles)
Aphids 108
Brown Rot **13**, 120
Brown Scale **58**, 120
Grey Mould **51**, 142
Iron Deficiency and Lime-induced Chlorosis **39**, 148

Jasmine (Jasminum)
Glasshouse Red Spider Mite **36**, 38, 141
Grey Mould **51**, 142
Viruses **38**, 42, 194

Jerusalem sage (Phlomis)
Leafhoppers 151

K

Kerria
Kerria Leaf and Twig Blight **44**, 149

L

Lavatera
Lavatera Leaf and Stem Rot 150
Leaf Spot (Fungal) **25**, 152
Rusts 180

Lavender (Lavandula)
Cuckoo Spit **56**, 61, 129
Grey Mould **51**, 142
Honey Fungus **50**, 145
Lavender Shab 150
Leaf Spot (Fungal) **25**, 152
Rosemary Beetle 180

Lilac (Syringa)
Honey Fungus **50**, 145
Leaf Spot (Fungal) **25**, 152
Lilac Blight **51**, 153
Lilac Leaf Miner **32**, 153
Powdery Mildews **15**, 173
Privet Thrips **36**, 173
Silver Leaf 182

M

Magnolia see Garden Trees 204–5

Mahonia
Berberis Sawfly 115
Mahonia Rust **25**, 39, 155
Powdery Mildews **15**, 173

Mexican orange (Choisya)
Carnation Tortrix Moth **34**, 123
Fluted Scale 137
Frost Damage **43**, 62, 137, 138
Glasshouse Red Spider Mite **36**, 38, 141
Scorch 181
Snails **45**, 184

Mock orange (Philadelphus)
Blackfly 116
Leaf Spot (Fungal) **25**, 152
Viruses **38**, 42, 194

Morning glory (Ipomoea)
Frost Damage (Foliage and Stems) **43**, 138
Glasshouse Red Spider Mite **36**, 38, 141
Magnesium Deficiency **37**, 154
Viruses **38**, 42, 194

N

New Zealand flax (Phormium)
Leaf Spot (Fungal) **25**, 152
Phormium Mealybug 167

P

Photinia
Apple Scab **28**, 109
Fireblight **51**, 135
Leaf Spot (Fungal) **25**, 152
Pear Scab **13**, 165

Pieris
Pieris Lacebug 168

Privet (Ligustrum)
Aphids 108
Honey Fungus **50**, 145
Leaf Spot (Fungal) **25**, 152
Lilac Leaf Miner **32**, 153
Privet Thrips **36**, 173
Viruses **38**, 42, 194

R

Rhododendron
Aphids 108
Azalea Whitefly 112
Bud, Dry 121
Caterpillars 124
Crown Gall **54**, 129
Frost Damage **43**, 62, 137, 138
Grey Mould **51**, 142
Honey Fungus **50**, 145
Iron Deficiency and Lime-induced Chlorosis **39**, 148
Leaf Scorch 151
Phyllosticta Leaf Spot and Dieback **44**, 167
Phytophthora 167
Pieris Lacebug 168
Rhododendron Bud Blast **59**, 176
Rhododendron Lacebug **35**, 176
Rhododendron Leafhopper 176
Rhododendron Petal Blight **60**, 177
Rhododendron Powdery Mildew **28**, 177
Rusts 180
Silver Leaf 182
Vine Weevil **45**, 193

Rock rose (Helianthemum)
Aphids 108
Leaf Spot (Fungal) **25**, 152
Powdery Mildews **15**, 173

Rose (Rosa)
Capsid Bugs **45**, 65, 123
Crown Gall **54**, 129
Deer **59**, 131
Dieback 132
Downy Mildew 132
Glasshouse Red Spider Mite **36**, 38, 141
Grey Mould **51**, 142
Honey Fungus **50**, 145
Large Rose Sawflies 150
Leaf-cutting Bees **46**, 151
Magnesium Deficiency **37**, 154
Nitrogen Deficiency 159
Potassium Deficiency **39**, 170
Rabbits 174
Robin's Pin Cushion **56**, 177
Rose Aphids **61**, 178
Rose Balling **64**, 178
Rose Black Spot **26**, 178
Rose Leaf-rolling Sawfly **41**, 178
Rose Leafhopper **35**, 179
Rose Powdery Mildew **33**, 179
Rose Root Aphid **51**, 179
Rose Rust **29**, 179
Rose Sickness/Replant Problems 179
Rose Slugworm **32**, 180
Viruses **38**, 42, 194
Weedkiller Damage – Broadleaved **42**, 196
Weedkiller Damage – Glyphosate 196
Winter Moth **47**, 198

Rosemary (Rosmarinus)
Brown Scale **58**, 120
Cuckoo Spit **56**, 61, 129
Frost Damage (Foliage and Stems) **43**, 138
Honey Fungus **50**, 145
Leafhoppers 151
Rosemary Beetle 180

Russian vine (Fallopia baldschuanica)
Leaf Miners 151

S

Skimmia
Horse Chestnut Scale **57**, 146
Iron Deficiency and Lime-induced Chlorosis **39**, 148
Magnesium Deficiency **37**, 154
Phytophthora 167
Vine Weevil **45**, 193
Waterlogging **50**, 195

Solanum
Flea Beetles **45**, 136
Grey Mould **51**, 142

Leaf Spot (Fungal) **25**, 152
Verticillium Wilts **44**, 192

Stranvaesia
Fireblight **51**, 135
Honey Fungus **50**, 145

Sumach (Rhus)
Frost Damage **43**, 62, 137, 138
Leaf Spot (Fungal) **25**, 152
Verticillium Wilts **44**, 192

Sun rose (Cistus)
Aphids 108

Sweet bay (Laurus) see Bay tree on facing page

T

Trumpet vine (Campsis)
Brown Scale **58**, 120
Glasshouse Red Spider Mite **36**, 38, 141

V

Viburnum
Blackfly 116
Capsid Bugs **45**, 65, 123
Carnation Tortrix Moth **34**, 123
Glasshouse Thrips 141
Grey Mould **51**, 142
Honey Fungus **50**, 145
Leaf Spot (Fungal) **25**, 152
Viburnum Beetle 192
Viburnum Scale 192
Viburnum Whitefly **27**, 193

W

Weigela
Brown Scale **58**, 120
Capsid Bugs **45**, 65, 123
Honey Fungus **50**, 145
Leaf and Bud Eelworm **57**, 151
Leaf Spot (Fungal) **25**, 152

Wisteria
Brown Scale **58**, 120
Frost Damage **43**, 62, 137, 138
Glasshouse Red Spider Mite **36**, 38, 141
Honey Fungus **50**, 145
Leaf Spot (Fungal) **25**, 152
Magnesium Deficiency **37**, 154
Powdery Mildews **15**, 173
Wisteria Scale 198

Y

Yucca
Blackfly 116
Snails **45**, 184
Yucca Leaf Spot **31**, 199

Herbaceous Perennials

Perennials do not suffer from many serious problems. The annual dying-back of foliage and stems gives an opportunity for the removal and disposal of plant tissue harbouring many diseases and pests. Regular division of established clumps also helps to keep plants healthy. A number of common problems may affect a wide range of unrelated plants (*see right*). Some, however, are more host-specific (*see below*). Resistant species or cultivars may be suggested in the entry for the problem in question in the *A–Z of Pests, Diseases, and Disorders*.

COMMON PROBLEMS

The most common non-host-specific ailments affecting herbaceous perennials are:

Aphids 108	Leaf Spots (Fungal) 152
Caterpillars 124	Powdery Mildews 173
Crown Rot 129	Rabbits 174
Deer 131	Slugs 183
Downy Mildew 132	Snails 184
Drought 133	Swift Moth 188
Frost Damage 138	Viruses 194
Grey Mould 142	Waterlogging 195

Page numbers in **bold** refer to colour illustrations

A

Acanthus
Powdery Mildews **37**, 173

Agapanthus
Agapanthus Gall Midges **60**, 194
Viruses **60**, 194

Alstroemeria
Viruses **60**, 194

Alyssum
Clubroot **21**, 127
Downy Mildew 132
Flea Beetles **45**, 136
Foot and Root Rots **21**, 137
Fusarium Wilt 138
Powdery Mildews **37**, 173
Sclerotinia **52**, 181
White Blister **27**, 197

Anchusa
Sclerotinia **52**, 181

Anemone *see also*
Japanese anemone
on facing page
Downy Mildew 132
Grey Mould **63**, 142
Plum Rust **36**, 169
Powdery Mildews **37**, 173
Smuts **51**, 184
Viruses **60**, 194

Astilbe
Powdery Mildews **37**, 173

Aubrieta
Aphids 108
Downy Mildew 132
Flea Beetles **45**, 136
Powdery Mildews **37**, 173
White Blister **27**, 196

Auricula
(*Primula auricula*)
Crown Rot 129
Glasshouse Leafhopper **34**, 141
Root Aphids **21**, 177
Vine Weevil **45**, 193

C

Californian poppy
(*Romneya*)
Caterpillars 124
Phytophthora 167
Verticillium Wilts **50**, 192

Campanula
Downy Mildew 132
Grey Mould **63**, 142
Leaf Spot (Fungal) **25**, 152
Phytophthora 167
Powdery Mildews **37**, 173
Rusts 180
Verticillium Wilts **50**, 192
Viruses **60**, 194

Candytuft (*Iberis*)
Clubroot **21**, 127
Downy Mildew 132
Leaf Spot (Fungal) **25**, 152
Powdery Mildews **37**, 173

Canna
Honey Fungus **50**, 145
Viruses **60**, 194

Carnation (*Dianthus*)
Aphids 108
Fusarium Wilt 138
Grey Mould **63**, 142
Leaf Spot (Fungal) **25**, 152
Leafy Gall **55**, 152
Powdery Mildews **37**, 173
Rusts 180
Slugs **45**, 183
Smuts **51**, 184
Snails **45**, 184

Verticillium Wilts **50**, 192
Viruses **60**, 194

Centaurea
Lettuce Downy Mildew **37**, 153
Powdery Mildews **37**, 173
Rusts 180
Sclerotinia **52**, 181

Chrysanthemum
Aphids 108, 155
Capsid Bugs **65**, 123
Caterpillars 124
Chrysanthemum Eelworm 126
Chrysanthemum Leaf Miner **37**, 126
Chrysanthemum White Rust **27**, 127
Crown Gall **54**, 129
Earwigs **65**, 133
Fusarium Wilt 138
Leaf Spot (Fungal) **25**, 152
Melon Cotton Aphid **61**, 155
Powdery Mildews **37**, 173
Rusts 180

Columbine (*Aquilegia*)
Aphids 108
Aquilegia Downy Mildew 110
Aquilegia Gall Midges 110
Aquilegia Leaf Miners 110
Aquilegia Sawfly 110
Leaf Spot (Fungal) **25**, 152
Powdery Mildews **37**, 173

Convolvulus
Frost Damage (Foliage and Stems) **43**, 138
Waterlogging **50**, 195

Cosmos
Frost Damage **43**, **62**, 138
Powdery Mildews **37**, 173

Cranesbill (*hardy geraniums*)
Capsid Bugs **65**, 123
Downy Mildew 132

Geranium Sawfly 139
Leafy Gall **55**, 152
Powdery Mildews **37**, 173
Vine Weevil **45**, 193

D

Daisy (*Bellis perennis*)
Fasciation **62**, 135
Powdery Mildews **37**, 173

Daylily (*Hemerocallis*)
Hemerocallis Gall Midge **63**, 143
Leaf Spot (Fungal) **25**, 152

Delphinium
Delphinium Leaf Miner **32**, 131
Delphinium Moth 131
Fasciation **62**, 135
Foot and Root Rots **21**, 137
Grey Mould **63**, 142
Leaf Spot (Bacterial) **26**, **44**, 151
Leaf Spot (Fungal) **25**, 152
Powdery Mildews **37**, 173
Sclerotinia **52**, 181
Slugs **45**, 183

E

Elephant's ears
(*Bergenia*)
Leaf and Bud Eelworms **37**, 151
Leaf Spot (Fungal) **25**, 152
Vine Weevils **45**, 193

Epimedium
Leaf-cutting Bees **46**, 151
Leaf Spot (Fungal) **25**, 152
Magnesium Deficiency **37**, 154
Vine Weevil **45**, 193

Euphorbia
Fasciation **62**, 135
Grey Mould **63**, 142

F

Foxglove (*Digitalis*)
Caterpillars 124
Downy Mildew 132
Glasshouse Leafhopper **34**, 141
Leaf Spot (Fungal) **25**, 152
Powdery Mildews **37**, 173

G

Gentian
Rusts 180

Geranium *see* **Cranesbill**

Geum
Aphids 108
Downy Mildew 132
Geum Sawflies 140

Globe thistle (*Echinops*)
Aphids 108
Leaf Miners 151

Goatsbeard (*Aruncus dioicus*)
Aruncus Sawfly **46**, 111
Powdery Mildews **37**, 173

Golden rod (*Solidago*)
Powdery Mildews **37**, 173

H

Hellebore (*Helleborus*)
Aphids 108
Downy Mildew 132
Hellebore Black Death **34**, 143
Hellebore Leaf Blotch **34**, 143
Slugs **45**, 183

Heuchera
Heuchera Rust **30**, 144
Leaf Spot (Fungal) **25**, 152
Leafy Gall **55**, 152
Rusts 180
Vine Weevil **45**, 193

Lawns

Many gardeners happily tolerate a certain quantity of weeds in the lawn, and indeed some would say that a crop of daisies adds to a lawn's appeal. A possible exception is moss, which can be pernicious, particularly on damp ground; there are several effective moss-killers available, often combined with a lawn-feed mixture. Eradication of severe moss infestation may also have to be combined with re-seeding.

Pests and diseases, however, are always disfiguring and controls are often necessary. In most cases, chemical control must be combined with cultural improvements if the lawn's health is to be improved in the long term.

COMMON PROBLEMS

The most common ailments affecting lawns are:

Bulbous Plants

The planting of bulbs, corms, tubers, and rhizomes in suitable site is crucial to their health. Many will not tolerate soils which become heavy or wet in winter, soon succumbing to disease. In most cases, it is essential for the foliage to be left in place for several weeks after flowering in order to allow the bulbs to build and store up energy reserves for the following year.

Division of established clumps allows for the removal of less vigorous or unhealthy-looking portions before problems can spread. Many bulbs are prone to virus infections, and regular inspection and removal of those showing symptoms is advised. Other problems are more host-specific; the particular problems affecting various plants are listed below. Resistant species or cultivars may be suggested in the entry for the problem in question in the *A–Z of Pests, Diseases, and Disorders*.

COMMON PROBLEMS

The most common non-host-specific ailments affecting bulbs are:

Slugs 183   Viruses 194

Page numbers in **bold** refer to colour illustrations

A

Acidanthera
Gladiolus Corm Rot **21**, 140
Gladiolus Dry Rot 140
Grey Mould **63**, 142
Leaf Spot (Fungal) **25**, 152

Autumn crocus (*Colchicum*)
Slugs **45**, **61**, 183
Smuts **51**, 184
Tulip Grey Bulb Rot 191

B

Bluebell (*Hyacinthoides*)
Bluebell Rust 117
Narcissus Bulb Flies **20**, 158
Viruses **60**, 194

C

Colchicum see **Autumn crocus**

Crinum
Narcissus Leaf Scorch **44**, 158

Crocosmia see **Montbretia**

Crocus
Blue Mould on Bulbs **20**, 117
Gladiolus Core Rot 140
Gladiolus Dry Rot 140
Grey Mould **63**, 142
Mice 156
Narcissus Basal Rot 158
Sparrows **65**, 185
Squirrels **18**, 186
Tulip Grey Bulb Rot 191
Violet Root Rot **23**, 193

Cyclamen, hardy
Bacterial Soft Rot 112
Foot and Root Rots **21**, 137
Fusarium Wilt 138
Grey Mould **63**, 142
Leaf Spot (Fungal) **25**, 152
Powdery Mildews **37**, 173
Slugs **45**, **61**, 183

D

Daffodil (*Narcissus*)
Blindness of Bulbs **59**, 117
Bulb Scale Mites 121
Narcissus Basal Rot 158
Narcissus Bulb Flies **20**, 158
Narcissus Eelworms **20**, 158
Narcissus Leaf Scorch **44**, 158
Narcissus Smoulder 158
Narcissus Viruses **38**, 159
Pollen Beetles **61**, 170
Slugs **45**, **61**, 183
Tulip Grey Bulb Rot 191
Weedkiller Damage – Contact 196

Dahlia
Blackflies 116
Capsid Bugs **45**, **65**, 123
Crown Gall **54**, 129
Dahlia Smut 130
Earwigs **65**, 133
Foot and Root Rots **21**, 137
Glasshouse Red Spider Mite **36**, **38**, 141
Grey Mould **63**, 142
Leaf Spot (Fungal) **25**, 152
Leafy Gall **55**, 152
Onion Thrips 161
Phytophthora 167
Powdery Mildews **37**, 173
Sclerotinia **52**, 181
Verticillium Wilts **44**, **50**, 192
Viruses **60**, 194

E

Erythronium
Grey Mould **63**, 142
Rusts 180

F

Freesia
Aphids 108
Blue Mould on Bulbs **20**, 117
Gladiolus Core Rot 140
Gladiolus Dry Rot 140
Gladiolus Hard Rot 140
Glasshouse Red Spider Mite **36**, **38**, 141
Grey Mould **63**, 142
Leaf Spot (Fungal) **25**, 152
Slugs **45**, **61**, 183
Viruses **60**, 194

Fritillary (*Fritillaria*)
Lily Beetle 153
Slugs **45**, **61**, 183
Viruses **60**, 194

G

Gladiolus
Aphids 108
Fusarium Bulb Rot 138
Gladiolus Core Rot 140
Gladiolus Corm Rot **21**, 140
Gladiolus Dry Rot 140
Gladiolus Scab and Neck Rot 140
Gladiolus Thrips **63**, 140
Leaf Spot (Fungal) **25**, 152
Slugs **45**, **61**, 183
Snails **45**, **61**, 184
Viruses **60**, 194

Grape hyacinth (*Muscari*)
Bluebell Rust 117
Smuts **51**, 184

H

Hyacinth (*Hyacinthus*)
Fusarium Bulb Rot 138
Grey Mould **63**, 142
Slugs **45**, **61**, 183
Tulip Grey Bulb Rot 191
Viruses **60**, 194

I

Iris
Aphids 108
Blue Mould on Bulbs **20**, 117
Iris Ink Disease 147
Iris Rhizome Rot **21**, 148
Iris Sawfly **45**, 148
Iris Scorch 148
Leaf Miners 151
Leaf Spot (Fungal) **25**, 152
Narcissus Basal Rot 158
Rusts 180
Slugs **45**, **61**, 183
Snails **45**, **61**, 184
Tulip Grey Bulb Rot 191
Viruses **60**, 194

L

Lily (*Lilium*)
Aphids 108
Lily Beetle 153
Lily Disease 154
Slugs **45**, **61**, 183
Vine Weevil **19**, **45**, 193
Viruses **60**, 194

M

Montbretia (*Crocosmia*)
Gladiolus Dry Rot 140
Glasshouse Red Spider Mite **36**, **38**, 141
Iris Ink Disease 147
Leaf Spot (Fungal) **25**, 152

O

Ornithogalum
Smuts **51**, 184

S

Snowdrop (*Galanthus*)
Gladiolus Dry Rot 140
Narcissus Bulb Fly **20**, 158
Narcissus Eelworm **20**, 158
Narcissus Leaf Scorch **44**, 158
Snowdrop Grey Mould 184

Solomon's seal (*Polygonatum*)
Slugs **45**, **61**, 183
Solomon's Seal Sawfly **47**, 185

T

Tulip (*Tulipa*)
Aphids 108
Blue Mould on Bulbs **20**, 117
Deer **59**, 131
Mice 156
Slugs **45**, **61**, 183
Squirrels **18**, 186
Tulip Fire **21**, **36**, 190
Tulip Grey Bulb Rot 191
Viruses **60**, 194

W

Winter aconite (*Erantbis*)
Smuts **51**, 184

Annuals and Biennials

Many problems affecting annuals and biennials can be avoided by careful initial choice of plants, good planting, and regular feeding and watering. To avoid the foot and root rots to which many are prone, do not grow the same, or closely related, plants on any site for more than one year. Some problems affect a wide range of unrelated plants (*see right*); others are more host-specific (*below*). Resistant cultivars may be suggested in the entry for the problem in question in the *A–Z of Pests, Diseases, and Disorders*.

Greenhouse and House Plants

A sheltered environment not only protects plants, but also pests and diseases. They may therefore build up very rapidly, and exotic species may flourish that would have little chance to cause serious problems out of doors. However, an enclosed environment also lends itself to effective and thorough treatment, and to the use of biological controls (*see pp.90–91*) to eradicate pests.

Some common problems may affect a wide range of unrelated plants (*see right*). Others are more host-specific (*see below*). Resistant species or cultivars may be suggested in the entry for the problem in question in the *A–Z of Pests, Diseases, and Disorders*.

Glossary

A

ABDOMEN The hind part of an insect's body.

ACARICIDE A pesticide effective against mites.

ACID (of soil) With a pH value of less than 7; *see also* Alkaline *and* Neutral.

ADVENTITIOUS (of roots) Arising above ground from a stem or leaf.

AERATE (of soil) 1. Loosen by mechanical means in order to allow air to enter. 2. Alter soil texture and structure by creating more air spaces.

ALKALINE (of soil) With a pH value of more than 7; *see also* Acid *and* Neutral.

ALTERNATE HOST One of two host plants needed for some pathogens or pests to complete their life cycles. The alternate host is often unrelated to the other host: most pathogens and pests are able to complete their life cycles using a single host plant.

ANAEROBIC (of an organism) Able to live and grow without oxygen.

ANNUAL A plant or fungus that completes its life cycle, from germination through to death, in one year.

ANTENNA (*pl.* antennae) A jointed sensory organ on the head of insects and some other arthropods.

ANTHER The part of a stamen that produces pollen; it is usually borne on a filament.

ANTHRACNOSE A term used to describe several unrelated fungal diseases that are seen as dark spots or lesions on foliage, stems, pods, or fruits.

ARTHROPODS Invertebrate animals that have jointed legs, e.g. insects, mites, woodlice, centipedes, and millipedes.

ASEXUAL REPRODUCTION A form of reproduction not involving fertilization.

AXIL The upper angle between a leaf and stem, between a main stem and a lateral branch, or between a stem and a bract; where an axillary bud develops.

B

BACTERIUM (*pl.* bacteria) A microscopic single-celled organism. Some are beneficial to plants, others may be pathogenic or cause secondary rotting.

BACTERICIDE A chemical capable of controlling bacteria.

BARK The surface layer of the trunk and branches of woody plants, protecting the tissues within; usually composed of dead corky cells.

BARK-RINGING The controlled removal of a partial ring of bark from the trunk or branches of certain fruit trees, to reduce vigorous growth and encourage fruit cropping. Complete removal or girdling of a trunk, as in damage by rabbits, squirrels, voles, and vine weevil grubs, can lead to the death of woody plants. *See also* Girdling.

BASAL PLATE The flattened base of a bulb, from which the roots grow.

BIENNIAL A plant that produces leafy growth in the first year, and then flowers, sets seed, and dies in the next year.

BIOLOGICAL CONTROL The control of pests, diseases, and weeds by the use of natural enemies such as predators, parasites, and pathogens.

BLEEDING The oozing of sap through a pruning cut or wound.

BLIGHT A term loosely used to describe a fungal, or occasionally an insect, attack.

BLIND SHOOT A shoot that does not form a terminal flower bud, or one where the growing point has been destroyed.

BOLT To produce flowers and seed prematurely, especially in vegetables.

BORDEAUX MIXTURE A fungicide mixture containing copper sulphate and lime.

BRACT A modified leaf at the base of a flower or flower cluster, sometimes resembling a flower petal or leaves. Bracts may also be reduced to scalelike structures.

BROADLEAVED Of trees or shrubs that bear broad, flat leaves, rather than needle-like foliage as in conifers.

BROAD-SPECTRUM Used to describe a chemical that has an effect on a wide range of often unrelated organisms.

BUD A condensed shoot containing an embryonic leaf, leaf cluster, or flower.

BULB A storage organ consisting mainly of fleshy scales and swollen, modified leaf-bases on a much reduced stem. Bulbs usually, but not always, grow from underground.

C

CALLUS The corky wound tissue formed by a plant in response to wounding. Callus frequently grows in from the edges of a wound created when a tree limb is removed, or at the base of a cutting.

CATERPILLAR The larval stage of moths and butterflies.

CERTIFIED STOCK Plants sold guaranteed to be of good quality and free from pests and diseases, especially virus infection, at the time of sale.

CHEMICAL In a garden situation, a fungicide, pesticide, weedkiller, repellent, or fertilizer used to control pathogens, pests, or weeds, or to increase the soil fertility.

CHESHUNT COMPOUND A fungicide mixture of ammonium carbonate and copper sulphate.

CHLOROPHYLL The green pigment in plants that absorbs light, providing the energy for photosynthesis.

CHLOROSIS Yellowing of plant tissue due to a deficiency or loss of chlorophyll (the green pigment). It may be brought about by a range of physiological disorders, nutrient deficiencies, pests, and diseases.

CHLOROTIC Showing the symptoms of chlorosis.

CHRYSALIS The resting stage between a fully-fed caterpillar and the adult moth or butterfly. *See also* Pupa.

COCOON A protective outer covering, usually of silk, produced by caterpillars and some other larvae before they pupate.

COLOUR-BREAKING A term used to describe a patterned change in flower colour, where the original colour is "broken" into elaborate feathered patterns. It is usually caused by virus infections.

COMPANION PLANTING Positioning plants together that are reputed to have a beneficial effect on neighbouring plants by discouraging pests and diseases or improving growth.

COMPLETE METAMORPHOSIS *See* Metamorphosis.

COMPOST 1. A potting medium comprising a mixture of loam, sand, peat, leaf mould, coir, bark, or other ingredients. 2. An organic material, rich in humus, formed by decomposed plant remains and other organic matter, including material from the kitchen and garden, used as a soil improver or mulch. 3. A growing medium available in different forms.

COMPOUND Made up of several or many parts, e.g. a leaf that is divided into two or more leaflets.

CONTACT Used to describe a chemical that remains on the outside of a plant and is absorbed by a pest or pathogen when it is sprayed or comes into contact with a treated surface.

CORM A bulb-like underground storage organ, consisting mainly of a swollen stem base and often surrounded by a papery tunic.

COTYLEDON A seed leaf; the first leaf or leaves to emerge from some seeds after germination, often markedly different from mature leaves. Flowering plants (angiosperms) are classified into monocotyledons (one) and dicotyledons (two) depending on how many cotyledons are contained in the mature seed. In gymnosperms (conifers) they are often produced in whorls.

CRUCIFEROUS Of the family Cruciferae, to which brassicas belong.

CULTIVAR A type of plant that has developed from the original species in cultivation (rather than in the wild), either by accident or by deliberate breeding; a contraction of "cultivated variety". *See also* Variety.

D

DECIDUOUS Plants that shed their leaves at the end of the growing season (in autumn and winter) and renew them in spring.

DEFOLIATION Loss of leaves.

DERRIS An insecticide derived from the roots of *Derris* and *Lonchocarpus* spp. Its active ingredient is rotenone. No longer used in the UK.

DIAPAUSE A resting period in the life cycle of invertebrate animals, the onset and duration of which is controlled by environmental factors such as day length and temperature. It enables them to survive adverse conditions, such as winter, and has similarities to hibernation in vertebrate animals.

DIEBACK The death of shoots, often spreading down from the tip of the stem, caused by damage or disease.

DORMANCY 1. The state of temporary cessation of growth in plants, and slowing down of other activities, usually during winter. 2. Seed dormancy: non-germination of seed when placed in conditions suitable for germination, due to physical, chemical, or other factors inherent in the seed.

DRUPES, DRUPELETS *See* Stone fruits.

DRENCH Pest or disease control applied as a liquid to the roots of a plant to control soil-borne pests or pathogens.

DUST Pest or disease control applied as a fine dust, usually to the plant, but occasionally to the soil round it.

E

ECDYSIS The process in invertebrate animals of shedding their outer skin as they grow and increase in size.

ECTOPARASITE A parasite that feeds on the outside of a plant or animal's body.

ELYTRON (*pl.* elytra) A beetle's wing case; when not in flight these are folded back and form a covering for the top of the abdomen.

ENDOPARASITE A parasite that feeds inside a plant or animal's body.

EYE On a stem or, more commonly, on a tuber, e.g. a potato, a dormant or latent growth bud that is visible. Of a fruit, e.g. apple, the end opposite the stalk.

F

f. [Lat.] *forma*; a variant within a species.

FAMILY A category in animal and plant classification, grouping together related genera. All these genera have characteristics that are all constant, or that are clearly different from those of other families.

FERTILIZATION The fusion of a pollen grain nucleus (male) with an ovule (female) to form a fertile seed.

FERTILIZER A plant food (usually in concentrated form), naturally or synthetically produced, applied to the soil or to plant foliage. A balanced fertilizer contains a balance of nitrogen, phosphate, and potash. A complete fertilizer contains both these and other nutrients.

FLEECE, HORTICULTURAL Synthetic material composed of very fine fibres woven or compressed together into a fabric, which can be draped over plants to protect them from weather extremes and certain pests.

FOLIAR FEED Liquid fertilizer formulated for application direct to healthy plant foliage. Nutrients applied in this manner are very rapidly taken up by the plant.

FRASS Insect excrement pellets, especially of caterpillars.

FREE-LIVING An organism, such as a nematode, which is associated with but not attached, to its host.

FROST HARDY A plant able to withstand temperatures down to −5°C (23°F).

FROST TENDER A plant vulnerable to frost damage.

FRUITING BODIES (of fungi) Spore-bearing structures of larger fungi, commonly known as toadstools, mushrooms, or bracket fungi. May be annual or perennial.

FULLY HARDY *See* Hardy.

FUNGICIDE A pesticide capable of controlling a fungus.

FUNGUS (*pl.* fungi) A very variable group of organisms, which obtain their food materials from living or dead organic matter (*see also* Parasite *and* Saprophyte). They vary greatly in size, from microscopic to clearly visible with the naked eye.

G

GALL An abnormal growth produced by a plant in response to chemicals secreted by an animal, fungus, or bacterium that lives within the galled tissue.

GENUS (*pl.* genera) A category in animal and plant classification, grouping related species. Denoted by the first part of the scientific name. *See also* Species.

GERMINATION The physical and chemical changes that take place as a seed starts to grow and develop into a plant.

GIRDLING The removal of bark all round a trunk, stem, or branch, by animals or physical injury. *See also* Bark-ringing.

H

HABIT 1. Characteristic, natural form of growth of a plant – upright, prostrate, weeping, etc. 2. The characteristic growth or general appearance of a plant.

HALF HARDY 1. Plants not able to tolerate frost, but generally able to withstand lower temperatures than frost-tender plants. 2. Not tolerating frost, but withstanding temperatures down to 0°C (32°F).

HARDY Plants tolerating year-round climatic conditions in temperate regions, including frost, without protection.

HAULM Used to describe the top-growth of plants such as potatoes and legumes.

HEAVY (of soil) Having a high proportion of clay, and often poorly drained.

HERBACEOUS 1. A non-woody plant in which the above-ground part dies down to the rootstock at the end of the growing season. 2. Dying down at the end of the growing season.

HERBICIDE A chemical used to control or kill weeds.

HERBIVORE A plant-feeding animal.

HERMAPHRODITE Animals with both female and male sexual organs, e.g. slugs, snails, and earthworms.

HONEYDEW A sugary liquid excreted by sap-feeding insects including many, but not all, aphids, whiteflies, suckers, mealybugs, and scale insects.

HOST The plant on which a pest or pathogen develops or feeds; also the insect or other animal in which parasitic animals develop.

HOST RANGE The range of host plants that a given pest or pathogen will use. The host range is often restricted to closely-related plants.

HYPHA (*pl.* hyphae) The minute thread-like growths formed by fungi.

I

IMMUNE A plant with characteristics that prevent a given pest or pathogen from attacking or colonizing it. A plant immune to one disease or pest may, however, be susceptible to others.

INCOMPLETE METAMORPHOSIS *See* Metamorphosis.

INORGANIC 1. Used generally, not of plant or animal origin, i.e. a mineral or synthetically-produced material. 2. Of a chemical compound, one that does not contain carbon. Many man-made pesticides contain carbon-based chemicals and so are organic compounds as defined by a chemist, but not in the sense in which gardeners use the word organic. Inorganic fertilizers are refined from naturally-occurring minerals or produced artificially. *See also* Organic.

INSECT Invertebrate animal whose body is divided into a head, thorax, and abdomen. Adult insects have three pairs of jointed legs on the thorax and usually two pairs of wings, e.g. moths, butterflies, beetles, ants, bees, wasps, sawflies, some aphids, whiteflies, and earwigs.

INSECTICIDAL SOAP A pesticide containing fatty acids, used against some pests, diseases, and weeds.

INSECTICIDE A pesticide capable of controlling insects.

INSTAR The stages between ecdysis in the development of larvae and nymphs of insects and mites.

INTEGRATED CONTROL A combination of chemical and biological controls with good cultivation techniques, aimed at keeping pests and diseases below the level at which damage occurs.

INVERTEBRATE Animals without backbones, e.g. insects, mites, woodlice, molluscs, nematodes, earthworms, and millipedes.

L

LARVA (*pl.* larvae) The immature feeding-stage in the life cycle of an insect undergoing complete metamorphosis, e.g. moth caterpillars, beetle grubs, fly maggots. *See also* Nymph.

LATENT INFECTION Infection by a pathogen that does not result in any visible symptoms on the plant.

LEACHING The loss of nutrients when they are washed down through the soil to areas out of reach of plant roots, often caused by excessive watering or rain.

LEGUME A plant of the family Leguminosae, to which peas and beans belong, or of one of its three sub-families, Caesalpiniaceae, Mimosaceae, and Papilionaceae.

LENTICEL A breathing pore found in the stems of woody plants. Often enlarged and clearly visible to the naked eye.

LESION Point or area of damage caused by pests, pathogens, or physical injury.

LIME Compounds of calcium; the amount of lime in soil determines whether it is alkaline, neutral, or acid.

M

MANDIBLES The biting mouthparts of insects, especially leaf-eating types.

METAMORPHOSIS The changes undergone during an insect or mite's development. Incomplete metamorphosis, as shown by aphids, capsid bugs, earwigs, and mites, involves little more than an increase in size and the gradual development of sexual organs and sometimes wings. Complete metamorphosis involves a dramatic change of form with the larva bearing no resemblance to the adult, e.g. in moths, flies, beetles, and sawflies.

MOISTURE SHELLS Layers of humid air that accumulate or build up round a plant or its leaves as they transpire and increase humidity locally.

MOLLUSCICIDE A pesticide capable of controlling slugs and snails.

MULCH 1. A material applied in a layer to the soil surface to suppress weeds, conserve moisture, and maintain a cool, even root temperature. In addition to organic materials such as manure, bark, and garden compost, which also enrich the soil, polythene, foil, and gravel may also be used. 2. A layer of organic matter applied to the soil over or around a plant to conserve moisture, protect the roots from frost, reduce the growth of weeds, and enrich the soil.

MULTI-STEMMED Tree or shrub with several main stems arising either directly from the ground or from a short main stem.

MYCELIUM A compacted mass of fungal hyphae, often developed into a distinct "sheet" of fungal growth.

MYCOPLASMA Microscopic organism closely related and with similar characteristics to a virus.

MYCORRHIZA (*pl.* mycorrhizae) Soil fungi that live in mutually beneficial association with plant roots, helping them to absorb nutrients from the soil.

N

NECROSIS The deterioration and death of plant tissues.

NECROTIC Dead, frequently brown or black, areas of plant tissue.

NECTAR A sweet, sugary liquid secreted by nectary glandular tissue, usually found in the flower, but sometimes on the leaves or stem.

NEMATICIDE A pesticide capable of controlling nematodes or eelworms.

NEMATODES Microscopic, unsegmented, worm-like animals, also known as eelworms. Some are used as pest controls, e.g. of slugs and vine weevils.

NEUTRAL (of soil) With a pH value of 7, the point at which soil is neither acid nor alkaline. *See also* Acid *and* Alkaline.

NITRATE A salt of nitric acid, having a high nitrogen content that is available to plants. Either produced by the activity of bacteria in the soil or manufactured.

NITRITE A salt of nitric acid in which the nitrogen is not readily available to plants.

NODE The point on a stem where one or more buds, leaves, shoots, or flowers are attached.

NODULE A small swelling.

NOSE (of a bulb or corm) The pointed upper end.

NOTIFIABLE A pest or disease is described as notifiable when it is obligatory to notify the Animal and Plant Health Agency (APHA) as soon as it is discovered.

NUTRIENTS Minerals (mineral ions) used to develop proteins and other compounds required for plant growth.

NYMPH The immature stage of an insect or mite that undergoes incomplete metamorphosis. *See also* Larva.

O

ORGANIC 1. Of garden chemicals, referring to compounds containing carbon derived from plant or animal organisms. 2. Loosely, mulches, composts, or similar materials derived from plant or animal-derived materials. 3. A system of crop production and gardening without using synthetic or non-organic materials.

OVICIDE A pesticide capable of controlling the egg stage of pests.

OVIPOSITOR The egg-laying organ of an insect.

P

PARASITE An organism that gets all or part of its food from another plant or animal over a period of time, often without killing its host. Parasitic insects that eventually kill the insect in which they have developed are more correctly termed parasitoids.

PARTHENOGENESIS The ability of some female pests to reproduce without fertilization. Many aphids have this ability, as have vine weevils, some sawflies, scale insects, and mealybugs. Males are rare or non-existent in these species.

PATHOGEN An organism that is parasitic and causes disease in another organism. Pathogens include certain bacteria, fungi, viruses, and mycoplasmas.

PATHOVAR A distinct strain or type of a named pathogen.

PEDICEL The stalk of an individual flower.

PERENNIAL Living for at least three seasons. In this book the term when used as a noun, and unless qualified, a herbaceous perennial. A woody-based perennial dies down only partially, leaving a woody stem at the base.

PERSISTENCE (of chemicals) The length of time a chemical persists in an active form either on the plant or in the soil or compost after it has been applied.

PESTICIDE A chemical substance; a term usually applied to chemicals that control pests, but in its broader definition also includes fungicides, herbicides, animal repellents, and wood preservatives.

PETIOLE The stalk of a leaf.

pH The scale by which the acidity or alkalinity of soil is measured. *See also* Acid, Alkaline, Neutral.

PHEROMONE Volatile chemicals produced by insects as a means of communicating with others of the same species. Males often locate virgin females by tracking down the source of the pheromone or scent released by the female.

PHLOEM Conducting elements within plant tissue, largely associated with the transport of food materials in solution, produced in the leaves and distributed to the rest of the plant.

PHOTOSYNTHESIS The synthesis of carbohydrates in green plants from carbon dioxide and water, using light energy from sunshine or artificial light absorbed by the green pigment chlorophyll.

PHYTOTOXIC Harmful to plants. Usually used to describe a chemical (usually an insecticide or fungicide) that can damage a plant onto which it has been sprayed.

POLLEN The male cells of a plant, formed in the anthers.

POLLINATION The transfer of pollen from the anthers to the stigma of the same or different flowers, often carried out by insects, especially bees, or by air movement, resulting in fertilization of the embryonic seeds in the ovary.

POLYPHAGOUS Pests that feed on a wide range of plants which may be botanically unrelated.

POME FRUIT A firm, fleshy fruit, with seeds enclosed in a central core, as in apples and pears.

PROBOSCIS Insect mouthparts, used for sucking up liquids, especially nectar.

PROLEGS The clasping legs on the abdomen of a caterpillar or sawfly larva.

PUPA (*pl*. pupae) The resting, non-feeding stage in the life cycle of an insect undergoing complete metamorphosis. Also known as a chrysalis.

PUPATION The period of time during which larval tissues are dissolved and reconstructed into the form of the adult insect.

PYCNIDIUM (*pl*. pycnidia) Minute raised, rounded fungal fruiting body, produced on the surface of infected plant tissue.

R

RACE (of fungi) A specific strain of a named fungus which may vary in its physical or pathogenic characteristics from other races of that fungus.

REPELLENTS Substances that deter pests without causing them harm.

RESISTANT A plant is described as resistant to a pest or disease when it is attacked by it, but shows no ill effects. A fungus may also develop resistance to a fungicide, or to a closely-related group of fungicides. Fungicide resistance occurs most commonly when a particular fungicide or closely-related ones have been used repeatedly for many years. Similar resistance problems also occur with some pests and weeds that are no longer controlled by insecticides and herbicides that were formerly effective.

RHIZOME A specialized, usually horizontally creeping, swollen, or slender underground stem that acts as a storage organ and produces aerial shoots at its apex and along its length.

RHIZOMORPH A cord-like fungal structure, which is very resilient and withstands extremes of temperature and moisture. Honey fungus rhizomorphs are dark and tough, and grow from host to host.

ROOT The part of a plant, normally underground, that anchors it and through which water and nutrients are absorbed.

ROTATION Planning where plants are grown, to ensure that different types of plants are grown on a given piece of land from season to season. By rotating plants some pest and disease outbreaks can be avoided. Rotation is a particularly effective method of avoiding problems with soil-borne pests and pathogens.

RUSSETING A discoloration, generally brown, of a plant surface (usually the skin of fruits) which is often roughened.

S

SAPROPHYTE An organism (most commonly a fungus) that lives on dead or decaying organic material but does not attack living material.

SCLEROTIUM (*pl*. sclerotia) The resilient fungal resting bodies formed by some fungi to allow them to overwinter successfully and withstand extremes of weather.

SHOOT The aerial part of a plant that bears leaves. A side-shoot arises from a bud along the length of a main shoot.

SPECIES A category in plant and animal classification, comprising individuals in the same genus that are alike and naturally breed with each other. Denoted by the second part of the scientific name. *See also* Genus *and* Family.

SPORE Microscopic reproductive structure of fungi and bacteria. Certain non-flowering plants, such as ferns, also produce spores.

SPUR A short fruiting branch on fruit bushes and trees, particularly apples and pears.

SSP. Subspecies; a higher category in plant classification than *forma* or variety.

STAMEN The male reproductive organ in a plant, comprising the pollen-producing anther and usually its supporting filament or stalk.

STEM The main axis of a plant, usually above ground and supporting leaves, flowers, and fruits.

STIGMA The female part of a flower that receives pollen during the process of pollination.

STOLON A horizontally spreading or arching stem, usually above ground, which roots at its tip to produce a new plant, e.g. strawberry.

STONE FRUITS Fruits, also known as "drupes", with one or more seeds ("stones") surrounded by fleshy, usually edible, tissue. They are common in the genus *Prunus* (e.g. apricots, plums, cherries).

STOOL A group of shoots emerging from the base of a single plant. In propagation the term refers to the roots and stem bases of a plant, such as a chrysanthemum, after it has finished flowering.

STRAIN (of bacteria, etc.) A distinct form of a named pathogen.

STYLETS The mouthparts of sap-feeding insects and mites, used for piercing into plant tissues.

SYMBIOSIS A relationship between two organisms, where neither harms the other and both derive benefit from the relationship, e.g. nitrogen-fixing bacteria in root nodules.

SYSTEMIC 1. Insecticides and fungicides that are absorbed into a plant's sap and translocated within the plant. Systemic fungicides are, however, rarely able to move great distances but may move from one leaf surface through the leaf to the lower surface. 2. Of diseases, some pathogens, in particular the viruses, that are systemic within the host plant, moving with the plant sap and being found in all or a large part of the host.

T

TENDER Vulnerable to low temperatures. Tender plants may be categorized as: cool-growing, down to a minimum temperature of 10°C (50°F); intermediate, minimum 13°C (55°F); or warm-growing, minimum 18°C (64°F). *See also* Frost tender.

THORAX The mid-part of an insect's body, between the head and abdomen.

TOLERANT Having the ability to tolerate an attack by a pest or a pathogen. A pest or pathogen may be described as tolerant if able to withstand the effects of a pesticide.

TRACE ELEMENTS Chemical elements needed by plants, but only in very small amounts, e.g. boron, copper, manganese, iron, molybdenum, and zinc.

TRANSPIRATION Loss of water by evaporation from the leaves and stems of plants.

TUBER A swollen, usually underground, organ derived from a stem or root, used for food storage.

V

VARIEGATED Irregularly marked with various colours; particularly of leaves patterned with markings in white, yellow, or other colours.

VARIETY A plant that has developed with a slight difference from a wild plant species. Plant varieties are often brought into cultivation once developed.

VECTOR *See* Virus vector.

VIRUS A sub-microscopic particle often responsible for causing plant virus diseases.

VIRUS VECTOR An insect, mite, or nematode that transmits plant viruses, usually on its mouthparts or in its saliva. Some fungal and bacterial diseases can also be spread by plant pests.

VIVIPAROUS Giving birth to live young. Most aphid reproduction is by this method.

W

WETTING AGENT A material used to lower the surface tension of liquids on a plant surface, so allowing a spray to form a film over the surface rather than forming into droplets. Many garden pesticide formulations contain wetting agents (also known as wetters or spreaders).

WIND SNAP Wind causing a tree to be broken off at ground level; this can be made more likely by decay round the base of the trunk.

WIND THROW Wind causing a tree to be uprooted. Decay in the tree roots can make the tree unstable and therefore susceptible to wind damage.

WORM Various animals with elongate, cylindrical bodies. Eelworms or nematodes have simple, unsegmented bodies; earthworms are more advanced animals with bodies divided into many segments.

X, Z

XYLEM Vascular tissue under bark taking water and nutrients up a stem.

ZOOSPORE A motile stage, capable of self-propulsion, in the life cycle of certain fungi.

Index

References to illustrations
and tables are in **bold**.
Subheadings are arranged
with symptoms first.

PHOTOGRAPHY CREDITS

The publisher would like to thank the following for their kind permission to reproduce their photographs:

key: t=top, b=bottom, c=centre, l=left, r=right

A–Z Botanical 56tc. DW Bevan: 12bc, 68tr. Martin Stankewitz: 75tr.
Alamy Images Nigel Cattlin: 31br. Brian Harris: 38cr.
Biofotos/Heather Angel 53bl, 54c, 55cl, 62tl, 62br, 67br, 69tl, 71tc, 71tr, 71cl, 80bl, 83tl, 104tl.
Bruce Coleman Jane Burton: 70bc. Kim Taylor: 71tl, 92tr.
DK Images Peter Anderson/RHS Hampton Court Flower Show: 72. Alan Buckingham: 10, 16tl. Mark Winwood/RHS Wisley: 2.
Dreamstime Isselee 4, 75br.
Forestry Commission 52bl, 54br, 55tl, 56bl, 59tc, 78cr. David Boles: 52tl. Glenn Brearley: 55tc. George Gate: 38tl. Ana Perez-Sierra: 24tc, 44tl, 49tr.
The Garden magazine Tim Sandall: 5c, 12bc, 13c, 14tl, 13bc, 14bc, 15tr, 16tl, 17tl, 17br, 18cl, 19tr, 19tc, 19c, 19bl, 20tl, 20tr, 20cr, 20br, 21cl, 21cr, 22bc, 23br, 24cr, 24br, 25tl, 25br, 26tr, 26bl, 26br, 27tc, 27c, 27bc, 28bl, 29tr, 29bc, 29tl, 29cl, 29bl, 29br, 30bl, 30cr, 30bc, 31cl, 31tr, 32tr, 33cr, 31bl, 32br, 33tl, 33br, 34cr, 34c, 35tl, 35tc, 35bl, 36c, 36cr, 36bl, 36br, 37tl, 37tc, 38bc, 38br, 38tr, 38bl, 39tc, 39tr, 39bl, 39br, 37cr, 37bl, 37bc, 40cl, 40bl, 41tl, 41cr, 41bl, 41bc, 41br, 42tr, 42bc, 42bl, 43tr, 45tl, 45tc, 45cr, 45bc, 45br, 46c, 46cr, 46bl, 47tl, 47tr, 48tr, 48tc, 48cl, 48c, 50tl, 50bl, 50tr, 51cl, 51cr, 52tr, 53tl, 54bl, 54bc, 55br, 57bl, 58tl, 58cr, 59bc, 61bl, 62tr, 63bl, 64tc, 65bl, 66bl, 66br, 83cr, 83cb. Derek St Romaine: 21tl, 27cr, 42br, 80bl.
The Garden Picture Library John Glover: 33bl, 82bl.
Garden/Wildlife Matters 13br, 14cl.
Andrew Halstead 15cr, 42tl, 56tl.
Jerry Harpur 97bl.
Harry Smith Collection 16cl.
Holt Studios John Adams: 43bl, 62cr. Gordon Roberts: 63ar. Inga Spence: 17cl. Nigel Cattlin: 6, 12bl, 13tr, 13cr, 15tl, 15bc, 16cr, 17tr, 18tl, 18br, 21tc, 21tr, 22tl, 22br, 22cr, 23cl, 24bl, 25tc, 25tr, 26bc, 26tl, 26cr, 31tc, 35br, 38cr, 39tl, 37c, 40tr, 40tl, 43tl, 43cr, 43bl, 43bc, 44cl, 44c, 44cr, 49bc, 51bl, 51c, 53br, 54tr, 56cr, 57cr, 60bc, 60tr, 61cr, 61tl, 62tc, 62bl, 62cr, 66bc, 67tc, 67cl, 68bl, 68bc, 69cl, 70tl, 80bl, 93tr, 104br.
Oxford Scientific Films 76cl. Marshall Black: 83trb. Scott Camazine: 80br. Harry Fox: 69tc. Harold Taylor: 68br, 82bc, 82br. Terry Heathcote: 78bl. Colin Milkins: 70br, 82tl. James Robinson: 83tr. Tony Tilford: 97tr.
Photos Horticultural 27bl, 31br, 43br, 59cl, 61tr, 64tr, 64bl, 65tc, 23bl, 70cr, 79br, 83crb, 83br, 92bl. BT: 12br, 50br, 57bc, 57br. RB: 14tr.
RHS Images 13tl, 13cl, 14br, 16br, 17tc, 18tc, 18bl, 19bc, 19cr, 20bc, 21bc, 21br, 24tr, 27tr, 28cl, 30br, 32c, 32bl, 32bc, 34tl, 34bc, 35cl, 36cl, 36c, 40c, 42bl, 46cl, 46br, 45c, 46bc, 47bl, 47bc, 48cr, 48bl, 48bc, 48br, 49tc, 49c, 49bl, 49br, 51bl, 54tl, 56tl, 57tl, 58cl, 58br, 59tr, 63tl, 63cl, 65tl, 65tc, 65tr, 65cl, 65bc, 66cr, 66cr, 69bl. Liz Beal: 55tr. Andrew Halstead: 17br, 32tl, 32cl, 45tr, 46tl, 47br, 48tl, 58t, 59cr, 64tl. Horticultural Science: 27br, 36tl, 41tr, 44bl, 50cr. Anne Kilty: 34tr. Joyce Maynard: 57tl. Pathology: 30tl, 44bc. Anna Platoni: 64br. Rebekah Robinson: 44br. Tim Sandall: 12tr, 16cl, 40br. Carol Sheppard: 47cr.
Carrie Thomas 28tr.
Jerry Young 71bl, 75br, 82tr.

Additional photography by **Peter Anderson**.

Every effort has been made to trace the copyright holders. Dorling Kindersley apologises for any unintentional omissions and would be pleased, in such cases, to add an acknowledgment in future editions.

Penguin Random House

DK UK
Senior Editor Alastair Laing
Art Editor Anne Fisher
Picture Research Myriam Megharbi
Senior Jacket Creative Nicola Powling
Pre-Production Producer Tony Phipps
Producer Stephanie McConnell
Managing Editor Stephanie Farrow
Managing Art Editor Christine Keilty
Art Director Maxine Pedliham
Publishing Director Mary-Clare Jerram

DK INDIA
Senior Editor Rupa Rao
Assistant Editor Tanya Singhal
Project Art Editor Vikas Chauhan
Managing Editor Soma B. Chowdhury
Managing Art Editor Arunesh Talapatra
Pre-Production Manager Sunil Sharma
DTP Designers Umesh Singh Rawat,
Rajdeep Singh, Satish Gaur

ROYAL HORTICULTURAL SOCIETY
Image Library Ewan Guilder
Publisher Rae Spencer-Jones
Head of Editorial Chris Young

This edition published in Great Britain in 2018
in association with The Royal Horticultural Society
by Dorling Kindersley Limited, 80 Strand,
London WC2R 0RL

First edition published in Great Britain in 1997
Copyright © 1997, 2003, 2007, 2009, 2018 Dorling
Kindersley Ltd; text copyright © 1997, 2003, 2007,
2009, 2018 Pippa Greenwood, Dorling Kindersley Ltd

A Penguin Random House company
10 9 8 7 6 5 4 3
003–307775–Mar/2018

A CIP catalogue record for this book is available
from the British Library.
ISBN 978-0-2413-1560-6

Printed and bound in China.

A WORLD OF IDEAS:
SEE ALL THERE IS TO KNOW

www.dk.com

Pippa Greenwood ran the Plant Pathology department at the RHS and is widely recognised as an authority on plant problems. She has been a regular gardening presenter on TV and radio for more than 25 years, most notably on the BBC for *Gardeners' World* and *Gardeners' Question Time*.

Andrew Halstead has more than 40 years' experience in the recognition, management, and control of garden pests, and was Principal Entomologist (insect specialist) for the RHS. He is now a gardening author, journalist, and lecturer.

RHS Plant Health
HEAD OF PLANT HEALTH Gerard Clover
PRINCIPAL ENTOMOLOGIST Andrew Salisbury
PRINCIPAL PLANT PATHOLOGIST Matthew Cromey

Authors' Acknowledgements

Pippa Greenwood would like to thank those colleagues past and present, and friends and family, who helped her to retrieve obscure but useful information and continued to put up with her whilst she was writing and later revising this book. In particular, Dr John Fletcher, her plant pathology "guru", who inspired her and taught her so much, and who really does share her fascination with diseased plants! She would also like to thank everyone at Dorling Kindersley who has been involved in this project for their tireless enthusiasm and patience, and for being such a pleasure to work with.

Andrew Halstead is indebted to Dorothy Gibson for her speedy and accurate conversion of hand-written pages to a word-processed script. He would also like to thank the companies and colleagues who have given assistance and encouragement in writing this book.

Publisher's Acknowledgements

For the first edition:
Project Editors Laura Langley, Martha Swift
Designer Rachael Parfitt
Managing Editor Louise Abbott
Managing Art Editor Lee Griffiths
Illustrations Sandra Pond and Will Giles

Dorling Kindersley would also like to thank Susanne Mitchell, Karen Wilson and Barbara Haynes at the Royal Horticultural Society and Ian Hodgson at *The Garden* magazine.